IMAGE SENSORS *and* SIGNAL PROCESSING *for* DIGITAL STILL CAMERAS

IMAGE SENSORS *and* SIGNAL PROCESSING *for* DIGITAL STILL CAMERAS

Edited by
Junichi Nakamura

Taylor & Francis
Taylor & Francis Group

Boca Raton London New York Singapore

A CRC title, part of the Taylor & Francis imprint, a member of the
Taylor & Francis Group, the academic division of T&F Informa plc.

Published in 2006 by
CRC Press
Taylor & Francis Group
6000 Broken Sound Parkway NW, Suite 300
Boca Raton, FL 33487-2742

© 2006 by Taylor & Francis Group, LLC
CRC Press is an imprint of Taylor & Francis Group

No claim to original U.S. Government works
Printed in the United States of America on acid-free paper
10 9 8 7 6 5 4 3 2 1

International Standard Book Number-10: 0-8493-3545-0 (Hardcover)
International Standard Book Number-13: 978-0-8493-3545-7 (Hardcover)
Library of Congress Card Number 2005041776

This book contains information obtained from authentic and highly regarded sources. Reprinted material is quoted with permission, and sources are indicated. A wide variety of references are listed. Reasonable efforts have been made to publish reliable data and information, but the author and the publisher cannot assume responsibility for the validity of all materials or for the consequences of their use.

No part of this book may be reprinted, reproduced, transmitted, or utilized in any form by any electronic, mechanical, or other means, now known or hereafter invented, including photocopying, microfilming, and recording, or in any information storage or retrieval system, without written permission from the publishers.

For permission to photocopy or use material electronically from this work, please access www.copyright.com (http://www.copyright.com/) or contact the Copyright Clearance Center, Inc. (CCC) 222 Rosewood Drive, Danvers, MA 01923, 978-750-8400. CCC is a not-for-profit organization that provides licenses and registration for a variety of users. For organizations that have been granted a photocopy license by the CCC, a separate system of payment has been arranged.

Trademark Notice: Product or corporate names may be trademarks or registered trademarks, and are used only for identification and explanation without intent to infringe.

Library of Congress Cataloging-in-Publication Data

Image sensors and signal processing for digital still cameras / edited by Junichi Nakamura.
    p. cm.
  Includes bibliographical references and index.
  ISBN 0-8493-3545-0 (alk. paper)
  1. Image processing—Digital techniques. 2. Signal processing—Digital techniques. 3. Digital cameras. I. Nakamura, Junichi.

TA1637.1448 2005
681'.418—dc22
                                                  2005041776

Taylor & Francis Group
is the Academic Division of T&F Informa plc.

Visit the Taylor & Francis Web site at
http://www.taylorandfrancis.com

and the CRC Press Web site at
http://www.crcpress.com

Preface

Since the introduction of the first consumer digital camera in 1995, the digital still camera (DSC) market has grown rapidly. The first DSC used a charge-coupled device (CCD) image sensor that had only 250,000 pixels. Ten years later, several models of consumer point-and-shoot DSCs with 8 million pixels are available, and professional digital single-lens reflex (DSLR) cameras are available with 17 million pixels. Unlike video camera applications in which the output is intended for a TV monitor and, thus, the vertical resolution of the image sensors is standardized, there is no standard output or resolution for DSCs, so the sensor pixel count continues to grow. Continuing improvements in sensor technology have allowed this ever increasing number of pixels to be fit onto smaller and smaller sensors until pixels measuring just 2.3 µm × 2.3 µm are now available in consumer cameras. Even with this dramatic shrinking of pixels, the sensitivity of sensors has improved to the point that shirt-pocket-sized consumer DSCs take good quality pictures with the exposure values equivalent to 400 ISO film.

Improvements in optics and electronics technologies have also been remarkable. As a result, the image quality of pictures taken by a DSC has become comparable to that normally expected of silver-halide film cameras. Consistent with these improvements in performance, the market for DSCs has grown to the point that shipments of DSCs in 2003 surpassed those of film cameras.

Image Sensors and Signal Processing for Digital Still Cameras focuses on image acquisition and signal-processing technologies in DSCs. From the perspective of the flow of the image information, a DSC consists of imaging optics, an image sensor, and a signal-processing block that receives a signal from the image sensor and generates digital data that are eventually compressed and stored on a memory device in the DSC. The image acquisition part of that flow includes the optics, the sensor, and front-end section of the signal-processing block that transforms photons into digital bits. The remainder of the signal-processing block is responsible for generating the image data stored on the memory device. Other technologies used in a DSC, such as mechanics, data compression, user interface, and the output-processing block that provides the signals to output devices, such as an LCD, a TV monitor, a printer, etc., are beyond the scope of this book.

Graduate students in electronics engineering and engineers working on DSCs should find this book especially valuable. However, I believe it offers interesting reading for technical professionals in the image sensor and signal-processing fields as well. The book consists of 11 chapters:

- Chapter 1, "Digital Still Cameras at a Glance," introduces the historical background and current status of DSCs. Readers will be able to understand

what DSCs are, how they have evolved, types of modern DSCs, their basic structure, and their applications.
- In Chapter 2, "Optics in Digital Still Cameras," a wide range of topics explains the imaging optics used in DSCs. It is obvious that high image quality cannot be obtained without a high-performance imaging optical system. It is also true that requirements for imaging optics have become higher as pixel counts have increased and pixel sizes have decreased.
- Reviews of image sensor technologies for DSC applications can be found in Chapter 3 through Chapter 6.
  - First, in Chapter 3, "Basics of Image Sensors," the functions and performance parameters common to CCD and complementary metal-oxide semiconductor (CMOS) image sensors are explained.
  - Chapter 4, "CCD Image Sensors," describes in detail the CCD image sensors widely used in imaging applications. The chapter ranges from a discussion of basic CCD operating principles to descriptions of CCD image sensors specifically designed for DSC applications.
  - Chapter 5, "CMOS Image Sensors," discusses the relatively new CMOS image sensor technology, whose predecessor, the MOS type of image sensor, was released into the market almost 40 years ago, even before the CCD image sensor.
  - Following these descriptions of image sensors, methods for evaluating image sensor performances relative to DSC requirements are presented in Chapter 6, "Evaluation of Image Sensors."
- Chapter 7 and Chapter 8 provide the basic knowledge needed to implement image processing algorithms.
  - The topics discussed in Chapter 7, "Color Theory and Its Application to Digital Still Cameras," could easily fill an entire book; however, the emphasis here is on how color theory affects the practical uses of DSC applications.
  - Chapter 8, "Image-Processing Algorithms," presents the algorithms utilized by the software or hardware in a DSC. Basic image-processing and camera control algorithms are provided along with advanced image-processing examples.
- This provides the framework for the description of the image-processing hardware engine in Chapter 9, "Image-Processing Engines." The required performance parameters for DSCs and digital video cameras are reviewed, followed by descriptions of the architectures of signal-processing engines. Examples of analog front-end and digital back-end designs are introduced.
- In Chapter 10, "Evaluation of Image Quality," readers learn how each component described in the previous chapters affects image quality. Image quality-related standards are also given.
- In Chapter 11, "Some Thoughts on Future Digital Still Cameras," Eric Fossum, the pioneer of CMOS image sensor technology, discusses future DSC image sensors with a linear extrapolation of current technology and then explores a new paradigm for image sensors. Future digital camera concepts are also addressed.

I would like to extend my gratitude to the contributors to this book, many of whom work actively in the industry, for their efforts and time given to making this book possible.

Also, I am grateful to all of my coworkers who reviewed draft copies of the manuscripts: Dan Morrow, Scott Smith, Roger Panicacci, Marty Agan, Gennnady Agranov, John Sasinowski, Graham Kirsch, Haruhisa Ando, Toshinori Otaka, Toshiki Suzuki, Shinichiro Matsuo, and Hidetoshi Fukuda.

Sincere thanks also go to Jim Lane, Deena Orton, Erin Willis, Cheryl Holman, Nicole Fredrichs, Nancy Fowler, Valerie Robertson, and John Waddell of the Mar-Com group at Micron Technology, Inc. for their efforts in reviewing manuscripts, appendices, and the table of contents; creating drawings for Chapter 3 and Chapter 5; and preparing the tables and figures for publication.

Junichi Nakamura, Ph.D.
Japan Imaging Design Center
Micron Japan, LTD

Editor

Junichi Nakamura received his B.S. and M.S. in electronics engineering from Tokyo Institute of Technology, Tokyo, Japan, in 1979 and 1981, respectively, and his Ph.D. in electronics engineering from the University of Tokyo, Tokyo, Japan, in 2000.

He joined Olympus Optical Co., Tokyo, in 1981. After working on optical image processing, he was involved in developments of active pixel sensors. From September 1993 to October 1996, he was resident at the NASA Jet Propulsion Laboratory, California Institute of Technology, as a distinguished visiting scientist. In 2000, he joined Photobit Corporation, Pasadena, CA, where he led several custom sensor developments. He has been with Japan Imaging Design Center, Micron Japan, Ltd. since November 2001 and is a Micron Fellow.

Dr. Nakamura served as technical program chairman for the 1995, 1999, and 2005 IEEE Workshop on Charge-Coupled Devices and Advanced Image Sensors and as a member of the Subcommittee on Detectors, Sensors and Displays for IEDM 2002 and 2003. He is a senior member of IEEE and a member of the Institute of Image Information and Television Engineers of Japan.

Contributors

Eric R. Fossum
Department of Electrical Engineering and Electrophysics
University of Southern California
Los Angeles, CA, USA

Po-Chieh Hung
Imaging System R&D Division
System Solution Technology R&D Laboratories
Konica Minolta Technology Center, Inc.
Tokyo, Japan

Takeshi Koyama
Lens Products Development Center, Canon, Inc.
Tochigi, Japan

Toyokazu Mizoguchi
Imager & Analog LSI Technology Department
Digital Platform Technology Division
Olympus Corporation
Tokyo, Japan

Junichi Nakamura
Japan Imaging Design Center, Micron Japan, Ltd.
Tokyo, Japan

Kazuhiro Sato
Image Processing System Group
NuCORE Technology Co., Ltd.
Ibaraki, Japan

Isao Takayanagi
Japan Imaging Design Center, Micron Japan, Ltd.
Tokyo, Japan

Kenji Toyoda
Department of Imaging Arts and Sciences
College of Art and Design
Musashino Art University
Tokyo, Japan

Seiichiro Watanabe
NuCORE Technology Inc.
Sunnyvale, CA, USA

Tetsuo Yamada
VLSI Design Department
Fujifilm Microdevices Co., Ltd.
Miyagi, Japan

Hideaki Yoshida
Standardization Strategy Section
Olympus Imaging Corp.
Tokyo, Japan

Table of Contents

Chapter 1
Digital Still Cameras at a Glance 1
Kenji Toyoda

Chapter 2
Optics in Digital Still Cameras 21
Takeshi Koyama

Chapter 3
Basics of Image Sensors.. 53
Junichi Nakamura

Chapter 4
CCD Image Sensors 95
Tetsuo Yamada

Chapter 5
CMOS Image Sensors 143
Isao Takayanagi

Chapter 6
Evaluation of Image Sensors.. 179
Toyokazu Mizoguchi

Chapter 7
Color Theory and Its Application to Digital Still Cameras...................... 205
Po-Chieh Hung

Chapter 8
Image-Processing Algorithms.. 223
Kazuhiro Sato

Chapter 9
Image-Processing Engines... 255
Seiichiro Watanabe

Chapter 10
Evaluation of Image Quality 277
Hideaki Yoshida

Chapter 11
Some Thoughts on Future Digital Still Cameras... 305
Eric R. Fossum

Appendix A
Number of Incident Photons per Lux with a Standard Light Source 315
Junichi Nakamura

Appendix B
Sensitivity and ISO Indication of an Imaging System 319
Hideaki Yoshida

Index 323

1 Digital Still Cameras at a Glance

Kenji Toyoda

CONTENTS

1.1 What Is a Digital Still Camera? 2
1.2 History of Digital Still Cameras 3
    1.2.1 Early Concepts 3
    1.2.2 Sony Mavica...4
    1.2.3 Still Video Cameras 5
    1.2.4 Why Did the Still Video System Fail? 7
    1.2.5 Dawn of Digital Still Cameras 8
    1.2.6 Casio QV-10...9
    1.2.7 The Pixel Number War 10
1.3 Variations of Digital Still Cameras 10
    1.3.1 Point-and-Shoot Camera Type 10
    1.3.2 SLR Type 13
    1.3.3 Camera Back Type 14
    1.3.4 Toy Cameras...14
    1.3.5 Cellular Phones with Cameras...15
1.4 Basic Structure of Digital Still Cameras 16
    1.4.1 Typical Block Diagram of a Digital Still Camera 16
    1.4.2 Optics 16
    1.4.3 Imaging Devices..17
    1.4.4 Analog Circuit 17
    1.4.5 Digital Circuit 17
    1.4.6 System Control..17
1.5 Applications of Digital Still Cameras 18
    1.5.1 Newspaper Photographs...18
    1.5.2 Printing Press 18
    1.5.3 Network Use 19
    1.5.4 Other Applications 19

In this chapter, the author describes briefly the basic concepts of digital still cameras and the history of digital and analog electronic cameras. This is followed by a discussion of the various types of digital still cameras and their basic construction. Descriptions of several key components of typical digital still cameras are given.

1.1 WHAT IS A DIGITAL STILL CAMERA?

An image can be described by "variation of light intensity or rate of reflection as a function of position on a plane." On the other hand, a camera is a piece of equipment that captures an image and records it, where "to capture" means to convert the information contained in an image to corresponding signals that can be stored in a reproducible way.

In a conventional silver halide photography system, image information is converted to chemical signals in photographic film and stored chemically at the same point where the conversion takes place. Thus, photographic film has the image storage function as well as the image capture function. Another method of image capture is to convert the image information to electronic signals. In this case, an image sensor serves as the conversion device. However, the image sensor used in the electronic photography system does not serve a storage function as does photographic film in the silver halide system. This is the most significant point in which the electronic system differs from the chemical silver halide system (Figure 1.1).

Naturally, the electronic photography system needs another device to store the image signals. Two primary methods have been adopted to perform this storage function: analog and digital. Analog electronic still cameras, which were once on the market, use a kind of floppy disk that electromagnetically records the image signals in the form of video signals. In digital still cameras, the image signals from the image sensor are converted to digital signals and stored in digital storage devices such as hard disks, optical disks, or semiconductor memories.

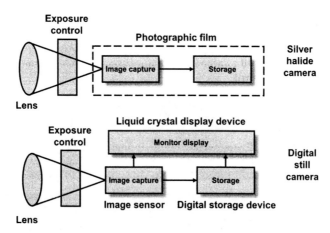

FIGURE 1.1 Difference between a silver halide photographic camera and an electronic still camera.

Digital Still Cameras at a Glance

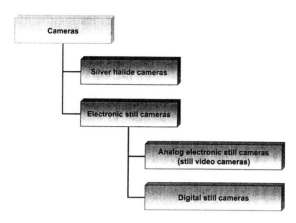

FIGURE 1.2 Classification of cameras.

Thus, still cameras are divided into two groups from the viewpoint of the image capture method: conventional silver halide cameras and electronic still cameras. Electronic still cameras are again divided into two groups: analog and digital (Figure 1.2). Thus, a digital still camera is defined as "a camera that has an image sensor for image capture and a digital storage device for storing the captured image signals."

1.2 HISTORY OF DIGITAL STILL CAMERAS

1.2.1 EARLY CONCEPTS

The idea of taking pictures electronically has existed for a long time. One of the earliest concepts was shown in a patent application filed by a famous semiconductor manufacturer, Texas Instruments Incorporated, in 1973 (Figure 1.3). In this embodiment drawing, the semiconductor image sensor (100) is located behind the retractable mirror (106). The captured image signals are transferred to the electromagnetic recording head (110) and recorded onto a removable ring-shaped magnetic drum. This concept did not materialize as a real product. In 1973, when this patent was filed, image sensor technology was still in its infancy, as was magnetic recording technology.

In another patent application filed in 1978, Polaroid Corp. suggested a more advanced concept (Figure 1.4). In this camera, the image signals are recorded on a magnetic cassette tape. Considering the amount of information an image contains, it must have taken a long time to record a single image. One of the advanced features of this camera was that it had a flat display panel (24) on the back to show the recorded image (Figure 1.5). Note that LCD panels at that time could only display simple numerals in a single color.

Another interesting item of note on Polaroid's concept is that this camera had a built-in color printer. Polaroid is famous for its instant camera products, so they must have thought that the output print was best done in the camera. The printing method was neither inkjet nor thermal transfer. Inkjet printers were not as popular then as they are now and the thermal dye transfer method did not even exist in 1978.

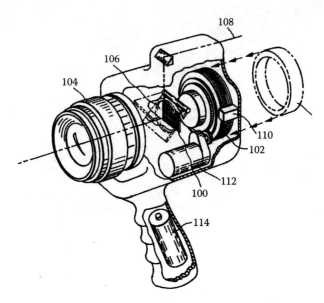

FIGURE 1.3 Early concept of electronic still cameras (USP 4,057,830).

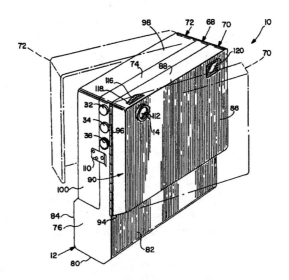

FIGURE 1.4 Another early concept of an electronic still camera (USP 4,262,301).

Actually, it was a wire dot impact printer. A paper cassette could be inserted into the camera body to supply ink ribbon and sheets of paper to the built-in printer.

Thus, several early digital still camera concepts were announced in the form of patent applications prior to 1981.

1.2.2 Sony Mavica

The year 1981 was a big one for camera manufacturers. Sony, the big name in the audio–visual equipment business, announced a prototype of an electronic still camera

Digital Still Cameras at a Glance

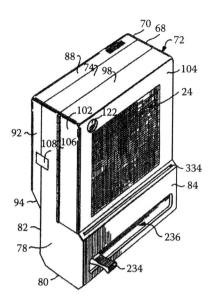

FIGURE 1.5 Rear view of the concept described in Figure 1.4.

FIGURE 1.6 Sony Mavica (prototype).

called "Mavica" (Figure 1.6). This name stands for "magnetic video camera." As its name suggests, this prototype camera recorded the image signals captured by the semiconductor image sensor on a magnetic floppy disk. This prototype camera had a single lens reflex finder, a CCD image sensor, a signal-processing circuit, and a floppy disk drive. Several interchangeable lenses, a clip-on type of electronic flash, and a floppy disk player to view the recorded images on an ordinary TV set were also prepared. The image signals recorded to the floppy disk were a form of modified video signals. Naturally, they were analog signals. Therefore, this was not a "digital" still camera. Nevertheless, it was the first feasible electronic still camera ever announced.

1.2.3 STILL VIDEO CAMERAS

The announcement of the Sony Mavica caused a big sensation throughout the camera business. Many people felt strong anxiety about the future of conventional silver halide photography. Some even said that silver halide photography would die before long.

Several camera manufacturers and electronic equipment makers started a consortium to advance Sony's idea. After a significant amount of negotiation, they came out with a set of standards for an electronic photography system. It was called "still video system," which included "still video camera" and "still video floppy." The still video floppy (Figure 1.7) was a flexible circular magnetic disk that measured 47 mm in diameter. It had 52 coaxial circular recording tracks. The image signals were recorded on tracks 1 through 50, with each track storing signals corresponding to one field — that is, one half of a frame image (Figure 1.8). Thus, a still video

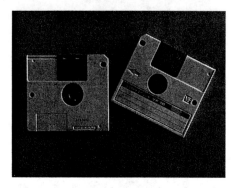

FIGURE 1.7 Still video floppy disk.

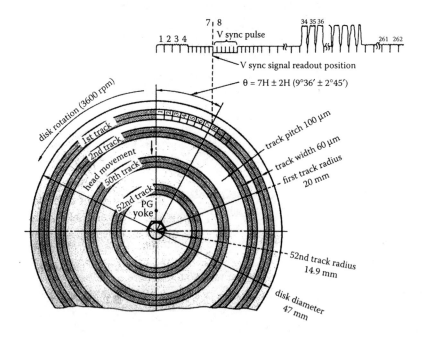

FIGURE 1.8 Track configuration of still video floppy.

Digital Still Cameras at a Glance

floppy disk could store 50 field images or 25 frame images. The image signals were analog video and based on the NTSC standard.

Once the standards were established, many manufacturers developed cameras, equipment, and recording media based on these standards. For example, Canon, Nikon, and Sony developed several SLR types of cameras with interchangeable lenses (Figure 1.9). Various point-and-shoot cameras were developed and put on sale by Canon, Fuji, Sony, Konica, Casio, and others (Figure 1.10); however, none of them achieved sufficient sales. These cameras had many advantages, such as instant play back and erasable and reusable recording media, but these were not sufficient to attract the attention of consumers.

1.2.4 Why Did the Still Video System Fail?

With all its advantages, why then did this new still video system fail? It is said that the main reason lies in its picture quality. This system is based on the NTSC video format. In other words, the image recorded in a still video floppy is a frame or a field cut out from a video motion picture. Thus, its image quality is limited by the number of scan lines (525). This means that the image resolution could not be finer than VGA quality (640×480 pixels). This is acceptable for motion pictures or even still pictures if they are viewed by a monitor display, but not for prints. Paper prints require high quality and one cannot get sufficient quality prints with VGA resolution, even if they are small type-C prints (approximately 3.5 in. $\times$ 5 in.). As a result, the

FIGURE 1.9 SLR type still video camera.

FIGURE 1.10 Point-and-shoot still video camera.

use of still video cameras was limited to applications in which paper prints were not necessary; these were quite limited.

Another major reason for failure was price. These cameras were expensive. Point-and-shoot cameras had a price tag of $1000 to $2500. SLR cameras were as expensive as $5000. Point-and-shoot models had a poor function equivalent to the $200 silver halide counterpart, with a no-zoom single focus length lens.

1.2.5 Dawn of Digital Still Cameras

The still video cameras can be categorized as analog electronic still cameras. It was natural that analog would transition to digital as digital technologies progressed. The first digital camera, Fuji DS-1P (Figure 1.11), was announced at the Photokina trade show in 1988. It recorded digital image signals on a static RAM card. The capacity of this RAM card was only 2 Mbytes and could store only five frames of video images because image compression technology was not yet applicable.

Though this Fuji model was not put on sale, the concept was modified and improved and several digital still cameras did debut in the camera market — for instance, Apple QuickTake 100 and Fuji DS-200F (Figure 1.12). However, they did not achieve good sales, the image quality was not sufficient yet for prints, and they were still fairly expensive. Apparently, they were not significantly different from still video cameras, except for digital storage function.

A rather unique model among them was the Kodak DCS-1 (Figure1.13), which was put on sale in 1991. This model was dedicated to newspaper photography. To

FIGURE 1.11 Fuji digital still camera DS-1P.

FIGURE 1.12 Early model of a digital still camera.

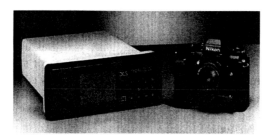

FIGURE 1.13 Kodak DCS-1.

FIGURE 1.14 Casio QV-10.

satisfy the requirements of newspaper photographers, Eastman Kodak attached a CCD image sensor that had 1.3 million pixels to the body of Nikon F3, the most popular SLR for these photographers. However, the image storage was rather awkward. A photographer had to carry a separate big box that connected to the camera body by a cable. This box contained a 200-Mbyte hard disk drive to store an adequate amount of image signals and a monochromatic monitor display. Nevertheless, this camera was welcomed by newspaper photographers because it could dramatically reduce the time between picture taking and picture transmitting.

1.2.6 CASIO QV-10

Casio announced its digital still camera, QV-10 (Figure 1.14), in 1994 and put it on sale the next year. Contrary to most expectations, this model had great success. Those who had had little interest in conventional cameras, such as female students, especially rushed to buy the QV-10.

Why then did this camera succeed while other similar ones did not? One thing is clear: it was not the picture quality because the image sensor of this camera had only 250,000 pixels and the recorded picture was no better than 240×320 pixels. The main point might be the LCD monitor display. Casio QV-10 was the first digital still camera with a built-in LCD monitor to view stored pictures. With this monitor display, users could see pictures immediately after they were taken. Furthermore, this camera created a new style of communication, which Casio named "visual communication." QV-10 was a portable image player, as well as a camera. Right after a picture was taken, the image could be shared and enjoyed on the spot among the friends. No other equipment was necessary, just the camera. Thus, this camera was enthusiastically welcomed by the young generation, even though its pictures did not have sufficient quality for prints.

Another key point was, again, the price. The QV-10 sold for 65,000 yen, approximately $600. To reduce the cost, Casio omitted many functions. It had no optical finder, with the LCD display working as a finder when taking pictures. It had no shutter blade or zoom lens. The semiconductor memory for storing image signals was fixed in the camera body and was not removable. In any case, the digital camera market began growing dramatically with this camera's debut.

1.2.7 THE PIXEL NUMBER WAR

Once the prospects for a growing market were apparent, many manufacturers initiated development of digital still cameras. Semiconductor makers also recognized the possibility of significant business and began to investigate development of image sensors dedicated to digital still cameras.

Although the Casio QV-10 created a "visual communication" usage that does not require paper prints, still cameras that cannot make prints were still less attractive. More than 1 million pixels were necessary to make fine type-C prints. Thus, the so-called "pixel number war" broke out. Prior to this point, most digital still camera makers had to utilize image sensors made for consumer video cameras. Subsequently, semiconductor manufacturers were more positive about the development of image sensors that had many more pixels.

In 1996, Olympus announced its C-800L (Figure 1.15) digital still camera, which had an approximately 800,000-pixel CCD image sensor. It was followed by the 1.3-Mpixel Fuji DS-300 and 1.4-Mpixel Olympus C-1400L the next year. In 1999, many firms rushed to announce 2-Mpixel models, and a similar situation was seen in 2000 for 3-Mpixel models. Thus, pixel numbers of digital still cameras used by consumers increased year by year up to 8 Mpixels for point-and-shoot cameras (Figure 1.16) and 16.7 Mpixels for SLR cameras in 2004. Figure 1.17 shows the increase in number of pixels of digital still cameras, plotted against their date of debut.

1.3 VARIATIONS OF DIGITAL STILL CAMERAS

The various digital still cameras sold at present are classified into several groups.

1.3.1 POINT-AND-SHOOT CAMERA TYPE

Most popular digital still cameras have a similar outfit to that of silver halide point-and-shoot cameras. For this type of camera, the LCD monitor display also works

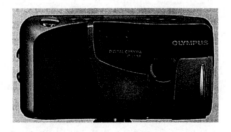

FIGURE 1.15 Olympus C-800L.

Digital Still Cameras at a Glance

FIGURE 1.16 Point-and-shoot digital still camera with 8 Mpixels.

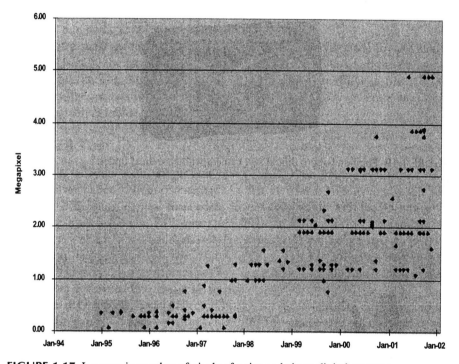

FIGURE 1.17 Increase in number of pixels of point-and-shoot digital cameras.

as the finder to show the range of pictures to be taken. Thus, an optical viewfinder is omitted in some models (Figure 1.18). However, the LCD monitor display has a disadvantage in that it is difficult to see under bright illumination such as direct sunlight. Many models in this category have optical viewfinders to compensate for this disadvantage (Figure 1.19). Most of these finders are the real image type of zoom finder. Figure 1.20 shows a typical optical arrangement of this type.

High-end models of the point-and-shoot digital still camera, which have rather high zoom ratio lenses, have electronic viewfinders (EVF) in place of optical ones (Figure 1.21). This type of viewfinder includes a small LCD panel that shows the output image of the image sensor. The eyepiece is a kind of magnifier to magnify this image to reasonable size. Most models that belong to this point-and-shoot

FIGURE 1.18 Point-and-shoot digital still camera with no optical finder.

FIGURE 1.19 Point-and-shoot digital still camera with an optical finder.

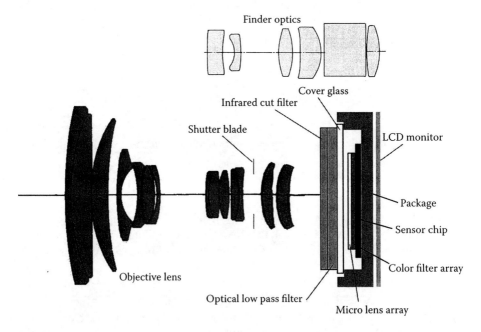

FIGURE 1.20 Typical arrangement of a point-and-shoot digital still camera.

Digital Still Cameras at a Glance

FIGURE 1.21 Point-and-shoot digital still camera with an EVF (electronic view finder).

category utilize the output signals of the image sensor as control signals for auto focus and automatic exposure.

Recently, most models of this type have included a movie mode to take motion pictures. Usually, the picture size is as small as a quarter VGA and the frame rate is not sufficient. The file format of these movies is motion JPEG or MPEG4; the duration of the movies is limited. However, this function may generate new applications that silver halide photography could not achieve.

1.3.2 SLR Type

Digital still cameras of the single lens reflex (SLR) type (Figure 1.22) are similar to their silver halide system counterparts in that they have an interchangeable lens system and that most models have a lens mount compatible with the 35-mm silver halide SLR system. Digital SLR cameras have instant return mirror mechanisms just like ordinary SLRs. The major difference is that they have image sensors in place of film and that they have LCD monitor displays. The LCD display built in SLR digital still cameras, however, does not work as a finder. It only functions as a playback viewer for stored images.

A few digital SLR models have an image sensor image size equivalent to the 35-mm film format, but most of them have smaller image size — one half or one

FIGURE 1.22 SLR digital still camera.

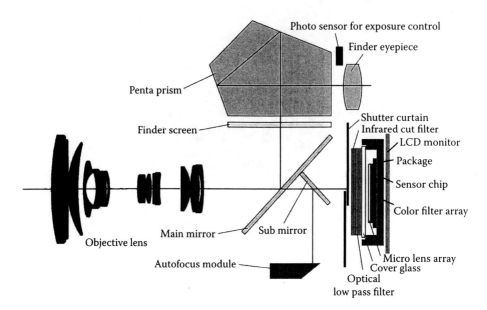

FIGURE 1.23 Typical arrangement of an SLR digital still camera.

fourth of the 35-mm full frame. As a result, the angle of view is smaller than that of 35-mm SLRs by 1/1.3 to 1/2 when an objective lens of the same focal length is attached. Recent cost reduction efforts have made this type of camera affordable to the average advanced amateur photographer. Figure 1.23 shows a typical optical arrangement of an SLR digital camera.

1.3.3 Camera Back Type

Camera back type digital still cameras are mainly for professional photographers to use in photo studios. This type of unit contains an image sensor; signal-processing circuit; control circuit; image storage memory; and, preferably, an LCD monitor display. One cannot take pictures with this unit alone; it must be attached to a medium-format SLR camera with an interchangeable film back or a large format view camera (Figure 1.24). In many cases, a computer is connected with a cable to control the camera and check the image. The image size is larger than point-and-shoot or SLR cameras, but not as large as medium-format silver halide cameras.

1.3.4 Toy Cameras

A group of very simple digital still cameras is sold for about $100 or less each (Figure 1.25). They are called "toy cameras" because manufacturers who do not make cameras, such as toy makers, sell them. To reduce the cost, the image sensor is limited to VGA class and they are equipped with fixed focus lenses and nondetachable image memory. LCD monitor displays are omitted and often no electronic flash is built in. The main use of this type of camera is for taking pictures for Web sites.

FIGURE 1.24 Camera back digital still camera.

FIGURE 1.25 Toy camera.

FIGURE 1.26 Cellular phone with a built-in camera.

1.3.5 CELLULAR PHONES WITH CAMERAS

Recently, cellular phones have begun to have built-in cameras (Figure 1.26). Their main purpose is to attach a captured image to an e-mail and send it to another person. Fine pictures that have many pixels tend to have large file sizes, thus increasing time and cost of communication. Therefore, the image size of these cameras was limited to approximately 100,000 pixels.

However, as this type of cellular phone became popular, people began to use them as digital still cameras that were always handy. To meet this requirement, cellular phone built-in cameras that have image sensors with more pixels — say, 300,000 or 800,000 — began to appear and the pixel number war started just as it did for ordinary digital cameras. In 2004, models debuted that can store an image as large as 3 million pixels. However, the camera functions are rather limited because the space in a cellular phone in which to incorporate a camera is very small.

1.4 BASIC STRUCTURE OF DIGITAL STILL CAMERAS

Digital cameras are thought to be a camera in which image sensors are used in place of film; therefore, its basic structure is not much different from that of silver halide cameras. However, some differing points will be discussed in this section.

1.4.1 TYPICAL BLOCK DIAGRAM OF A DIGITAL STILL CAMERA

Figure 1.27 shows a typical block diagram of a digital still camera. Usually, a digital camera includes an optical and mechanical subsystem, an image sensor, and an electronic subsystem. The electronic subsystem includes analog, digital–processing, and system control parts. An LCD display, a memory card socket, and connectors to communicate with other equipment are also included in most digital cameras. Each component will be discussed in detail later in this book.

1.4.2 OPTICS

Fundamental optics of digital still cameras are equivalent to those of silver halide cameras, except for the fact that the focal length is much shorter because of the smaller image size in most models. However, some additional optical elements are required (Figure 1.20 and Figure 1.23). A filter to attenuate infrared rays is attached

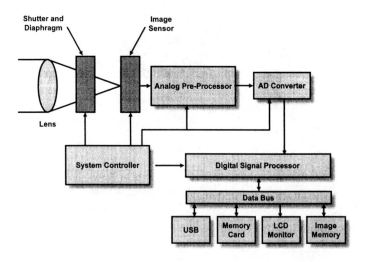

FIGURE 1.27 Typical block diagram of a digital still camera.

in front of the image sensor because the sensor has significant sensitivity in infrared range, which affects the image quality. Also arranged in front of the image sensor is an optical low pass filter (OLPF) to prevent moiré artifacts.

1.4.3 Imaging Devices

Charge-coupled devices (CCDs) are the most popular image sensors for digital still cameras. However, CMOS sensors and other x–y address types of image sensors are beginning to be used in SLR cameras and toy cameras for various reasons. Arranged on the image receiving surface of these imaging devices are a mosaic filter for sensing color information and a microlens array to condense the incident light on each pixel (Figure 1.20 and Figure 1.23).

1.4.4 Analog Circuit

The output signals of the image sensor are analog. These signals are processed in an analog preprocessor. Sample-and-hold, color separation, AGC (automatic gain control), level clamp, tone adjustment, and other signal processing are applied in the analog preprocessor; then they are converted to digital signals by an analog-to-digital (A/D) converter. Usually, this conversion is performed to more than 8-b accuracy — say, 12 or 14 b — to prepare for subsequent digital processing.

1.4.5 Digital Circuit

The output signals of the A/D converter are processed by the digital circuits, usually digital signal processors (DSPs) and/or microprocessors. Various processing is applied, such as tone adjustment; RGB to YCC color conversion; white balance; and image compression/decompression. Image signals to be used for automatic exposure control (AE), auto focus (AF), and automatic white balance (AWB) are also generated in these digital circuits.

1.4.6 System Control

The system control circuits control the sequence of the camera operation: automatic exposure control (AE), auto focus (AF), etc. In most point-and-shoot digital still cameras, the image sensor also works as AE sensor and AF sensor. Before taking a picture, the control circuit quickly reads sequential image signals from the image sensor, while adjusting the exposure parameters and focus. If the signal levels settle appropriately in a certain range, the circuit judges that proper exposure has been accomplished. For auto focus, the control circuit analyzes the image contrast, i.e., the difference between maximum and minimum signal levels. The circuit adjusts the focus to maximize this image contrast. However, this is not the case in SLR digital cameras. These models have a separate photo sensor for exposure control and another sensor for auto focus, just like the silver halide SLRs (Figure 1.23).

1.5 APPLICATIONS OF DIGITAL STILL CAMERAS

1.5.1 Newspaper Photographs

Among the various users of cameras, newspaper photographers have showed the greatest interest in digital still cameras since early days when their ancestors, still video cameras, barely made their debut. It is always their most serious concern to send the picture of an event to their headquarters at the earliest possible time from the place at which the event has taken place. Until 1983, they used drum picture transmitters, which required making a print and wrapping it around the drum of the machine. When the film direct transmitter that could send an image directly from the negative film (Figure 1.28) was developed, the time to send pictures was dramatically reduced because making prints was no longer necessary. However, they still had to develop a negative film.

Using electronic still cameras could eliminate this film development process and the output signals could be sent directly through the telephone line. This would result in a significant reduction of time. Many newspaper photographers tested the still video cameras as soon as they were announced, but the result was negative. The image quality of these cameras was too low even for rather coarse newspaper pictures. They had to wait until the mega-pixel digital still cameras appeared (Figure 1.29).

Newspaper technologies have seen major innovation during years in which the Olympic Games are held. Still video cameras were tested at the Seoul Olympic Games in 1988, but few pictures actually were used. Mega-pixel digital still cameras were tested in Barcelona in 1992. In 1996, approximately half the cameras that newspaper photographers used in Atlanta were digital still cameras. This percentage became almost 100% at the Sydney Games in 2000.

1.5.2 Printing Press

Printing press technologies were computerized fairly early under the names of CEPS (color electronic prepress system), DTP (desktop publishing), and so on. Only the

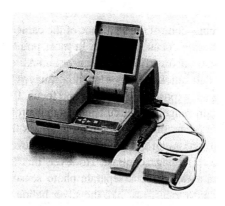

FIGURE 1.28 Film direct transmitter.

FIGURE 1.29 Mega-pixel digital still camera for newspaper photography (Nikon D1).

picture input device, i.e., the camera, was not digital. Therefore, as the technology of digital still cameras progressed, they were gradually incorporated into the system.

At first, digital still cameras were used for printing flyers or information magazines for used cars. The pictures that they use are relatively small and do not require very high resolution; they benefit greatly from the instant availability of pictures that digital cameras provide. They became popular among commercial photographers who take pictures for ordinary brochures, catalogues, or magazines because the picture quality grew higher. Many photographers nowadays are beginning to use digital still cameras in place of silver halide cameras.

1.5.3 Network Use

One of the new applications of digital still cameras in which conventional silver halide cameras are difficult to use is for network use. People attach pictures taken by digital still cameras to e-mails and send them to their friends or upload pictures to Web sites. Thus, the "visual communication" that Casio QV-10 once suggested has expanded to various communication methods. This expansion also gave birth to the cellular phone built-in cameras.

1.5.4 Other Applications

Digital still cameras opened a new world to various photographic fields. For instance, astrophotographers could increase the number of captured stars using cameras with cooled CCDs. In the medical arena, endoscopes have drastically changed. They no longer have the expensive image guide made of bundled optical fibers because the combination of an image sensor and a video monitor can easily show an image from inside the human body. Thus, digital still cameras have changed various applications related to photography, and will continue changing them.

2 Optics in Digital Still Cameras

Takeshi Koyama

CONTENTS

2.1 Optical System Fundamentals and Standards for Evaluating Optical Performance 22
    2.1.1 Optical System Fundamentals 22
    2.1.2 Modulation Transfer Function (MTF) and Resolution 27
    2.1.3 Aberration and Spot Diagrams 30
2.2 Characteristics of DSC Imaging Optics 31
    2.2.1 Configuration of DSC Imaging Optics 32
    2.2.2 Depth of Field and Depth of Focus 32
    2.2.3 Optical Low-Pass Filters 34
    2.2.4 The Effects of Diffraction 36
2.3 Important Aspects of Imaging Optics Design for DSCs 38
    2.3.1 Sample Design Process 38
    2.3.2 Freedom of Choice in Glass Materials 40
    2.3.3 Making Effective Use of Aspherical Lenses 41
    2.3.4 Coatings 43
    2.3.5 Suppressing Fluctuations in the Angle of Light Exiting from Zoom Lenses 46
    2.3.6 Considerations of the Mass-Production Process in Design 47
2.4 DSC Imaging Lens Zoom Types and Their Applications 49
    2.4.1 Video Zoom Type 49
    2.4.2 Multigroup Moving Zooms 50
    2.4.3 Short Zooms 50
2.5 Conclusion 51
References 51

In recent years, the quality of images produced by digital still cameras (DSCs) has improved dramatically to the point that they are now every bit as good as those produced by conventional 35-mm film cameras. This improvement is due primarily to advances in semiconductor fabrication technology, making it possible to reduce the pixel pitch in the imaging elements and thereby raising the total number of pixels in each image.

However, other important factors are involved. These include the development of higher performance imaging optics to keep pace with the lower pixel pitches, as well as improvements to the image-processing technology used to convert large amounts of digital image data to a form visible to the human eye. These three factors — imaging optics, imaging elements, and image-processing technology — correspond to the eye, retina, and brain in the human body. All three must perform well if we are to obtain adequate image quality.

In this chapter, I explain the imaging optics used in a DSC, or its "eyes" in human terms, focusing primarily on the nature of the optical elements used and on the key design issues. Because this book is not intended for specialists in the field of optics, I make my descriptions as simple as possible, omitting any superfluous parameters rather than giving a strictly scientific explanation. For a more rigorous discussion of the respective topics, I refer the reader to the specialist literature. In this chapter, lenses in an optical system that provide the image are called *imaging lenses*, and the entire optical system, including any optical filters, is called the *imaging optics*. Imaging elements such as CCDs do not form part of the imaging optics.

2.1 OPTICAL SYSTEM FUNDAMENTALS AND STANDARDS FOR EVALUATING OPTICAL PERFORMANCE

The text begins by briefly touching on a few prerequisites that are indispensable to understanding the DSC imaging optics that will be discussed later. The first is a basic understanding of optical systems and the second is a familiarity with the key terms used in discussions of the performance of those systems.

2.1.1 Optical System Fundamentals

First, I explain basic terminology such as focal length and F-number. Figure 2.1 shows the image formed of a very distant object by a very thin lens. Light striking the thin lens from the object side (literature on optics normally shows the object

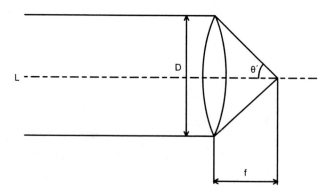

FIGURE 2.1 Schematic diagram of single lens.

side to the left and the imaging element side to the right) enters the lens as parallel beams when the object is very distant. The light beams are refracted by the lens and focused at a single point that is at a distance f (the focal length) from the lens. Because Figure 2.1 is a simplified diagram, it only shows the image-forming light beams above the lens's optical axis L.

The focal length f for a single very thin lens can be calculated using a relational expression such as that shown in Equation 2.1, where the radius of curvature for the object side of the lens is R_1; the radius of curvature for the image side of the lens is R_2; and the refractive index of the lens material is n. In this equation, the radius of lens surface curvature is defined as a positive value when it is convex relative to the object and as a negative value when it is convex relative to the imaging side.

$$\frac{1}{f} = (n-1) \cdot \left(\frac{1}{R_1} - \frac{1}{R_2} \right) \qquad (2.1)$$

For example, for a very thin lens that is convex on both sides and has a refractive index of 1.5 and values of 10 and −10, respectively, for R_1 and R_2, the focal length is 10 mm. From this equation, we can also see that a direct correlation exists between the focal length of the lens and its refractive index minus 1. Thus, simply by increasing the refractive index from 1.5 to 2.0, we halve the focal length of the lens.

Of course, actual lenses also have thickness. Here, if we express the thickness of a single lens as d, we get the following relational expression:

$$\frac{1}{f} = \frac{n-1}{R_1} + \frac{1-n}{R_2} + \frac{d(n-1)^2}{nR_1R_2} \qquad (2.2)$$

From this equation, we can see that for a lens that is convex on both sides, the thicker the lens is, the longer the focal length will be. This equation can also be used to calculate the combined focal length of a lens that is made up of two very thin lenses. Thus, if we take the focal length of the first lens as f_1, the focal length of the second lens as f_2, and the gap between the lenses as d, the expression that corresponds to Equation 2.2 is shown in Equation 2.3.

$$\frac{1}{f} = \frac{1}{f_1} + \frac{1}{f_2} - \frac{d}{f_1 f_2} \qquad (2.3)$$

The inverse of the focal length is what is called the refractive power or, simply, the power of a lens. Saying that a lens has a high power means that it is strongly refractive, just as we talk about the lenses in a pair of spectacles as being strong. These terms will be used again in the discussion of zoom types in Section 2.4.

To get a clearer grasp of the direct correlation between multiple lens configurations and focal length, see Figure 2.2 and Figure 2.3. In Figure 2.2, where the first lens is concave and the second lens is convex, the effect is equivalent to having a

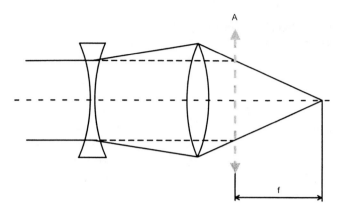

FIGURE 2.2 Schematic diagram of retrofocus lens.

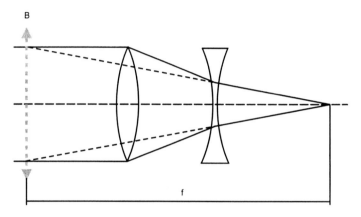

FIGURE 2.3 Schematic diagram of telephoto lens.

single lens in position A. Therefore, the overall lens is rather long in comparison with its focal length. This configuration is called a retrofocus type and is often used in wide-angle lenses and compact DSC zoom lenses. (The position of the equivalent lens is called the principal point (or more correctly, the rear principal point.)

Figure 2.3 shows the reverse situation, in which the first lens is convex and the second lens is concave; the effect is equivalent to having a single lens in position B. This configuration makes it possible for the overall lens to be short in comparison with its focal length. This configuration is called a telephoto type and is widely used in telephoto lenses and in zoom lenses for compact film cameras. However, regardless of the effectiveness of this configuration in reducing the total lens length, the telephoto configuration used in zoom lenses for compact film cameras is not used in compact DSCs. The reason for this is explained in Section 2.3.5.

The principal methods used to bend light, other than refraction as discussed earlier, are reflection and diffraction. For example, mirror lenses mostly use reflection to bend light. Lenses for some of the latest single-lens reflex (SLR) cameras on the market now make partial use of diffraction. Even in DSCs in which diffraction is not actively used for image formation, the effects of diffraction are sometimes unavoidable; this is discussed further in Section 2.2.4.

The F-number of a lens (F) is expressed as half the opening angle θ' for the focused light beams in Figure 2.1, as shown as in Equation 2.4.

$$F = \frac{1}{2 \sin \theta'} \qquad (2.4)$$

In reality, because this depends on the cross-sectional area of the light beams, the brightness of the lens (the image plane brightness) is inversely proportional to the square of this F-number. This means that the larger the F-number is, the less light passes through the lens and the darker it becomes as a result. The preceding equation also shows that the theoretical minimum (brightest) value for F is 0.5. In fact, the brightest photographic lens on the market has an F-number of around 1.0. This is due to issues around the correction of various aberrations, which are discussed later. The brightest lenses used in compact DSCs have an F-number of around 2.0. When the value of θ' is very small, it can be approximated using the following equation in which the diameter of the incident light beams is taken as D. This equation is used in many books.

$$F = \frac{f}{D} \qquad (2.5)$$

However, this equation tends to give rise to the erroneous notion that we can make the F-number as small as we like simply by increasing the size of the lens. Therefore, it is important to understand that Equation 2.4 is the defining equation and Equation 2.5 should only be used as an approximation.

In addition, because actual lens performance is also affected by reflection from the lens surfaces and internal absorption of light by the optical materials used, lens brightness cannot be expressed by the F-number alone. For this reason, brightness may also be discussed in terms of the T-number, which is a value that corresponds to the F-number and takes into account the transparency (T) of the imaging optics. To make the imaging optics as transparent as possible, it is necessary to suppress surface reflections from optical system elements such as the lenses. This is achieved through the use of coatings, which are discussed in Section 2.3.4.

Generally speaking, the transparency of imaging lenses to the wavelengths of light typically used by compact DSCs (around 550 nm) is between 90 and 95% for most compact cameras with relatively few lenses. In cameras with high-magnification zooms that use ten or more lenses, the figure is generally around 80%. An infrared cut filter and optical low-pass filter have a combined transparency in the 85 to 95% range. This means that, given these transparency levels for lenses and filters, when the transparency of the imaging lens and filters is, for example, 90% in each case, no more than around 80% of the light entering the camera actually reaches the imaging element. Indeed, to be strictly correct, we should also include losses caused by the glass plate covering the imaging element and losses in the element.

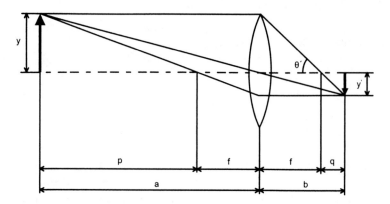

FIGURE 2.4 Schematic diagram of single lens.

Next, I discuss the image formation correlations for objects at limited distances from the camera, as shown in Figure 2.4. Compare this figure with Figure 2.1. Where the object size is y and its distance from the lens is a, an image with a size of y' is formed in a position only a distance of q from the focal length f of the lens (distance b from the lens). Distance p is obtained by subtracting the focal length f from distance a. Here, for the sake of simplicity, we will express all these symbols as absolute values rather than giving them positive or negative values. This gives us the following simple geometrical relationship where m is the imaging magnification:

$$m = \frac{y'}{y} = \frac{q}{f} = \frac{f}{p} = \frac{b}{a} \tag{2.6}$$

By substituting $a - f$ for p, we come to the following well-known equation:

$$\frac{1}{f} = \frac{1}{a} + \frac{1}{b} \tag{2.7}$$

In situations like this, in which the distance to the object is limited, the image is formed farther away from the lens than is the case for very far-off objects, so the brightness at the image plane (the surface of the imaging element) is lower. This is similar to the way things get darker as one moves farther away from a window. In this situation, the value that corresponds to the F-number is called the "effective F-number" and can be approximated as F' using the same approach as Equation 2.5:

$$F' = \frac{f+q}{D} = \left(1 + \frac{q}{f}\right) \cdot \frac{f}{D} = (1+m) F \tag{2.8}$$

Thus, an F2.8 lens that produces a life-size image will actually be as dark as an F5.6 lens. Because the F-number acts as a square, as discussed earlier, the brightness (image plane brightness) falls to one fourth that of an image of a very distant object.

Optics in Digital Still Cameras

The image plane brightness E_i can be expressed using Equation 2.9, where E_o is the object brightness (luminance). However, essentially, we can think of the brightness as inversely proportional to the square of the effective F-number $(1 + m)F$ and directly proportional to the transparency of the optical system (T).

$$E_i = \frac{\pi}{4} E_o T \left(\frac{1}{(1+m)F} \right)^2 \qquad (2.9)$$

This equation gives us the brightness at the center of the image, but the brightness at the periphery of the image is generally lower than this. This is known as edge illumination fall-off. For a very thin lens with absolutely no vignetting or distortion, edge illumination falls in proportion to $\cos^4\theta$, where θ is the angle diverging from the optical axis facing towards the photographed area on the object side (half the field of view). This is called the cosine fourth law. However, depending on the lens configuration, the actual amount of light at the edge of an image may be greater than the theoretical value, or the effects of distortion may be not inconsiderable. The former can be seen, for example, when we look from the front into a wide-angle lens with a high degree of retrofocus; in this case the lens opening (the image of the aperture) appears larger when we look from an angle than it does when we look from directly in front. With wide-angle lenses, which tend toward negative distortion (barrel distortion), this corresponds to compression of the periphery of the photographed image, which has a beneficial effect on the edge illumination. Ultimately, matching this to the imaging element is an unavoidable problem for camera design, as I discuss further in Section 2.3.5.

In imaging optics, the field of view is twice the angle θ (half the field of view) discussed previously. Where the radius of the imaging element's recording area is y', the relationship between y' and θ can be expressed by the following equation:

$$y' = f \tan \theta \qquad (2.10)$$

There is no clear definition for the terms "wide-angle lens" and "telephoto lens"; lenses with a field of view of 65° or more are generally regarded as wide-angle and those with a field of view of 25° or less are called telephoto.

2.1.2 MODULATION TRANSFER FUNCTION (MTF) AND RESOLUTION

Modulation transfer function (MTF) is frequently used as a standard for evaluating the imaging performance of any lens, not merely those used in DSCs. MTF is a transfer function for spatial frequency and is one of the measures used to show how faithfully object patterns are transferred to the image. The graph in Figure 2.5 shows an example of MTF spatial frequency characteristics, with the vertical axis showing the MTF as a percentage and the horizontal axis showing the spatial frequency (line-pairs/mm). The unit for the horizontal axis shows the number of stripes of light and

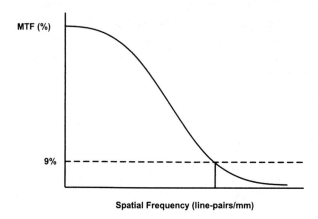

FIGURE 2.5 Example of MTF spatial frequency characteristics.

dark (line-pairs) photographed per millimeter in the image plane. The graph normally shows a line falling away to the right as it does in the figure because the higher the frequency (the more detailed the object pattern), the greater the decline in object reproducibility. These graphs are best seen as a gauge for reproducing contrast through a range from object areas with coarse patterning to areas with fine patterning. The term "resolution" has also often been used in this context. This is a measure for evaluating the level of detail that can be captured, and line-pairs/mm is also widely used as a unit of resolution.

TV lines are also used as a unit of resolution when lenses are used for electronic imaging, such as in DSC and, particularly, video lenses. This is based on the concept of scanning lines in TVs and is calculated as line-pairs × 2 × the vertical measurement of the imaging element recording plane. For example, for a Type 1/1.8 CCD, a resolution of 100 line-pairs/mm equates to roughly 1000 TV lines. In general, a "high-resolution" lens refers to a sharp lens that gives excellent reproduction of finely detailed patterns.

Thus, a very close correlation exists between MTF and resolution. This is discussed in more detail in Section 2.2.4, but the theoretical limit on the number of imaging lines is reached at the point at which the MTF is roughly 9%. In an actual lens, the resolution limit is generally reached at MTF levels between 10 and 20% due to the effects of factors such as aberration, which will be discussed later. The simplest approach is to think of a high MTF at high frequencies as indicating sharpness, and a high MTF at medium frequencies as indicating the reproducibility of contrast in ordinary objects.

Figure 2.6 shows three patterns of MTF spatial frequency characteristics. In this figure, the MTF for C is low in the medium frequencies and is retained even at high frequencies, indicating that when detailed patterns are photographed, contrast is insufficient overall, with no modulation between light and dark. By contrast, in pattern B, the MTF is high in the medium frequencies but low for high frequencies. This equates to high contrast but low resolution, which will produce images that appear to be out of focus. Pattern A has a high MTF for high and medium frequencies

Optics in Digital Still Cameras

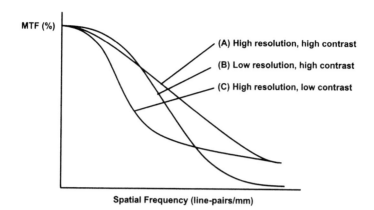

FIGURE 2.6 (See Color Figure 2.6 following page 178.) Three patterns of MTF spatial frequency characteristics.[1]

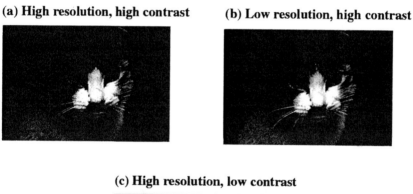

FIGURE 2.7 (See Color Figure 2.7 following page 178.) Images for three patterns of MTF spatial frequency characteristics: high resolution, high contrast; low resolution, high contrast; high resolution, low contrast.

and will give images with good contrast and resolution. Figure 2.7 shows the images for three patterns of MTF spatial frequency characteristics.

However, the MTF is by no means an all-purpose evaluation tool; care is essential because it frequently leads to misunderstandings. For example, even when the MTF is high, differences in the actual images produced can arise in which the MTF is limited by chromatic aberration or by comatic flare. Distortion has no direct impact on MTF. These types of aberration are discussed in more detail in the next section.

2.1.3 Aberration and Spot Diagrams

Ideal image formation by a lens, expressed in simple terms, meets the following conditions:

- Points appear as points.
- Planes appear as planes.
- The subject and its image are the same shape.

The reasons for a lens's failure to meet these conditions are the aberrations in that lens. Lens aberrations were classified mathematically by Seidel in 1856. When polynomials are used to approximate spherical surfaces (lens surfaces), aberrations that apply as far as third-order areas are called third-order aberrations; my explanation will be limited to these aberrations, even though consideration of higher order aberrations is indispensable in actual lens design.

According to Seidel's classifications, five basic third-order aberrations affect monochromatic light, collectively referred to as the Seidel aberrations (there are also nine fifth-order aberrations, known as the Schwarzschild aberrations):

1. Spherical aberration is an aberration caused by the fact that the lens surface is spherical. This means that an image formed of a point light source on the optical axis cannot be focused to one point. This can be corrected by reducing the aperture size.
2. Comatic aberration is an aberration that produces a comet-like flare with a tail for a point light source that is off the optical axis. This can usually be corrected by reducing the aperture size.
3. Astigmatism is an aberration that causes a point light source to be projected as a line or ellipse rather than as a point. The shape of the line changes by 90°, depending on the focal point (e.g., a vertical line becomes horizontal). Reducing the aperture reduces the effects of this aberration.
4. Curvature of field is an aberration that causes the focal plane to curve in the shape of a bowl so that the periphery of the image is out of focus, when the object is a flat surface. Reducing the aperture size also reduces the effects of this aberration because it increases the depth of focus.
5. Distortion refers to aberration that distorts the image. The previously mentioned barrel distortion sometimes found in wide-angle lenses is an example of distortion expanding middle of the image relative to the top and bottom, reminiscent of the shape of a barrel. This aberration by itself has no effect on the MTF and is not corrected by reducing the aperture size.

Some aberrations are collectively referred to as chromatic aberration:

1. Axial or longitudinal chromatic aberration refers to the fact that different colors (wavelengths of light) have different focal points. Reducing the aperture size remedies the effects of this aberration.
2. Chromatic difference of magnification or lateral chromatic aberration refers to the fact that different colors have different rates of magnification. Accordingly, point-symmetrical color bleeding can be seen from the center of the image towards the margins of the image. Reducing the aperture size does not correct this aberration.

These aberrations can be recorded in a figure known as an aberration chart, but that alone does not give the viewer a complete understanding of the nature of the lens. Spot diagrams, which show the images of a point light source as a number of spots, are often used for this. Figure 2.8 shows an example of a color spot diagram. The function that describes light and shade in a spot diagram is called the point spread function. When converted to its real number component using Fourier transformation, this function gives the MTF discussed earlier.

2.2 CHARACTERISTICS OF DSC IMAGING OPTICS

In this section, I examine the characteristics of the imaging optics used in DSCs, focusing particularly on how they differ from the optical systems used in conventional film cameras. Section 2.3 deals with information on some important design aspects of DSC imaging optics; this section will be confined to the external characteristics.

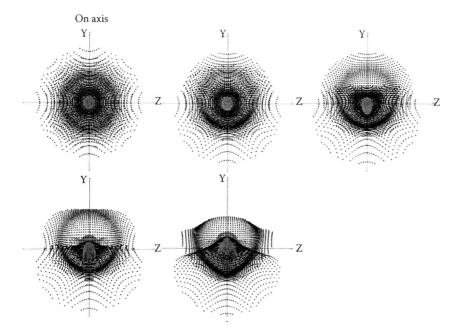

FIGURE 2.8 (See Color Figure 2.8 following page 178.) Example of a color spot diagram.

2.2.1 CONFIGURATION OF DSC IMAGING OPTICS

Figure 2.9 shows an example of a typical configuration for imaging optics used in a DSC. DSC imaging optics normally consist of the imaging lenses, an infrared cut filter, and an optical low-pass filter (OLPF). Added to this is an imaging element, such as a charge-couple device (CCD). As mentioned earlier, in this chapter I refer to that part of the optical system that contributes to forming the image as the imaging lens and to the total optical system, including any filters, as the imaging optics. Imaging elements such as CCDs are not included in the imaging optics.

Imaging lenses such as zoom lenses are made up of several groups of lenses, and zooming is achieved by varying the spacing between these lens groups. This, together with the differences between DSCs and film cameras, is explained in more detail in Section 2.3.5 and Section 2.4.

Infrared cut filters are, as the name suggests, filters that cut out unwanted infrared light. These filters are needed because imaging elements such as CCDs are by their nature highly sensitive to unwanted infrared light. The infrared cut filter is generally positioned behind the imaging lens (between the lens and the imaging element), but can in some cases be fitted in front (on the object side). Most infrared cut filters are absorption-type filters, but some are reflective, using a thin film deposited on the filter by evaporation, and some combine both methods.

Optical low-pass filters (OLPFs) are normally positioned in the closest part of the imaging lens to the imaging element. OLPFs are discussed in more detail in Section 2.2.3.

2.2.2 DEPTH OF FIELD AND DEPTH OF FOCUS

One of the key characteristics of DSCs, particularly compact DSCs, is that they combine a large depth of field with a small depth of focus. This section looks at this characteristic in detail.

Depth of field — in other words, the area (depth) within which the object is in focus — is proportional to the square of the focal length of the imaging optics.

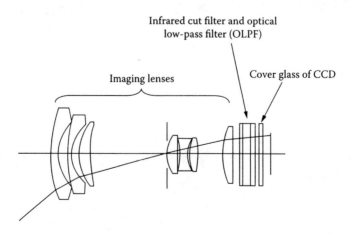

FIGURE 2.9 Example of a typical configuration for imaging optics used in a DSC.

Optics in Digital Still Cameras

Where δ is the circle of least confusion, which is the area that is allowed to be out of focus on the imaging element, the hyperfocal distance, D_h (the distance which, when used to set the focus, effectively sets the focus to infinity), is given by:

$$D_h = \frac{f^2}{F\delta} \quad (2.11)$$

As this equation shows, the depth of field at the object grows rapidly as the focal length is shortened. Whereas the normal image size for 35-mm film is 43.27 mm measured diagonally, the diagonal measurement of the CCD imaging area in current compact DSCs is generally between 5 and 11 mm. Given lenses with the same field of view, the focal length f is proportional to this diagonal length. For instance, if we were to compare the image quality of prints of the same size, given a sufficiently small pixel pitch in the DSC imaging element, the circle of least confusion δ is also proportional to the diagonal length. Consequently, provided the F-number of the lens is the same, δ is in effect proportional to the focal length f and the hyperfocal distance becomes longer. When the DSC is an SLR-type camera, the imaging element is also larger; this means that the depth of field is smaller, which is ideal for shots such as portraits when the background should be out of focus. With a compact DSC, it is difficult to soften the background in this way, but images with a large depth of field can be obtained without having to reduce the aperture size.

The circle of least confusion δ is generally said to be around 35 μm on a 35-mm camera. As regards the adequacy of this figure, there is a view that examples exist in which its adequacy is indicated in calculations from factors such as the camera's ability to discriminate and resolve two points based on observations of a final printed image.

The depth of focus Δ, which refers to the depth on the imaging element side, is expressed in simple terms by Equation 2.12.

$$\Delta = F\delta \quad (2.12)$$

Because the depth of focus Δ is directly proportional to the circle of least confusion δ and the F-number, compact DSCs that have imaging elements with a short diagonal also have a small depth of focus.

Table 2.1 shows specific examples of calculations for the depth of field and depth of focus in 35-mm film cameras and compact DSC cameras. The examples in the table compare the hyperfocal distance and depth of focus for different image sizes for the equivalent of a 38-mm F2.8 lens when converted to 35-mm film format.

For example, the hyperfocal distance for a 38-mm F2.8 lens for a 35-mm film camera is 14.7 m, but the focal length of a lens with the same field of view using a type 1/2.7 imaging element is 5.8 mm. Given a 3-Mpixel imaging element and the same F-number of F2.8, the hyperfocal distance for the lens is roughly 2 m. In other words, by setting the focus to 2 m, an image can be obtained that is in focus from a distance of 1 m from the camera to infinity, enabling the camera to be used in what is virtually pan-focus mode.

TABLE 2.1
Comparison of Depth between 35-mm Film Camera and DSCs (F2.8)

	f	No. pixels	Pixel pitch	Image circle	δ	Dh	Δ
35-mm film	38 mm			ø 43.27 mm	35 μm	14.7 m	98 μm
Type 1/1.8	7.8 mm	4 Mpixels	3.125 μm	ø 8.9 mm	6.3–7.5 μm	3.4–2.9 m	17–21 μm
Type 1/2.7	5.8 mm	3 Mpixels	2.575 μm	ø 6.6 mm	5.2–6.2 μm	2.3–1.9 m	14–17 μm

The corollary of this is that the depth of focus is extremely small. For a 35-mm film camera, the circle of least confusion δ is around 35 μm, as stated earlier, which gives a depth of focus of 98 μm when calculated for an F2.8 lens. In other words, the image is focused through a range 98 μm in size in front of and behind the film. There are various approaches to determining the circle of least confusion in DSCs, but it is generally taken to be between 2 and 2.4 units of the imaging element's pixel pitch. For example, for the type 1/2.7 imaging element mentioned earlier, the circle of least confusion is between 5.2 and 6.2 μm, so that the depth of focus for an F2.8 lens is between 12 and 20 μm, or around one sixth the depth of the depth on a 35-mm film camera. This figure for the depth of focus is generally of little interest to users, but requires extremely precise focus control from the manufacturers. It also demands lens performance that gives a very high level of field flatness with minimal curvature of field.

2.2.3 OPTICAL LOW-PASS FILTERS

As discussed earlier, optical low-pass filters (OLPFs) are currently an indispensable part of DSCs and camcorders. They are normally made using thin, birefringent plates made from liquid crystal or lithium niobate, but diffractive optical elements or special aspherical surfaces may sometimes be used instead. Birefringency refers to the property of a material to have different refractive indexes depending on the direction in which light is polarized. Put simply, a birefringent material is a single material with two refractive indexes. Many readers will be familiar with the experience of holding a calcite plate over text and seeing the text appear in two places. If we take the two refractive indexes as n_e and n_o and the thickness of the OLPF as t, the distance separating the two points (e.g., the amount by which the text is displaced) is given by Equation 2.13.

$$S = t \cdot \frac{n_e^2 - n_o^2}{2n_e n_o} \qquad (2.13)$$

The reason for inserting an OLPF is to mitigate the moiré effect caused by the interaction of patterns in the object and the pattern on the imaging element caused by the arrangement of imaging element pixels at a fixed pitch. To prevent moiré (false colors) caused by the high-frequency component included in the object, frequencies that are higher than the Nyquist frequency should be eliminated. The details of this process are the subject of another chapter, but its impact on optical performance is briefly discussed here.

Figure 2.10 shows an example of the MTF for an imaging lens and the MTF frequency characteristics for a typical OLPF. By combining these two sets of characteristics, we see the final optical performance of the imaging optics as a whole. Accordingly, it is essentially not necessary for lens performance at frequencies above the OLPF cut-off frequency.[2] For example, if we set the Nyquist frequency as the OLPF cut-off frequency and use a CCD with a pixel pitch of 3.125 μm (equivalent to a type 1/1.8, 4-Mpixel CCD), a simple calculation shows that MTF above 160 line-pairs/mm is no longer relevant. Therefore, if our premise is that we use this sort of CCD, even with an imaging lens capable of resolutions of 200 line-pairs/mm or above, that resolving power is no longer needed.

However, because lens performance is not something that suddenly disappears at high frequencies, in order to ensure a high MTF at frequencies below the Nyquist frequency, a lens with high resolving power is normally essential. The key goal here is to guarantee performance at frequencies below those cut out by the OLPF, and the important thing is the level of the MTF at the range of frequencies between 30 and 80% of the Nyquist frequency. The cut-off frequency actually used in products varies slightly depending on the manufacturer, but the norm is to set it above the Nyquist frequency.

OLPFs made from materials such as liquid crystal are constructed in various different configurations, including composite filters composed of multiple liquid-crystal plates; filters with phase plates sandwiched between other layers; and more simple filters that comprise a single liquid-crystal plate. Product specifications also vary from manufacturer to manufacturer. In an actual DSC, other factors besides the imaging optics MTF affect the final overall image quality. These include MTF deterioration due to the imaging element (involving characteristics such as pixel aperture ratio and color filter positioning) and variations in the MTF caused by image processing (edge sharpening, etc.).

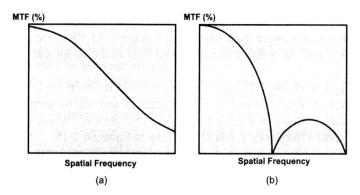

FIGURE 2.10 Example of the MTF for (a) an imaging lens and (b) a typical OLPF.

2.2.4 THE EFFECTS OF DIFFRACTION

The recent progress in the miniaturization of the pixel pitch in the imaging elements used in DSCs discussed at the start of this chapter has resulted in reductions in the size of the circle of least confusion. This progress is particularly marked in compact DSCs. We have now reached the level at which the pixel pitch is only five or six times the wavelength of light, and it is inevitable that the effects of diffraction will be increasingly felt. It is believed that further reduction in the pixel pitch will occur in the future, so this problem is becoming more pressing.

Light has the characteristics of rays and of wave motion. The imaging optics used in conventional film cameras have been designed in the so-called geometric areas of refraction and reflection, but the design of modern DSCs with their tiny pixel pitch must be handled using so-called wave optics that also allows for diffraction. Diffraction refers to the way light bends around objects. One can still listen to the radio when standing between two buildings because the signals from the broadcasting station bend around the buildings into the spaces between them. Light, which can be thought of simply as ultrahigh-frequency radio waves, is also bent around in the same way into microscopically small areas.

In a geometrical approach that treats light simply as straight lines, the image formed by an aberration-free lens of a point body (an object that is a point) will be capable of being focused to a perfect point. However, in wave optics, which takes wave characteristics into consideration, the image is not focused to a point. For example, given an aberration-free lens with a circular aperture, a bright disc surrounded by concentric dark circles is formed. This is known as the Airy disc, and the brightness of the first ring (the primary diffracted light) is just 1.75% of the brightness at the center; although it is very dim, it does exist. The radius of the first dark ring r is shown by Equation 2.14, where λ is the wavelength and F is the F-number.

$$r = 1.22\lambda F \tag{2.14}$$

This shows that the radius largely depends on the F-number.

Rayleigh took this distance to the first dark ring to be the criterion of resolving power for a two-point image. This is known as the Rayleigh limit, and the intensity at the median area between these two points is roughly 73.5% of the peak intensity for the two points. As a dimension, this correlates to the reciprocal of the spatial frequency.

The graph in Figure 2.11 shows the relationship between the F-number of an ideal (aberration-free) lens with a circular aperture and the monochromatic MTF frequency characteristics for the helium d-line (587.56 nm). Where the horizontal axis is the spatial frequency v, this is shown as in Equation 2.15.

$$MTF(v) = \frac{2}{\pi} \cdot \left(\cos^{-1}(\lambda F v) - \lambda F v \sqrt{1 - (\lambda F v)^2} \right) \tag{2.15}$$

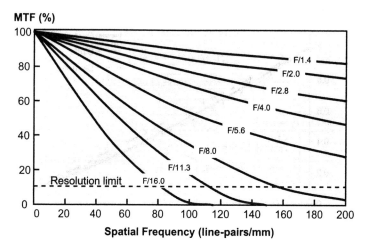

FIGURE 2.11 Relationship between the F-number and the MTF.

If we substitute $1/r$ from Equation 2.14 for v in this equation, we see that the MTF value is 8.94%. Thus, with the Rayleigh limit as our premise, in ideal conditions the point at which the MTF is roughly 9% corresponds to the number of spatial resolution lines at the Rayleigh limit.

In this way, if we take the number of resolution lines at the Rayleigh limit in ideal conditions as the number of lines at which the MTF is roughly 9%, then F11, for instance, equates to roughly 123 line-pairs/mm. This is lower than the Nyquist frequency of 160 line-pairs/mm for the type 1/1.8, 4-Mpixels CCD mentioned earlier. This leads us to conclude that small apertures should be avoided as far as possible with lenses for high-megapixel imaging elements. The usual methods for avoiding the use of small apertures include increasing the shutter speed and using an ND filter.

There is also an approach to considering resolution in a situation in which absolutely no intensity is lost in the area between the two points — that is, where the two points are perfectly linked. This is called Sparrow resolution and is shown in Equation 2.16. However, this is hardly ever used in imaging optics.[3]

$$r' = 0.947\lambda F \qquad (2.16)$$

In reality, due to the effects of aberrations, the resolution limit in most situations is reached at an MTF of between 10 and 20%.

Figure 2.11 shows the MTF for an ideal lens with a circular aperture; however, the effects of diffraction also vary slightly depending on the shape of the aperture. This is shown in the graph in Figure 2.12, which shows the respective monochromatic MTF frequency characteristics for an ideal F2.8 lens with a circular, rectangular (square), and diamond-shaped aperture. Also, if the goal is simply to increase the resolving power, a degree of control is possible by inserting a filter with a distributed density into the imaging optics (e.g., using an apodization filter or "super resolution").

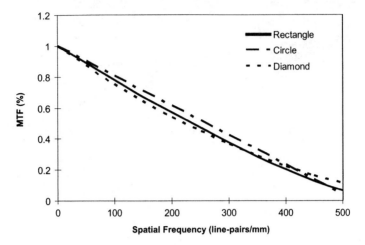

FIGURE 2.12 MTF for an ideal F2.8 lens with some aperture shape.

2.3 IMPORTANT ASPECTS OF IMAGING OPTICS DESIGN FOR DSCS

In this section, the design process used for designing imaging optics for DSCs is discussed as well as the key issues that must be considered in that design process.

2.3.1 Sample Design Process

We begin by describing an ordinary design process, which is not limited specifically to imaging optics for DSCs. Figure 2.13 shows a typical optical system design process.

- The first step is to formulate the target design specifications, as well as optical specifications, such as the focal length, F-number, and zoom ratio. This includes physical specifications such as the optical performance targets and the total length and diameter.
- Next, we select the lens type. As described in Section 2.4, because the optimum lens type differs depending on the specifications, efficient lens design requires that a number of lens configurations suited to the specifications be selected in advance.
- Once the lens type has been selected, the next step is to build a prototype that will act as the basis for the design. For a simple lens system, mathematical methods may be used to determine the initial lens shape analytically. However, currently, the starting point is determined in most cases based on the database of existing lenses and on experience.
- Next, we perform simulations of light passing through the lens (light tracing) and evaluate the results in terms of how well the optical performance correlates with the target specifications. Based on the results, we change parameters such as the curvature, lens thickness, lens spacing, and

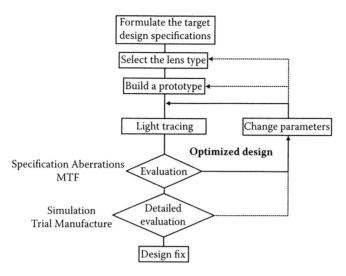

FIGURE 2.13 Typical optical system design process.

the type of glass. Then we go back and repeat the light tracing evaluation. This repetition or loop is normally performed using a computer, which makes small incremental changes to the parameters while searching for the optimal balance of factors such as optical performance. For this reason, it is referred to as optimized design. Particularly recently, because aspherical lens elements are used in almost all zoom lenses, the number of factors to be evaluated and the number of parameters that can be varied have grown enormously, and high-speed computers have become an indispensable tool. It is interesting to recall that until 50 years ago, design work followed a program in which a pair of women known as "computers" turned the handles on hand-wound calculators in accordance with calculations written by designers, and their calculations were then checked to see whether they matched before the design process moved on to the next stage. In those days there were no zoom lenses or aspherical lenses and probably no way to deal with them. Evaluation at this point mainly involves factors such as the amount of various types of aberration and the lens sizes for a range of object distances, focal lengths, and image heights (location in the image). However, once design has reached a certain point, factors such as MTF, peripheral brightness (the brightness around the edge of the image), and any suspected ghosting are also simulated. The lens configuration is then varied based on the results of these simulations and the testing loop is repeated.
- The final stage of testing naturally includes detailed design performance evaluation, as well as simulations of problems such as the effects of aberrations in actual manufactured models. If problems are discovered even at this late stage, the lens configuration may still be modified. Of course, it goes without saying that throughout the design process, close

collaboration with the design of the lens barrel that will hold the lenses must take place.

2.3.2 Freedom of Choice in Glass Materials

DSC imaging optics have progressed to a level at which they are far more compact, more powerful, and more highly specified than was the case several years ago. Advances in design techniques have obviously played a part in this; however, a very significant role has also been played by expansion in the range of available glass types and improvements in aspherical lens technology, which will be discussed later.

Figure 2.14 maps the types of glass that can be used in optical design. The vertical axis shows the refractive index, and the horizontal axis shows the Abbe number v_d. The Abbe number shows the dispersion and is given by Equation 2.17, where the refractive indexes for the helium d-line (587.6 nm), F-line (486.1 nm), and C-line (656.3 nm) are n_d, n_F, and n_C, respectively.

$$v_d = \frac{n_d - 1}{n_F - n_C} \qquad (2.17)$$

The Abbe number is an index that shows the extent of the difference in the refractive index caused by the wavelength of the light and can be thought of simply as describing the relationship between the amplitudes of the seven colors of the spectrum into which white light splits when it passes through a prism. The smaller the Abbe number is, the larger the amplitude of the seven colors of the spectrum is.

In Figure 2.14, the green portion indicates the existence area of the optical glasses, and the yellow portion indicates the most frequently used types of glass.

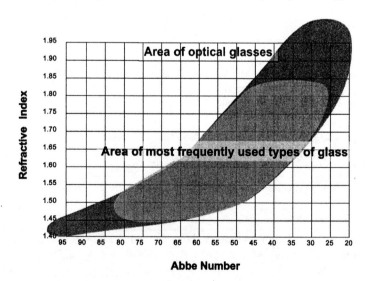

FIGURE 2.14 (See Color Figure 2.14 following page 178.) Schematic view of the glass map.

The number of types of glass used in optical systems is more than 100 just from a single glass manufacturer; if we count the total number of lens glass types from all the manufacturers, there are several hundred or more. From all these, individual glass types are chosen based on a wide range of considerations that include optical performance, durability, environmental friendliness, delivery, and cost. Recently, in addition to these, glasses that show any special dispersion tendencies or that have an ultrahigh refractive index have come to be used in DSCs. The former are effective in reducing chromatic aberration, and the latter are extremely useful in helping produce more compact lenses that offer higher performance.

Fluorite is well known as a glass material that shows special dispersion tendencies. It is a crystalline substance rather than an amorphous substance like glass. It not only has an extremely high Abbe number (very low dispersion), but also has abnormal dispersal properties and splits light into the seven colors of the spectrum in a different way from glass. This makes it very effective in correcting chromatic aberration, particularly in telephoto lenses and high-magnification zoom lenses. On the other hand, it is soft and requires particular care when it is handled. Its surface is also very delicate and machining fluorite requires considerable expertise.

Glass with an ultrahigh refractive index has in the past presented problems such as deterioration in its spectral transmission characteristics when exposed to strong sunlight (solarization) and staining. However, dramatic improvements in quality have occurred in recent years and it is already used in some compact DSCs to give them a refractive index of more than 2.0. The use of new materials of this sort provides greater benefits for electronic imaging devices such as DSCs than for film cameras. For instance, staining in glass materials is less of a problem for DSCs because, as long as the spectral transmission characteristics do not change over time, even a little staining from the outset can be corrected by electronically setting the white balance.

2.3.3 Making Effective Use of Aspherical Lenses

Aspherical lenses, as well as correcting the problem of spherical aberration described in Section 2.1.3, are effective in correcting other nonchromatic aberrations. Depending on where the aspherical elements are used, the dampening effect on the respective aberrations differs. An aspherical lens is said to have the effect of two or three spherical lenses, but when minor improvements in performance are taken into account, it is not unusual for them to be even more effective than this.

Because the onus is on size reductions to provide greater convenience and portability in a compact DSC, the key requirements are a combination of greater compactness and improved performance to take advantage of the smaller pixel pitches provided by the imaging element. When the only requirement is better performance, the simple solution is to increase the number of lenses, but this is counterproductive from the viewpoint of size reductions. On the other hand, simply reducing the number of lenses means that no option remains except to reduce the power (the reciprocal of the focal length) of each lens group in order to maintain the same performance. This often results in the lens getting larger. In short, there is no alternative but to include an efficient aspherical lens.

Aspherical lenses are generally classified into the following for types according to how they are made and the materials used:

- *Ground and polished glass aspherical lenses* are expensive to produce because they must be individually ground and polished. Altough they are unsuited to mass production, the glass allows a great deal of design flexibility and they can also be used in large diameters.
- *Glass molded (GMo) aspherical lenses* are formed by inserting glass with a low melting point into a mold and applying pressure. These lenses provide excellent precision and durability. Small-diameter GMo lenses are ideally suited to mass production. However, they are very difficult to make in large diameters and this gives rise to productivity problems.
- *Composite aspherical lenses* are spherical lenses to which a resin film is applied to make them aspherical. Factors such as temperature and humidity must be taken into account when considering how they are to be used. These lenses are relatively easy to use in large diameters.
- *Plastic aspherical lenses* are produced by inserting plastic into a mold, usually by injection molding. They are ideally suited to mass production and are the least costly type of aspherical lens to produce. However, care is needed because they are susceptible to variations in temperature and humidity.

Of these four, all but the first type are used in the compact DSCs currently on the market. However, with the increasingly tiny pixel pitches in imaging elements in recent years, demand for glass-molded aspherical lenses has increased enormously due to their excellent precision, reliability, and suitability for mass production. The problem with this is that the materials used in glass-molded lenses are special and often differ from those used in ordinary spherical lenses. This is because the glass must have a low melting point and be suitable for molding. Few types of glass offer these characteristics, so the range of optical characteristics available is far narrower than that offered by the materials used in ordinary spherical lenses.

Figure 2.15 shows an example of the effect of aspherical lenses in the PowerShot S100 DIGITAL ELPH (Canon's compact digital still camera using the type 1/2.7CCD). In order to make the imaging optics smaller, it is vital to improve areas of the optics that have a significant impact on the optical system volume. In the case of DSCs like the PowerShot S100 DIGITAL ELPH, where the goal is greater compactness, the largest component in the optical system is usually the first lens group, so the aim should be to slim down that group somehow. Here, the technological genesis for the reduction of the group 1 size is the technology for mass producing concave meniscus-shaped aspherical GMo lenses with a high refractive index. (Meniscus-shaped lenses are convex on one side and concave on the other.)

In Figure 2.15, the group 1 configuration in a conventional lens is shown on the left and the group 1 configuration in the PowerShot S100 DIGITAL ELPH is shown on the right. The conventional lens also uses an aspherical lens element, but by using glass with a higher refractive index ($n_d > 1.8$), the current two concave lenses (one of which is aspherical) can be successfully replaced by a single concave aspherical

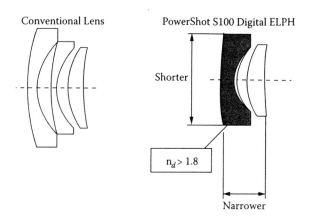

FIGURE 2.15 Example of the effect of aspherical lenses.

lens. This is an excellent example of an improvement that makes the lens shorter and narrower while at the same time making it higher performance.[4]

The level of precision required in the surfaces of the current aspherical lenses is so strict that if the surface of a small-diameter aspherical lens to be used in a compact DSC were a baseball field, a bulge the size of the pitcher's mound would be unacceptable, and even errors in the length to which the grass was cut would ordinarily be cause for rejection.

2.3.4 COATINGS

Like conventional film cameras, DSCs are susceptible to problems involving unwanted light in the image. This is called ghosting and is caused by light reflecting off the surfaces of the lens elements inside the lens. However, DSCs differ significantly from conventional film cameras in that they include elements such as the infrared cut filters discussed earlier that have surface coatings that reflect some of the infrared light. They also include an imaging element, such as a CCD, that has relatively high surface reflectance. To minimize the negative impact of these elements, it is vital to have appropriate coatings on the lens element surfaces.

Coatings are applied in a variety of ways. The most widely used method is vacuum deposition, in which the coating material is heated to a vapor in a vacuum and deposited as a thin film on the surface of the lens. Other methods, which are not discussed in detail here, include sputtering and dipping, which involves immersing the lens in a fluid.

The history of lens coatings is relatively short, with the first practical applications occurring in Germany in the 1940s. In those days, the initial aim appears to have been to improve the transparency of periscopes. Periscopes use at least 20 lenses, so the drastic loss of transparency due to lens surface reflections was a major stumbling block.[5] The technology was later applied to photographic lenses, starting with single-layer coatings and gradually progressing to the current situation in which a range of different multilayer coatings have been developed. However, it has been known for quite a long time that applying a thin, clear coating has the effect of reducing surface reflections. The inhabitants of some South Pacific islands knew to spread oils such as coconut oil on the surface of the water when they were hunting for fish from boats because it made it easier to see the fish. This was, in effect, one type of coating applied to the ocean's surface. With lenses also, people over the years have noticed that the surface reflectance of a lens was lower when a very thin layer of the lens surface was altered. This alteration was called tarnish, which is one type of defect in lens glass. This accidental discovery is tied up with the development of coating technology.

The surface reflectance for perpendicular incident light on an uncoated lens varies depending on the refractive index of the glass used, as shown in Equation 2.18. In the equation, n_g is the refractive index of the glass.

$$R = \left(\frac{n_g - 1}{n_g + 1}\right)^2 \quad (2.18)$$

For example, the reflectance for perpendicular incident light on glass with a refractive index of 1.5 is 4%. However, for glass with a refractive index of 2.0, the reflectance rises to 11%. This shows why coatings are particularly essential for glass that has a high refractive index.

The surface reflectance for perpendicular incident light when a single-layer coating is applied is given by Equation 2.19, where n_f is the refractive index of the coating and the product of the coating's refractive index and the coating thickness is one fourth the wavelength of the light.

$$R' = \left(\frac{n_g - n_f^2}{n_g + n_f^2}\right)^2 \quad (2.19)$$

Given the preceding equation, the solution in which there is no reflection is given by Equation 2.20.

$$n_f = \sqrt{n_g} \quad (2.20)$$

From this equation, for glass with a refractive index of 1.5, we can see that we must use a coating material with a refractive index of 1.22 to eliminate the reflection of light completely with a specific wavelength. However, the coatings with the lowest refractive indexes are still in the 1.33 (Na_3AlF_6) to 1.38 (MgF_2) range, so no solution offers zero reflectance using a single coating. For glass with a refractive index of 1.9, reflections of some wavelengths of light can be completely eliminated even with a single-layer coating, but, of course, reflectance still exists for other wavelengths of light.

The reflectance when a two-layer coating is applied is given by Equation 2.21, where n_2 is the refractive index of the inner coating applied directly to the glass; n_1

Optics in Digital Still Cameras

is the refractive index of the outer coating exposed to the air; and the product of the coatings' refractive indexes and the coating thickness is one fourth the wavelength of the light.

$$R'' = \left(\frac{n_1^2 n_g - n_2^2}{n_1^2 n_g + n_2^2}\right)^2 \qquad (2.21)$$

From this equation, the solution in which there is no reflection is given by:

$$\frac{n_2}{n_1} = \sqrt{n_g} \qquad (2.22)$$

This shows that it is possible to eliminate reflections in at least the specific wavelengths from glass with a refractive index of 1.5 by using two coatings with refractive indexes that differ by a factor of 1.22. This sort of combination of coating materials offers a highly practicable solution. Also, by using multilayer coatings in which very thin layers of vaporized material are deposited in alternating layers, techniques have been developed for suppressing reflectance across a wide range of wavelengths for glass materials with widely divergent refractive indexes. Almost all the current DSC imaging lenses include lens element surfaces on which multilayer coatings are used.

Figure 2.16 shows the correlation between the refractive index of glass and its reflectance, along with an example of the reflectance when a single-layer coating with a refractive index of 1.38 is applied. From the graph it is clear that, as the refractive index of the glass increases, its reflectance rises rapidly if no coating is used.

Figure 2.17 shows the wavelength dependency of reflectance for single-layer and multilayer coatings. From this, we can see that although reflectance is lowest

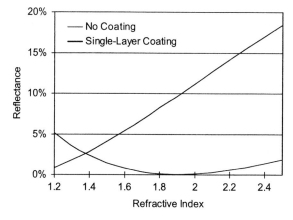

FIGURE 2.16 (See Color Figure 2.16 following page 178.) Correlation between the refractive index of glass and its reflectance.

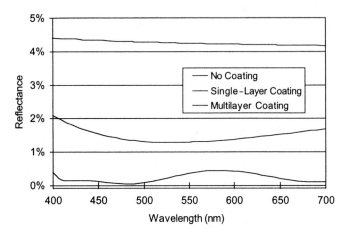

FIGURE 2.17 (See Color Figure 2.17 following page 178.) Wavelength dependency of reflectance for coatings.

for a single-layer coating for only one wavelength, there are several for the multilayer coating. It is also clear that the multilayer coating suppresses reflectance across a wider range of wavelengths than the single-layer coating does.

My discussion so far has looked at the contribution that coatings make to reducing reflections from a lens surface so as to increase the transparency of the imaging optics and prevent ghosting. However, coatings are also used in a number of other roles, such as preventing degeneration of the glass. A typical instance of this degeneration is the tarnishing mentioned earlier in this section. In another example, as discussed in Section 2.2.1, the purpose of the film deposited on some infrared cut filters is to actively reflect some of the infrared light.

2.3.5 Suppressing Fluctuations in the Angle of Light Exiting from Zoom Lenses

The image side (imaging element side) of a DSC imaging lens has an OLPF made of a material such as liquid crystal, as discussed earlier, and an imaging element fitted with microlenses for each pixel to ensure that the image is sufficiently bright. If zooming or focusing makes significant changes to the angles of the light exiting from the imaging lens, this can cause problems such as vignetting in the imaging element microlenses and changes in the low-pass effect. The vignetting in the imaging element is particularly undesirable because it leads to shading of the peripheral light in the resulting image. No matter how much thought has been given in the design stages to the provision of ample peripheral light, large variations in the exiting light angle due to zooming will result in a final image that is dark around the edges. This is a major restriction in terms of optical design and imposes limits on the types of zoom that can be used in the design.

Figure 2.18 compares the variations in the angle of the exit light during zooming in a zoom lens for a compact film camera and a zoom lens for a compact DSC. From this figure, it is clear that in most cases, the zoom lens group configurations

Optics in Digital Still Cameras

Zoom lens for a compact film camera

Diverges markedly on the image side

Wide-angle

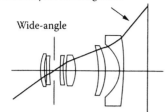

Zoom lens for a compact DSC

Nearly parallel to the optical axis

Wide-angle

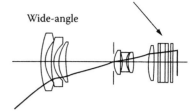

Marked difference between the wide-angle and telephoto settings

Telephoto

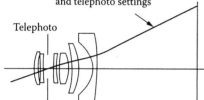

No marked difference between the wide-angle and telephoto settings

Telephoto

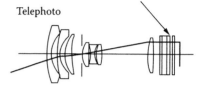

FIGURE 2.18 Comparison of the variations in the angle of the exit light during zooming.

are reversed front to back in zoom lenses for compact film cameras compared with zoom lenses for compact DSCs. In zoom lenses for compact film cameras, the convex lens group is at the object end of the lens while the concave lens group is at the film end of the lens. This is the retrofocus type configuration described in Section 2.1.1. The result of this is that in zoom lenses for compact film cameras, light beams passing through the center of the aperture at the wide-angle setting, diverge markedly on the image (film) side. In the zoom lens for a compact DSC, there is no marked difference between the wide-angle and telephoto settings, and the angle of the light is always nearly parallel to the optical axis. This is called a telecentric lens.[6]

Given that most imaging elements are now fitted with microlenses to ensure sufficient light levels directly in front of the imaging elements, this telecentric requirement means that it is basically impossible to transfer zoom lenses for compact film cameras directly to DSCs. However, in the interchangeable lenses used on SLR cameras, the distance from the rearmost lens to the film plane (the back focus) is long to avoid any interference with the quick-return mirror; the result is that the light beams are necessarily almost telecentric. Consequently, many of the lenses developed for SLR film cameras can be used on DSCs without causing problems.

2.3.6 Considerations of the Mass-Production Process in Design

As has already been mentioned in Section 2.2.2., the depth of focus in imaging optics for DSCs is extremely small due to the very small pixel pitch of the imaging elements. Also, because image-forming performance is required at very high spatial frequencies, the level of difficulty in the manufacturing process is generally extremely high. For this reason, the conversion to mass production not only requires higher standards of manufacturing precision at the individual component level, but also means that efficient and highly precise adjustment must also form a key part of the assembly process.

Even at the optical design stage, designers must consider how to reduce the likelihood that errors in the final manufacturing and assembly stages will affect the product. Normally, greater compactness is achieved by increasing the power (the reciprocal of the focal length) of the lens groups. However, the use of this means alone makes achieving basic performance more difficult and also increases the susceptibility of the lens to optical system manufacturing errors. Then problems such as eccentricity caused by very tiny manufacturing errors have significant adverse effects on the final optical performance. This is why every effort is made to eliminate such factors through measures introduced at the design stage.

Figure 2.19 shows an example of one such measure taken at the design stage. This figure compares the configurations of the second lens group in the PowerShot A5 Zoom (also Canon's compact digital still camera using the type1/2.7CCD) and the PowerShot S100 DIGITAL ELPH, along with their respective vulnerabilities to eccentricity within the group. The vertical axis shows the amount of image plane curvature in the meridional image plane for an image with a height that is 70% of the diagonal measurement of the imaging plane. In other words, the length of this bar graph effectively indicates the likelihood that blurring will occur around the edges of the image.

In the PowerShot A5 Zoom, groups G4 to G6 are made up of an independent three-lens configuration in which one concave lens is sandwiched between two convex lenses in what is known as a triplet-type configuration. Although this is not an unusual configuration, any shift in the relative positions of the respective lens elements will have a major impact on optical performance, as the bar graph in Figure

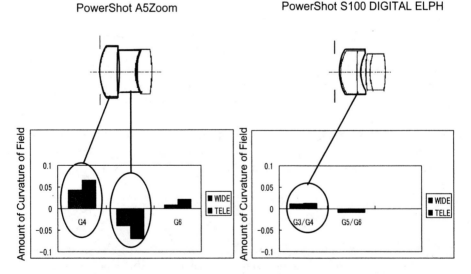

FIGURE 2.19 (See Color Figure 2.19 following page 178.) Comparison of decenter sensitivity.

2.19 shows. The configuration in the PowerShot S100 DIGITAL ELPH uses two pairs of joined lens elements (G3-G4 and G5-G6). This configuration takes the triplet-type configuration used in the PowerShot A5 Zoom and links the convex and concave lenses together. The measure introduced here was to reduce the vulnerability of the lenses that were particularly sensitive to changes in their relative positions by connecting together the lenses involved. Accordingly, we can see that the second lens group in the PowerShot S100 DIGITAL ELPH has far lower vulnerability to eccentricity than previous models and has very highly stable manufacturing characteristics.

Among the current DSCs, the cameras with the smallest pixel pitch in the imaging elements have now reached the 2-μm level, so even very slight manufacturing errors will have a major impact on the final optical performance. For this reason, from the design stage on down, it is vital that careful thought be given to manufacturing quality issues such as mass-production performance.

2.4 DSC IMAGING LENS ZOOM TYPES AND THEIR APPLICATIONS

At present, the zoom lenses for DSCs are not divided into wide-angle zooms and telephoto zooms, as is the case for the interchangeable lenses for SLR cameras; all fall into the standard zoom range. Nevertheless, a wide range of options is available, with number of pixels supported ranging from the submegapixel level through to several megapixels, and the zoom magnifications also ranging from two times up to ten times or higher. Thus, it is necessary to choose the zoom type that best matches the camera specifications. The illustration in Figure 2.20 shows the correlation between zoom types and the zoom magnifications and their supported pixel counts.

Bear in mind that the classifications used here are for the sake of convenience and are not intended to be strictly applied. The following sections will look at the characteristics of the respective zoom types.[7]

2.4.1 VIDEO ZOOM TYPE

In this type of zoom, the first lens group is a relatively high-powered convex group, and almost all the magnification during zooming is normally done by the relatively powerful concave second group. In total, the optics most often consist of four lens groups with the final convex group used for focusing, although this type of lens can also be configured with five or six lens groups.

This zoom type shows little variation in the angle at which light exits from the lens during zooming and is the type best suited to high-magnification applications. Also, by remaining fairly flexible in the allocation of power in the groups and increasing the number of lens elements used in the configuration, these lenses can be designed so that their optical performance can cope with large numbers of pixels. However, because it is difficult to reduce the total number of lens elements in the configuration and the diameter of the first group is large, it is difficult to reduce the lens size; they have a tendency to increase in cost. All of the 10× zoom lenses currently on the market are of this type.

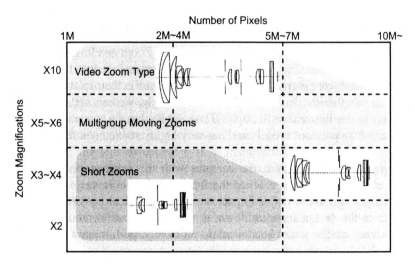

FIGURE 2.20 (See Color Figure 2.20 following page 178.) The correlation between zoom types and the zoom magnifications and their supported pixel counts.

2.4.2 MULTIGROUP MOVING ZOOMS

These are zoom lenses with four or five lens groups that do not fall into the preceding category. In these lenses, the first group may be convex or concave, depending on the specifications and aims of the lens. Lenses of this type can be miniaturized to some extent; however, they are also capable of achieving the twin goals of zoom magnifications at the four-times or five-times level and high performance capable of supporting large numbers of pixels. If the first group is a relatively weak convex group, they can be designed with a relatively bright F-number (around F2). If the first lens group is concave, they are ideal for zoom lenses that begin at a wide-angle setting. However, these lenses are not suitable for ultracompact applications, and in terms of cost they are looked upon as being towards the high end of the scale. Of the zoom lenses currently on the market, those at around F2 with a magnification of between 3× and 4× are probably of this type.

2.4.3 SHORT ZOOMS

These are the type of zoom that is currently most widely used as the imaging optics for compact DSCs. They are generally composed of a concave first group, a convex second group that takes care of most of the changes in magnification, and a convex third group that handles the focusing. Short zooms with a two-group configuration have also been around for a while. Recently, lenses have appeared on the market that consist of four lens groups and fall somewhere between short zooms and the multigroup moving zooms discussed previously.

An aspherical lens is normally included in the first group to correct distortion, but when the first group is composed entirely of spherical lens elements, the first lens (the lens closest to the object) is made convex. At least one aspherical lens element is normally used in the second group. This is the zoom type most suited to ultracompact applications because it is the most amenable to reductions in the number of component lens elements. However, this type of zoom is not capable of particularly high magnifications due to the large variations in the angle of light exiting the lens during zooming and the large changes in the F-number. The current limit on the magnification for short zooms is around 4×.

2.5 CONCLUSION

In this chapter, I have discussed imaging optics which, if we look at the components of a DSC in human terms, would represent the eyes of the camera. Of course, as far as the quality of the image finally produced by a DSC goes, the characteristics of the imaging element, which acts like a net to catch the light, and the image processor, which is the brain of the camera, make a huge difference in terms of the MTF, the amount of light close to the optical axis (center brightness), and the peripheral brightness. Thus, even when the same imaging optics are used, the quality and character of the final image may be very different. Nonetheless, just as a person who is near or far sighted is unable to see the details of an object, there is no doubt that the capacity of the imaging optics that provide the initial input is crucially important to the quality of the final image on a DSC.

In recent years, increasing miniaturization of the pixel pitch in imaging elements has brought us to the threshold of 2-µm resolutions. However, as discussed in Section 2.2.4, as long as we give sufficient thought to the size of the F-number (and do not reduce the aperture size too much), we can probably build optical systems that will cope with imaging elements that have a pixel pitch of as little as 2 µm or thereabouts.

In the future, with the emergence of new optical materials and the effective use of new technologies such as diffraction, we can look forward to further advances in DSC imaging optics. Cameras are irreplaceable tools for capturing memorable moments in life. As optical engineers, it is our job to actively cultivate new ideas that will make DSCs a more integral part of our lives and allow us to continue providing the world with new optical systems.

REFERENCES

1. EF LENS WORK III (Canon Inc.), 205–206, 2003.
2. T. Koyama, The lens for digital cameras, *ITE 99 Proc.*, 392–395, 1999.
3. K. Murata, *Optics* (Saiensu-sha), 178–179, 1979.
4. M. Sekita, Optical design of IXY DIGITAL, *J. Opt. Design Res. Group*, 23, 51–56, 2001.
5. I. Ogura, *The Story of Camera Development in Japan* (Asahi Sensho 684), 72–77, 2001.
6. T. Koyama, *J. Inst. Image Inf. TV Eng.*, 54(10), 1406–1407, 2000.
7. T. Koyama, Optical systems for camera (3), Optronics, (11), 185–190, 2002.

3 Basics of Image Sensors

Junichi Nakamura

CONTENTS

3.1 Functions of an Image Sensor 55
    3.1.1 Photoconversion 55
    3.1.2 Charge Collection and Accumulation 56
    3.1.3 Scanning of an Imaging Array 56
        3.1.3.1 Charge Transfer and X–Y Address 56
        3.1.3.2 Interlaced Scan and Progressive Scan 59
    3.1.4 Charge Detection 59
        3.1.4.1 Conversion Gain 60
3.2 Photodetector in a Pixel 61
    3.2.1 Fill Factor 61
    3.2.2 Color Filter Array 62
    3.2.3 Microlens Array 63
    3.2.4 Reflection at the SiO_2/Si Interface 64
    3.2.5 Charge Collection Efficiency 65
    3.2.6 Full-Well Capacity 66
3.3 Noise 66
    3.3.1 Noise in Image Sensors 66
    3.3.2 FPN 67
        3.3.2.1 Dark Current 68
        3.3.2.2 Shading 72
    3.3.3 Temporal Noise 72
        3.3.3.1 Thermal Noise 73
        3.3.3.2 Shot Noise 74
        3.3.3.3 1/f Noise 74
        3.3.3.4 Temporal Noise in Image Sensors 74
        3.3.3.5 Input Referred Noise and Output Referred Noise 77
    3.3.4 Smear and Blooming 77
    3.3.5 Image Lag 77
3.4 Photoconversion Characteristics 78
    3.4.1 Quantum Efficiency and Responsivity 78
    3.4.2 Mechanics of Photoconversion Characteristics 79
        3.4.2.1 Dynamic Range and Signal-to-Noise Ratio 79
        3.4.2.2 Estimation of Quantum Efficiency 81
        3.4.2.3 Estimation of Conversion Gain 81

		3.4.2.4	Estimation of Full-Well Capacity	82
		3.4.2.5	Noise Equivalent Exposure	83
		3.4.2.6	Linearity	83
		3.4.2.7	Crosstalk	83
	3.4.3		Sensitivity and SNR	84
	3.4.4		How to Increase Signal-to-Noise Ratio	84
3.5	Array Performance			85
	3.5.1		Modulation Transfer Function (MTF)	85
	3.5.2		MTF of Image Sensors, MTF_{Imager}	86
	3.5.3		Optical Black Pixels and Dummy Pixels	87
3.6	Optical Format and Pixel Size			88
	3.6.1		Optical Format	88
	3.6.2		Pixel Size Considerations	89
3.7	CCD Image Sensor vs. CMOS Image Sensor			90
References				90

A solid-state image sensor, also called an "imager," is a semiconductor device that converts an optical image that is formed by an imaging lens into electronic signals, as illustrated in Figure 3.1. An image sensor can detect light within a wide spectral range, from x-rays to infrared wavelength regions, by tuning its detector structures and/or by employing material that is sensitive to the wavelength region of interest. Moreover, some image sensors can reproduce an "image" generated by charged particles, such as ions or electrons. However, the focus of this chapter is on "visible" imaging, corresponding to the spectral response of the human eye (from 380 to 780 nm). Silicon, the most widely used material for very large-scale integrated circuits (VLSIs), is also suitable for visible-image sensors because the band gap energy of silicon matches the energy of visible wavelength photons.

To reproduce an image with acceptable resolution, a sufficient number of picture elements or "pixels" must be arranged in rows and columns. These pixels convert

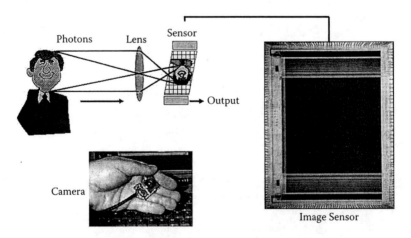

FIGURE 3.1 (See Color Figure 3.1 following page 178.) Image sensor.

Basics of Image Sensors

the incoming light into a signal charge (electrons or holes, depending on the pixel's structure).

The number of applications using image sensors is growing rapidly. For example, they are now found in state-of-the-art cellular phones, making the phones serious competition for the low-end digital still camera (DSC) market. It is also expected that they will make serious inroads into the automotive industry, which will become a major market for electronic cameras. In fact, some more expensive vehicles already have built-in cameras.

Until recently, the charged-coupled device (CCD) image sensor was the technology of choice for DSC applications. However, complementary metal-oxide semiconductor (CMOS) image sensors are rapidly replacing CCDs in low-end camera markets such as toy cameras and PC cameras. In addition, large-format CMOS image sensors have been used for higher end digital single lens reflex (DSLR) cameras.

Obviously, image sensors for DSC applications must produce the highest possible image quality. This is achieved through high resolution, high sensitivity, a wide dynamic range, good linearity for color processing, and very low noise. Also, special operation modes such as fast readout for auto exposure, auto focus, and auto white balance are needed, as well as viewfinder and video modes. CCD and CMOS image sensors are discussed in detail in Chapter 4 and Chapter 5, respectively; therefore, this chapter describes overall image sensor functions and performance parameters for both technologies.

3.1 FUNCTIONS OF AN IMAGE SENSOR

3.1.1 PHOTOCONVERSION

When a flux of photons enters a semiconductor at energy levels that exceed the semiconductor's band gap energy, E_g, such that

$$E_{photon} = h \cdot \nu = \frac{h \cdot c}{\lambda} \geq E_g \tag{3.1}$$

where h, c, ν, and λ are Planck's constant, the speed of light, the frequency of light, and the wavelength of light, respectively, the number of photons absorbed in a region with thickness dx is proportional to the intensity of the photon flux $\Phi(x)$, where x denotes the distance from the semiconductor surface. Because the band gap energy of silicon is 1.1 eV, light with wavelengths shorter than 1100 nm is absorbed and photon-to-signal charge conversion takes place. On the other hand, silicon is essentially transparent to photons with wavelengths longer than 1100 nm.

The continuity of photon flux absorption yields the following relationship[1]:

$$\frac{d\Phi(x)}{dx} = -\alpha \cdot \Phi(x) \tag{3.2}$$

where α is the absorption coefficient and depends on wavelength. Solving this equation with the boundary condition $\Phi(x = 0) = \Phi_0$, one can obtain

$$\Phi(x) = \Phi_0 \cdot \exp(-\alpha x) \tag{3.3}$$

Thus, the photon flux decays exponentially with the distance from the surface. The absorbed photons generate electron-hole pairs in the semiconductor with densities that follow Equation 3.3.

Figure 3.2 shows the absorption coefficient of silicon[2] and how the flux is absorbed. The figure shows that the penetration depth, $1/\alpha$, the depth at which the flux decays to $1/e$, for blue light ($\lambda = 450$ nm) is only 0.42 µm, while that of red light ($\lambda = 600$ nm) reaches 2.44 µm.

3.1.2 Charge Collection and Accumulation

This section outlines how the generated signal charge is collected at a charge accumulation area inside a pixel. Figure 3.3 illustrates a simple photodiode as a charge collection device. In this example, the p region is grounded and the n+ region is first reset at a positive voltage, V_R. It then becomes electrically floating with the reverse bias condition being held. Electrons excited by photons tend to collect at the n+ region, reducing this region's potential; holes flow to the ground terminal. In this case, the electrons are the signal charge. All CCD and CMOS image sensors for DSC applications operate in this charge-integrating mode first proposed by G. Weckler in 1967.[3]

Figure 3.4 shows another photoelement, a metal-oxide semiconductor (MOS) diode. When a positive voltage is applied to the gate electrode, the energy bands bend downward, and the majority of carriers (holes) are depleted. The depletion region is now ready to collect free charge. As described in Chapter 4, the MOS diode is a building block of the surface CCD. A buried channel MOS diode is used in the pixels of a frame transfer CCD (see Section 4.1.3 and Section 4.2.1).

In both cases (a reverse-biased photodiode and MOS diode), electrons generated inside the depletion region are fully utilized as signal charge. However, only a fraction of the electrons generated in the neutral region deep in the bulk can reach the depletion region through diffusion because no electric field exists at the neutral region; some of the electrons are lost by the recombination process before reaching the depletion region. This issue is revisited in Section 3.2.5.

3.1.3 Scanning of an Imaging Array

3.1.3.1 Charge Transfer and X-Y Address

The accumulated charge or the corresponding signal voltage or current must be read out from a pixel in an image sensor chip to the outside world. The signals distributed in two-dimensional space should be transformed to a time-sequential signal. This is

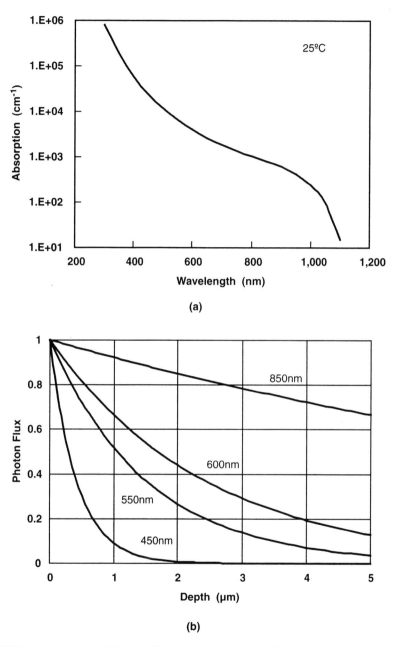

FIGURE 3.2 Absorption of light in silicon: (a) absorption coefficient; (b) intensity vs. depth.

called "scanning," and an image sensor should have this capability. Figure 3.5 shows two types of scanning schemes.

Several CCD readout architectures, such as the full-frame transfer (FFT), interline transfer (IT), frame transfer (FT), and frame-interline transfer (FIT) architectures, are

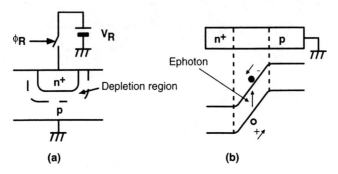

FIGURE 3.3 Reverse biased photodiode: (a) cross-sectional view; (b) energy band diagram.

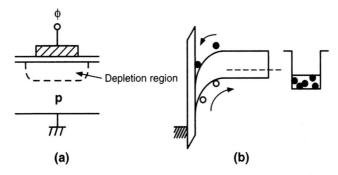

FIGURE 3.4 Reverse biased MOS diode: (a) cross-sectional view; (b) energy band diagram.

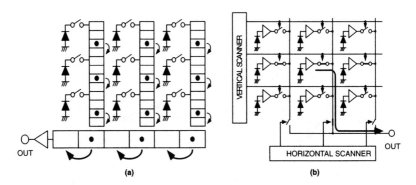

FIGURE 3.5 Imaging array scanning scheme: (a) charge transfer scheme; (b) X–Y address scheme.

discussed in Chapter 4. Figure 3.5(a) illustrates the IT CCD charge transfer scheme, in which each signal charge stored at a pixel's photodiode is shifted to a vertical CCD (V-CCD) simultaneously over the entire imaging array and is then transferred from the V-CCD to the horizontal CCD (H-CCD). The charge within the H-CCD is transferred to the output amplifier, which converts it into a voltage signal. The charge transfer readout scheme requires almost perfect charge transfer efficiency, which, in turn, requires highly tuned semiconductor structures and process technology.

Basics of Image Sensors

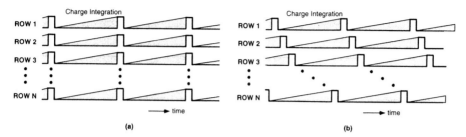

FIGURE 3.6 Operational timing of CCD and CMOS image sensors: (a) CCD image sensor; (b) CMOS image sensor.

Figure 3.5(b) shows an X–Y addressing scheme used in CMOS image sensors. In most of them, the signal charge is converted to a voltage or a current by an active transistor inside a pixel. As the X–Y address name suggests, a pixel signal addressed by a vertical scanner (a shift register or a decoder) selects a row (Y) to be read out, and a horizontal scanner selects a column (X) to be read out. As is apparent when comparing the two diagrams, the X–Y addressing scheme is much more flexible in realizing several readout modes than the charge transfer scheme.

Because CCD and CMOS image sensors are charge-integrating types of sensors, the signal charge on a pixel should be initialized or reset before starting the charge integration. The difference in scanning schemes results in operational timing differences, as shown in Figure 3.6. In the CCD image sensor, the charge reset is done by transferring the charge from a photodiode to a V-CCD. This action occurs at the same time over the entire pixel array. Alternately, the charge reset and the signal readout occur on a row-by-row basis in most CMOS image sensors.

3.1.3.2 Interlaced Scan and Progressive Scan

In conventional color television systems, such as National Television Systems Committee (NTSC), Phase Alternating Line (PAL), and Sequential Couleur Avec Memoire (SECAM), the interlace scanning mode is used in which half of the total lines (rows) are scanned in one vertical scan and the other half are scanned in a second vertical scan. Each vertical scan forms a "field" image and a set of two fields forms a single "frame" image, as shown in Figure 3.7(a).

Figure 3.7(b) shows a progressive scanning mode, which matches the scanning scheme of a PC monitor. Although progressive scan is preferable for DSC applications, the V-CCD structure becomes complicated, and thus it is more difficult to keep the photodiode area sufficiently large in CCDs. (This issue is addressed in Section 4.3.2 in Chapter 4.) CMOS image sensors for DSC applications operate in the progressive scanning mode.

3.1.4 CHARGE DETECTION

The charge detection principle is basically identical for CCD image sensors and most CMOS image sensors. As illustrated in Figure 3.5, CCD image sensors perform charge detection at an output amplifier, and CMOS image sensors accomplish it inside the pixel. Figure 3.8 shows a conceptual diagram of the charge detection

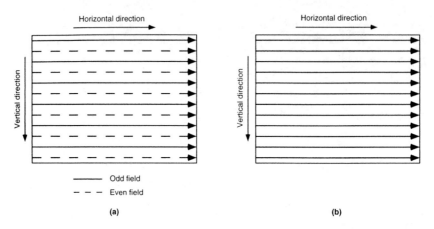

FIGURE 3.7 Interlaced scan and progressive scan: (a) interlaced scan; (b) progressive scan.

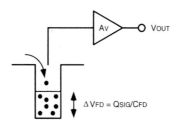

FIGURE 3.8 Charge detection scheme.

principle. Signal charge, Q_{sig}, is fed into a potential well, which is monitored by a voltage buffer. The potential change, ΔV_{FD}, caused by the charge is given by

$$\Delta V_{FD} = \frac{Q_{sig}}{C_{FD}} \tag{3.4}$$

where C_{FD} is the capacitance connecting to the potential well and acts as the charge-to-voltage conversion capacitance. The output voltage change is given by

$$\Delta V_{OUT} = A_V \cdot \Delta V_{FD} \tag{3.5}$$

where A_V represents the voltage gain of the voltage buffer.

3.1.4.1 Conversion Gain

Conversion gain (μV/e$^-$) expresses how much voltage change is obtained by one electron at the charge detection node. From Equation 3.4 conversion gain is obtained as

$$C.G. = \frac{q}{C_{FD}} \quad [\mu\text{V/electron}] \tag{3.6}$$

Basics of Image Sensors

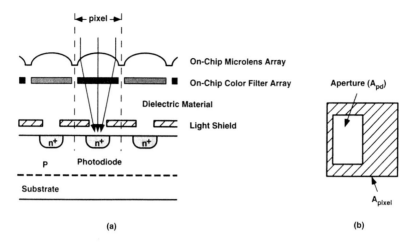

FIGURE 3.9 A simplified pixel structure: (a) cross-sectional view; (b) plane view.

where q is the elementary charge (1.60218×10^{19} C). Obviously, Equation 3.6 represents the "input referred" conversion gain, and it is not measured directly. The "output referred" conversion gain is obtained by multiplying the voltage gain from the charge detection node to the output and is given by

$$C.G._{output\_referred} = A_V \cdot \frac{q}{C_{FD}} \qquad (3.7)$$

The most commonly used charge detection scheme is floating diffusion charge detection.[4] The charge detection is performed in a CCD image sensor by a floating diffusion structure located at the end of the horizontal CCD register; in CMOS active-pixel sensors (APSs), it is performed inside a pixel. In combination with correlated double sampling (CDS) noise reduction,[5] extremely low-noise charge detection is possible. These schemes in CCD image sensors are addressed in Section 4.1.5 in Chapter 4, and those in CMOS image sensors are described in Section 5.1.2.1 and Section 5.3.1 in Chapter 5.

3.2 PHOTODETECTOR IN A PIXEL

A simplified pixel structure is shown in Figure 3.9. The following chapters provide details of pixel structures in CCD and CMOS image sensors.

3.2.1 FILL FACTOR

Fill factor (FF) is defined as the ratio of the photosensitive area inside a pixel, A_{pd}, to the pixel area, A_{pix}. That is,

$$\text{Fill factor} = (A_{pd} / A_{pix}) \times 100 \ [\%] \qquad (3.8)$$

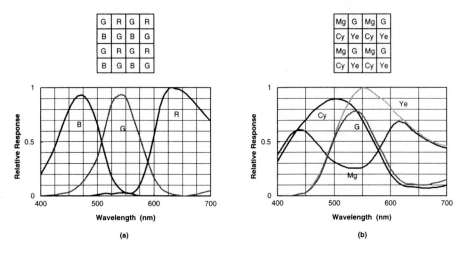

FIGURE 3.10 Color filter arrangement and spectral transmittance: (a) Bayer primary color filter pattern and its responses; (b) CMY complementary color filter pattern and its responses.

Without an on-chip microlens, it is defined by the aperture area not covered with a light shield in typical IT and FIT CCDs. In IT and FIT CCDs, the portion of the pixel covered with the light shield includes the area that holds a transfer gate, a channel stop region that isolates pixels, and a V-CCD shift register. The FF of an FT CCD is determined by the nonphotosensitive channel-stop region separating the V-CCD transfer channels and the CCD gate clocking.

In active-pixel CMOS image sensors, at least three transistors (a reset transistor, a source follower transistor, and a row select transistor) are needed and are covered by a light shield. If more transistors are used, the FF degrades accordingly. The area required for them depends on the design rules (feature sizes) of the process technology used.

The microlens condenses light onto the photodiode and effectively increases the FF. The microlens plays a very important role in improving light sensitivity on CCD and CMOS image sensors.

3.2.2 Color Filter Array

An image sensor is basically a monochrome sensor responding to light energies that are within its sensitive wavelength range. Thus, a method for separating colors must be implemented in an image sensor to reproduce an image of a color scene. For consumer DSC applications, an on-chip color filter array (CFA) built above the photodiode array provides a cost-effective solution for separating color information and meeting the small size requirements of DSCs.* Figure 3.10 shows two types of color filter arrays and their spectral transmittances.

---

*In some high-end video cameras, three (sometimes four) separate image sensors are used. They are attached on a dichroic prism with each image sensor detecting a primary color. This configuration is usually used for high-end video applications.

Basics of Image Sensors

DSC applications mainly use the red, green, and blue (RGB) primary color filter array. RGB CFAs have superior color reproduction and higher color signal-to-noise ratio (SNR), due to their superior wavelength selectivity properties.

The most commonly used primary color filter pattern is the "Bayer" pattern, as shown in Figure 3.10(a). Proposed by B.E. Bayer,[6] this pattern configuration has twice as many green filters as blue or red filters. This is because the human visual system derives image details primarily from the green portion of the spectrum. That is, luminance differences are associated with green whereas color perception is associated with red and blue.

Figure 3.10(b) shows the CMY complementary color filter pattern consisting of cyan, magenta, and yellow color filters. Each color is represented in the following equation:

$$Ye = R + G = W - B$$

$$Mg = R + B = W - G \quad (3.9)$$

$$Cy = G + B = W - R$$

$$G = G$$

The transmittance range of each complementary color filter is broad, and higher sensitivity can be obtained compared to RGB primary color filters. However, converting complementary color components to RGB for display can cause a reduction in S/N due to subtraction operations. Also, color reproduction quality is usually not as accurate as that found in RGB primary filters.

Material for on-chip color filter arrays falls into two categories: pigment and dye. Pigment-based CFAs have become the dominant option because they offer higher heat resistance and light resistance compared to dye-based CFAs. In either case, thicknesses ranging from submicron to 1 µm are readily available.

3.2.3 MICROLENS ARRAY

An on-chip microlense collimates incident light to the photodiode. The on-chip microlens array (OMA) was first introduced on an IT CCD in 1983.[7] Its fabrication process is as follows: first, the surface of the color filter layer is planarized by a transparent resin. Next, the microlens resin layer is spin-coated on the planarization layer. Last, photolithographic patterning is applied to the resin layer, which is eventually shaped into a dome-like microlens by wafer baking.

Recent progress in reducing pixel size and increasing the total number of pixels has been remarkable. However, sensitivity becomes poorer as a pixel shrinks. This can be countered by a simple on-chip microlens array, but it produces shading due to the positional dependence of incident light angles from an imaging lens to the image sensor, as illustrated in Figure 3.11.

Decreasing the distance between the microlens and the photodiode surface reduces this angular response dependency.[8,9] A technique has also been introduced

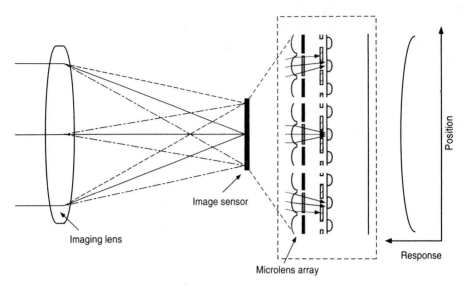

FIGURE 3.11 Shading caused by positional dependence of incident light angle.

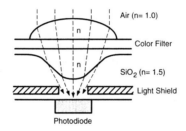

FIGURE 3.12 Double-layer microlens.

in which the positions of the microlenses at the periphery of the imaging array are shifted to correct shading.[10,11] FT CCDs provide a wider angular response than IT CCDs because the former have inherently larger fill factors than the latter.[12]

To increase the light-collection efficiency even further, the gap between each microlens has been reduced.[13,14] Also, a double-layer microlens structure, which has an additional "inner" microlens beneath the conventional "surface" microlens, as shown in Figure 3.12, has been developed.[15] The inner microlens improves the angular response, especially when smaller lens F-numbers are used, as well as smaller pixel sizes.[16] In addition to enhancing sensitivity, the microlens helps reduce smear in CCD image sensors and crosstalk between pixels caused by minority carrier diffusion in CCD and CMOS image sensors.[11,17]

3.2.4 REFLECTION AT THE SiO₂/Si INTERFACE

Incident light is reflected at the interface of two materials when the refractive indices are different. Reflectivity (R) of light rays that are incident perpendicular to the materials is given by[18]

$$R = \left(\frac{n_1 - n_2}{n_1 + n_2}\right)^2 \quad (3.10)$$

Thus, with the refractive indices of 1.45 for SiO_2 and 3 to 5 for Si, more than 20 to 30% of the incident light is reflected at the silicon surface in the visible light range (400 to 700 nm). To reduce the reflection at the SiO_2/Si interface, antireflective films formed above the photodiode have been introduced. A 30% increase in photosensitivity was reportedly obtained with an antireflective film consisting of optimized $SiO_2/Si_3N_4/SiO_2$ layers.[19]

3.2.5 CHARGE COLLECTION EFFICIENCY

Although the upper structure above a detector has been the focus of this chapter, photoconversion at the detector must also be examined. A simplified pixel structure that uses a p+-substrate and a potential profile along the surface to the substrate is shown in Figure 3.13(a). This structure corresponds to the photodiode structure shown in Figure 3.3(a). Because the p+-substrate wafer is widely used for CMOS VLSI or memory devices such as dynamic random access memory (DRAM), most CMOS image sensors, in which signal-processing circuits can be integrated on-chip, use the p+-substrate. In this structure, the depth of the p-region and the minority carrier lifetime (or the diffusion length) of the p-region and the p+-substrate affect the response in long wavelength regions of the spectrum. In general, the response from red to NIR is much higher with this structure than that with an n-type substrate.

Figure 3.13(b) shows another detector structure, in which an n-substrate is used. The n-type substrate is biased at a positive voltage, and the p-region is grounded. This structure is commonly used for IT CCDs and is also an option for CMOS image sensors. In it, electrons generated deep below a depth, x_p, are swept away to the n-

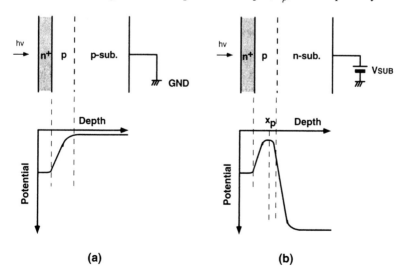

FIGURE 3.13 n+-p photodiode on different substrate types: (a) P-type substrate; (b) N-type substrate.

substrate and do not contribute to the signal. Thus, the spectral response at long (red to NIR) wavelengths is reduced.

From the preceding discussion, the charge collection efficiency, $\eta(\lambda)$, is defined by

$$\eta(\lambda) = \frac{Signal\ charge}{Photo-generated\ charge} \qquad (3.11)$$

Charge collection efficiency is determined by the substrate type, impurity profile, minority carrier lifetime in the bulk, and how the photodiode is biased. The p-substrate and n-substrate structures are discussed in Section 4.2.2 in Chapter 4 and Section 5.2.2 in Chapter 5.

3.2.6 Full-Well Capacity

The photodiode operates in the charge-integrating mode, as described in Section 3.1, and, therefore, has a limited charge handling capacity. The maximum amount of charge that can be accumulated on a photodiode capacitance is called "full-well capacity" or "saturation charge" and is given by

$$N_{sat} = \frac{1}{q} \int_{V_{reset}}^{V_{max}} C_{PD}(V) \cdot dV \quad [\text{electrons}] \qquad (3.12)$$

where C_{PD} and q are the photodiode capacitance and the charge of an electron, respectively. The initial and maximum voltages, V_{reset} and V_{max}, depend on photodiode structures and the operating conditions.

3.3 NOISE

3.3.1 Noise in Image Sensors

Table 3.1 summarizes noise components in image sensors. Noise deteriorates imaging performance and determines the sensitivity of an image sensor. Therefore, the term "noise" in image sensors may be defined as any signal variation that deteriorates an image or "signal."

An image sensor for still pictures reproduces two-dimensional image (spatial) information. Noise appearing in a reproduced image, which is "fixed" at certain spatial positions, is referred to as fixed-pattern noise (FPN). Because it is fixed in space, FPN at dark can be removed, in principle, by signal processing. Noise fluctuating over time is referred to as "random" or "temporal" noise. In this book, we use "temporal" when referring to noise fluctuating over time because "random" can also be associated with FPN; for example, "pixel-random" FPN is seen randomly in two-dimensional space.

Temporal noise is "frozen" as spatial noise when a snapshot is taken by a DSC so its peak-to-peak value can appear in a reproduced image. Although temporal noise is fixed spatially in a particular shot, it will vary in sequential shots. Temporal noise in

TABLE 3.1
Noise in Image Sensors

		Dark	Illuminated	
			Below saturation	Above saturation

Fixed Pattern Noise (FPN)		Dark signal nonuniformity
		  Pixel random
		  Shading
			Photo-response nonuniformity
			  Pixel random
			  Shading	

		Dark current nonuniformity (Pixel-wise FPN) (Row-wise FPN) (Column-wise FPN)		
		Defects		
Temporal Noise		Dark current shot noise	Photon shot noise	

		Read noise (Noise floor)
		  Amplifier noise, etc.
		  (Reset noise)		

				Smear, Blooming
		Image Lag		

video images, on the other hand, is more or less filtered out by the human eye, which cannot respond accurately within a field time (1/60 sec) or a frame time (1/30 sec).

Table 3.1 shows noise under dark conditions and under illumination. Under illumination, the noise components seen at dark still exist. The magnitude of FPN at dark and under illumination is evaluated as dark signal nonuniformity (DSNU) and photoresponse nonuniformity (PRNU). Also, smear and blooming are seen beyond the saturation level.

3.3.2 FPN

FPN at dark can be treated as an offset variation in the output signal and is evaluated as DSNU. FPN is also seen under illuminated conditions, where it is evaluated as PRNU. If the magnitude of the FPN is proportional to exposure, it is observed as sensitivity nonuniformity or gain variation.

The primary FPN component in a CCD image sensor is dark current nonuniformity. Although it is barely noticeable in normal modes of operation, it can be seen in images that have long exposure times or that were taken at high temperatures. If the dark current of each pixel is not uniform over the whole pixel array, the nonuniformity is seen as FPN because the correlated double sampling (CDS) cannot remove this noise component. In CMOS image sensors, the main sources of FPN are dark current nonuniformity and performance variations of an active transistor inside a pixel.

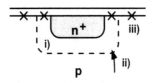

i) Generation in the depletion region
ii) Diffusion from a neutral bulk
iii) Generation at the surface of Si

FIGURE 3.14 Dark current component in a pixel.

In this section, only the dark current is discussed as a source of FPN. Other sources of FPN in CCD and CMOS image sensors are described in Section 4.2.4 in Chapter 4 and Section 5.1.2.3, Section 5.3.1, and Section 5.3.3 in Chapter 5.

3.3.2.1 Dark Current

Observed when the subject image is not illuminated, dark current is an undesirable current that is integrated as dark charge at a charge storage node inside a pixel. The amount of dark charge is proportional to the integration time and is represented by

$$N_{dark} = \frac{Q_{dark}}{q} = \frac{I_{dark} \cdot t_{INT}}{q} \qquad (3.13)$$

and is also a function of temperature.

The dark charge reduces the imager's useable dynamic range because the full well capacity is limited. It also changes the output level that corresponds to "dark" (no illumination). Therefore, the dark level should be clamped to provide a reference value for a reproduced image. Figure 3.14 illustrates three primary dark current components. Each will be examined and mechanisms of dark current generation discussed.

3.3.2.1.1 Generation Current in the Depletion Region

Silicon is an indirect-band-gap semiconductor in which the bottom of the conduction band and the top of the valence band do not occur at the same position along the momentum axis in an energy-momentum space. It is known that the dominant generation-recombination process is an indirect transition through localized energy states in the forbidden energy gap.[20] In the depletion region formed at the interface of a reverse-biased p–n junction, the minority carriers are depleted, and the generation processes (electron and hole emissions) become the dominant processes for restoring the system to equilibrium.*

---

* In a forward-biased diode, the minority carrier density is above the level of the equilibrium, so the recombination processes take place. In the equilibrium condition (zero bias diode), the recombination and the generation are balanced to maintain the relationship of $p \cdot n = n_i^2$, where p, n, and n_i denote the electron density, hole density, and intrinsic carrier density, respectively.

From the Shockley–Read–Hall theory,[21,22] the rate of electron–hole pair generation under the reverse bias condition can be expressed as

$$G = \left\{ \frac{\sigma_p \sigma_n v_{th} N_t}{\left[\sigma_n \exp\left(\frac{E_t - E_i}{kT} \right) + \sigma_p \exp\left(\frac{E_i - E_t}{kT} \right) \right]} \right\} n_i \quad (3.14)$$

where

σ_n = electron capture cross section
σ_p = hole capture cross section
v_{th} = thermal velocity
N_t = concentration of generation centers
E_t = energy level of the center
E_i = intrinsic Fermi level
k = Boltzmann's constant
T = absolute temperature

Assuming $\sigma_n = \sigma_p = \sigma_0$, Equation 3.14 can be rewritten as

$$G = \frac{(v_{th}\sigma_0 N_t) \cdot n_i}{2\cosh\left(\frac{E_t - E_i}{kT} \right)} = \frac{n_i}{\tau_g} \quad (3.15)$$

This equation indicates that only the energy states near midgap contribute to the generation rate, and the generation rate reaches a maximum value at $E_t = E_i$ and falls off exponentially as E_t moves from E_i. The generation lifetime and the generation current are thus given by

$$\tau_g = \frac{2\cosh\left(\frac{E_t - E_i}{kT} \right)}{v_{th}\sigma_0 N_t} \quad (3.16)$$

$$J_{gen} = \int_0^W qG\,dx \approx qGW = \frac{qn_i W}{\tau_g} \quad (3.17)$$

3.3.2.1.2 Diffusion Current

At the edges of the depletion regions, the minority carrier density is lower than that of equilibrium and it approaches the equilibrium density of n_{p0} in the neutral bulk region by the diffusion process. Here, our interest is the behavior of the minority carriers (electrons) in the p region. The continuity equation in the neutral region is given by

$$\frac{d^2 n_p}{dx^2} - \frac{n_p - n_{p0}}{D_n \tau_n} = 0 \qquad (3.18)$$

where D_n and τ_n denote the diffusion coefficient and the minority carrier lifetime, respectively. Solving this equation with a boundary condition of $n_p (x = \text{infinite}) = n_{p0}$ and $n(0) = 0$, yields the diffusion current as

$$J_{diff} = \frac{q D_n n_{p0}}{L_n} = q \sqrt{\frac{D_n}{\tau_n}} \cdot \frac{n_i^2}{N_A} \qquad (3.19)$$

3.3.2.1.3 Surface Generation

Because of the abrupt discontinuity of the lattice structure at the surface, a much higher number of energy states or generation centers tends to be created. A discussion similar to the one on generation current can be applied to the surface generation current, which is expressed as

$$J_{surf} = \frac{q S_0 n_i}{2} \qquad (3.20)$$

where S_0 is the surface generation velocity.[23]

3.3.2.1.4 Total Dark Current

Based on previous discussion, the dark current, J_d, is expressed as

$$J_d = \frac{q n_i W}{\tau_g} + q \sqrt{\frac{D_n}{\tau_n}} \cdot \frac{n_i^2}{N_A} + \frac{q S_0 n_i}{2} \quad [\text{A/cm}^2] \qquad (3.21)$$

Among these three major components, using typical values at room temperature for the parameters, it can be shown that $J_{surf} \gg J_{gen} \gg J_{diff}$.

However, the surface component can be suppressed by making an inversion layer at the surface of the n region. The inversion empties the midgap levels of electrons through recombination with the holes, thus reducing the chance for electrons trapped in the midgap levels to emit to the conduction band. This is accomplished by introducing a pinned photodiode structure[24] that is used in most IT and FIT CCD and CMOS image sensors (see Section 4.2.3 in Chapter 4 and Section 5.2.2 in Chapter 5).

3.3.2.1.5 Temperature Dependence

As seen in Equation 3.17, Equation 3.19, and Equation 3.20, the generation current and the surface generation current are proportional to n_i, the intrinsic carrier density, and the diffusion current is proportional to n_i^2. Because

$$n_i^2 \propto T^3 \cdot \exp\left(-\frac{E_g}{kT}\right) \qquad (3.22)$$

the temperature dependence of dark current is expressed as

$$I_d = A_{d,gen} \cdot T^{3/2} \cdot \exp\left(-\frac{E_g}{2kT}\right) + B_{d,diff} \cdot T^3 \cdot \exp\left(-\frac{E_g}{kT}\right) \qquad (3.23)$$

where $A_{d,gen}$ and $B_{d,diff}$ are coefficients. Figure 3.15 illustrates a typical temperature dependence of the dark current. In real devices, the temperature dependence of total dark current varies, depending on the magnitude of the coefficients, $A_{d,gen}$ and $B_{d,diff}$. Also, the temperature dependence is expressed as $\exp(-E_g/nkT)$, where n is between 1 and 2 and E_g/n corresponds to the activation energy of the dark current.

3.3.2.1.6 White Spot Defects

As design and process technologies have progressed, dark currents have decreased to very low levels. Therefore, pixels that have extremely high dark currents with an extra generation center become visible as a white spot defect. These defects determine the quality of the image sensor. The causes of white spot defects include

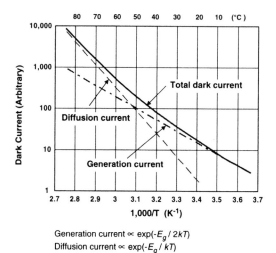

FIGURE 3.15 Temperature dependence of dark current.

contamination by heavy metals, such as gold, nickel, cobalt, etc., and crystal defects induced by stress during fabrication.[25]

3.3.2.1.7 Dark Current from CCD Registers

So far, we have focused on the dark currents generated inside a pixel, but dark currents are also generated in the CCD transfer channels of CCD image sensors. As described in Section 4.2.3 in Chapter 4, a negative voltage is applied momentarily to appropriate CCD gates to reduce surface-oriented dark currents. The technique is called "valence band pinning"; a negative voltage inverts the surface for a short period of time, creating a hole layer. Valence band pinning empties the generation centers at the surface, and it takes some time for the centers to start to generate again.[26] A detailed description of the process is found in Section 4.1.3.

3.3.1.2.8 Dark Current from a Transistor in an Active Pixel of a CMOS Image Sensor

In CMOS image sensors, it is reported that an additional dark current component originates at the active transistor inside a pixel. This is due to hot-carrier effects in the high electric field region near the drain end of an amplification transistor.[27,28] A careful pixel layout and proper transistor length and bias setting are required to suppress this dark current component.

3.3.2.2 Shading

Shading is a slowly varying or low spatial frequency output variation seen in a reproduced image. The main sources of shading in CCD/CMOS image sensors include:

- Dark-current-oriented shading: if a local heat source exists, the resultant thermal distribution in an imaging array produces dark current gradients.
- Microlens-oriented shading: if the light collection efficiency of a microlens at the periphery of an imaging array is reduced due to an inclined light ray angle, the output of pixels at the periphery decreases (see Figure 3.11).
- Electrical-oriented shading: in CCD image sensors, the amplitude of driving pulses to V-CCDs may change spatially due to resistance of the poly-Si gates that carry the driving pulses. This may cause local degradation of charge transfer efficiency, thus yielding shading.

In CMOS image sensors, nonuniform biasing and grounding may cause shading.

3.3.3 TEMPORAL NOISE

Temporal noise is a random variation in the signal that fluctuates over time. When a signal of interest fluctuates around its average, which is assumed to be constant, the variance is defined as

Basics of Image Sensors

$$\text{Variance} = <(N - <N>)^2> = <N^2> - <N>^2 \quad (3.24)^*$$

where $<>$ expresses the ensemble average or statistical average, which is an average of a quantity at time, t, from a set of samples. When the system is "ergodic" or stationary, an average over time from one sample is considered equal to the ensemble average.

The variance of a signal corresponds to the total noise power of the signal.** When several noise sources exist that are uncorrelated, the total noise power is given by

$$<n_{total}^2> = <\sum_{i=1}^{N} n_i^2> \quad (3.25)$$

From the central limit theorem, the probability distribution of a sum of independent, random variables tends to become Gaussian as the number of random variables being summed increases without limit. The Gaussian distribution is represented by

$$p(x) = \frac{1}{\sqrt{2\pi}\sigma} \exp\left[-\frac{(x-m)^2}{2\sigma^2}\right] \quad (3.26)$$

where m is the mean or average value and σ is the standard deviation or root mean square (rms) value of the variable x. In this case, the standard deviation, σ, can be used as a measure of temporal noise.

Three types of fundamental temporal noise mechanisms exist in optical and electronic systems: thermal noise, shot noise, and flicker noise. All are observed in CCD and CMOS image sensors.

3.3.3.1 Thermal Noise

Thermal noise comes from thermal agitation of electrons within a resistance. It is also referred to as Johnson noise because J.B. Johnson discovered the noise in 1928. Nyquist described the noise voltage mathematically using thermodynamic reasoning the same year. The power spectral density of the thermal noise in a voltage representation is given by

$$S_V(f) = 4kTR \quad [\text{V}^2/\text{Hz}] \quad (3.27)$$

where k is Boltzmann's constant; T is the absolute temperature; and R is the resistance.

---

* Hereafter, we will use an upper case N (or V) for an average and a lower case n (or v) for temporal noise.
** The square root of variance is the deviation.

3.3.3.2 Shot Noise

Shot noise is generated when a current flows across a potential barrier. It is observed in a thermionic vacuum tube and semiconductor devices, such as pn diodes, bipolar transistors, and subthreshold currents in a metal-oxide semiconductor (MOS) transistor. In CCD and CMOS image sensors, shot noise is associated with incident photons and dark current. A study of the statistical properties of shot noise shows that the probability that N particles, such as photons and electrons, are emitted during a certain time interval is given by the Poisson probability distribution, which is represented as

$$P_N = \frac{(\bar{N})^N \cdot e^{-\bar{N}}}{N!} \tag{3.28}$$

where N and $\bar{N}$ are the number of particles and the average, respectively. The Poisson distribution has an interesting property that the variance is equal to the average value, or

$$n_{shot}^2 = <(N-\bar{N})^2> = \bar{N} \tag{3.29}$$

The power spectral densities of thermal noise and shot noise are constant over all frequencies. This type of noise is called "white noise," an analogy to the white light that has a flat power distribution in the optical band.

3.3.3.3 1/f Noise

The power spectral density of $1/f$ (one-over-f) noise is proportional to $1/f^\gamma$, where γ is around unity. Obviously, the average over time of $1/f$ noise may not be constant. The output amplifier of CCD image sensors and the amplifier in a CMOS image sensor pixel suffer from $1/f$ noise at low frequencies. However, $1/f$ noise is mostly suppressed by correlated double sampling (CDS) as long as the CDS operation is performed in such a way that the interval between the two samples is short enough that the $1/f$ noise is considered an offset. Discussion of noise in the CDS process can be found in Section 5.3.3.1 in Chapter 5.

3.3.3.4 Temporal Noise in Image Sensors

3.3.3.4.1 Reset Noise or kTC Noise

When a floating diffusion capacitance is reset, noise called "reset" or "kTC" noise appears at the capacitance node when the MOS switch is turned OFF. This noise comes from the thermal noise of the MOS switch. Figure 3.16 shows an equivalent circuit of the reset operation. An MOS transistor is considered resistance during the ON period, and thermal noise appears, as shown in Equation 3.27. This noise is sampled and held by a capacitor. The resulting noise power can be calculated by integrating the thermal noise power over all frequencies, with R in Equation 3.27

Basics of Image Sensors

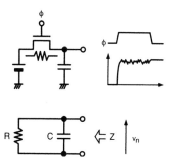

FIGURE 3.16 kTC noise.

replaced by the real part of the complex impedance of the RC low-pass filter, as follows:

$$v_n^2 = \int_0^\infty 4kT \cdot \frac{R}{1+(2\pi fRC)^2} \cdot df = \frac{kT}{C} \quad (3.30)$$

The noise charge is given by

$$q_n^2 = C^2 \cdot v_n^2 = kTC \quad (3.31)$$

It can be concluded that the noise is a function only of the temperature and the capacitance value, and thus is called kTC noise.

The kTC noise that appears in the floating diffusion amplifier in CCD image sensors can be suppressed by a CDS circuit. In CMOS image sensors, the kTC noise appears at the reset of the charge-detecting node. Suppressing kTC noise through CDS in CMOS sensors depends on the pixel's configuration, as described in Section 5.2 in Chapter 5.

3.3.3.4.2 Read Noise

Read noise, or noise floor, is defined as noise that comes from the readout electronics. Noise generated in a detector is not included. In CCD image sensors, the noise floor is determined by the noise generated by the output amplifier, assuming that the charge transfer in the CCD shift registers is complete. In CMOS image sensors, the noise floor is determined by the noise generated by readout electronics, including the amplifier inside a pixel.

In the noise model of an MOS transistor shown in Figure 3.17, two voltage noise sources, thermal noise and $1/f$ noise, are modeled in series with the gate. The thermal noise is represented by

$$\overline{v_{eq}^2} = \frac{4kT\alpha}{g_m} \cdot \Delta f \quad [V^2] \quad (3.32)$$

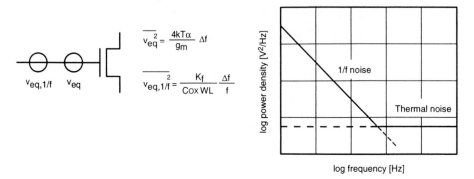

FIGURE 3.17 Noise in MOS transistor.

where g_m is the transconductance of an MOS transistor and α is a coefficient that depends on the modes of MOS transistor operation. The value of α is equal to 2/3 for long-channel transistors and a larger value for submicron transistors.

The 1/f noise is modeled as

$$\overline{v_{eq,1/f}^2} = \frac{K_f}{C_{ox}WL} \cdot \frac{\Delta f}{f} \quad [V^2] \tag{3.33}$$

where K_f is a process-dependent constant and C_{ox}, W, and L denote the gate capacitance per unit area, width, and length of the gate, respectively.[29]

The noise floor of an image sensor can be estimated using Equation 3.32 or Equation 3.33, depending on the specific amplifier configurations. If an image sensor has additional circuits, such as the gain amplifier and/or FPN suppression circuit often seen in CMOS image sensors, noise generated by those circuits should be added (using Equation 3.25). If the kTC noise mentioned earlier cannot be suppressed by the CDS process, this component should also be included in the read noise.

3.3.3.4.3 Dark Current Shot Noise and Photon Shot Noise

Referencing Equation 3.29, dark current shot noise and photon shot noise are given by

$$n_{dark}^2 = N_{dark} \tag{3.34}$$

$$n_{photon}^2 = N_{sig} \tag{3.35}$$

where N_{dark} and N_{sig} are the average of the dark charge given by Equation 3.13 and the amount of signal charge given by Equation 3.11, respectively.

Referencing Equation 3.25, the total shot noise under illumination is given by

Basics of Image Sensors

$$n_{shot\_total}^2 = N_{dark} + N_{sig} \tag{3.36}$$

3.3.3.5 Input Referred Noise and Output Referred Noise

Obviously, the previous discussion refers to "input referred" noise at the charge detection node, which is obtained from a measured "output referred" noise, in the manner discussed in Section 3.1.4, so that

$$v_{n,output}^2 = (A_V \cdot C.G.)^2 \cdot n_{n,pix}^2 + v_{n,sig\_chain}^2 \tag{3.37}$$

$$n_{n,input}^2 = n_{n,pix}^2 + \frac{v_{n,sig\_chain}^2}{(A_V \cdot C.G.)^2} \tag{3.38}$$

where $n_{n,pix}$ and v_{n,sig_chain} are noise generated at a pixel and noise voltage generated in a signal chain, respectively.

3.3.4 SMEAR AND BLOOMING

These phenomena occur when a very strong light illuminates a sensor device. Smear, which appears as white vertical stripes, occurs when stray light impinges on a V-CCD register or a charge generated deep in the silicon bulk diffuses into the V-CCD. Blooming occurs when the photogenerated charge exceeds a pixel's full-well capacity and spills over to neighboring pixels and /or to the V-CCD. To suppress blooming, an overflow drain should be implemented in a pixel.

Examples of smear are shown in Figure 4.28 in Chapter 4. For CCD image sensors, see Section 4.2.4.2 in Chapter 4, and for CMOS image sensors, see Section 5.3.3.4 in Chapter 5. Because smear noise can be considered signal chain noise (v_{n,sig_chain} in Equation 3.38) in CMOS image sensors, its contribution is effectively reduced by a factor of (C.G.·A_V). In CCD image sensors, smear noise directly deteriorates CCD image quality because A_v·C.G. in equations 3.37 and 3.38 is equal to 1. (See the architectural differences shown in Figure 3.5.)

3.3.5 IMAGE LAG

Image lag is a phenomenon in which a residual image remains in the following frames after the light intensity suddenly changes. Lag can occur if the charge transfer from the photodiode to the V-CCD in an IT CCD is not complete. In a CMOS image sensor with a four-transistor pixel (see Section 5.2.2), this can be caused by an incomplete charge transfer from the photodiode to the floating diffusion. In a CMOS sensor with a three-transistor pixel (see Section 5.2.1), its origin is the soft reset mode when the photodiode reset is performed in a subthreshold mode of an MOS transistor.

3.4 PHOTOCONVERSION CHARACTERISTICS

3.4.1 Quantum Efficiency and Responsivity

Overall quantum efficiency (QE) is given by

$$QE(\lambda) = N_{sig}(\lambda) / N_{ph}(\lambda) \qquad (3.39)$$

where N_{sig} and N_{ph} are the generated signal charge per pixel and the number of incident photons per pixel, respectively.

As described earlier, part of the incident photons are absorbed or reflected by upper structures above the photodiode. The microlens and photodiode structure (from the surface to the bulk) determine the effective FF and the charge collection efficiency, respectively. Thus, Equation 3.39 can be expressed as

$$QE(\lambda) = T(\lambda) \cdot FF \cdot \eta(\lambda) \qquad (3.40)$$

where $T(\lambda)$, FF, and $\eta(\lambda)$ are the transmittance of light above a detector, the effective FF, and the charge collection efficiency of the photodiode, respectively. N_{sig} and N_{ph} are represented by

$$N_{sig} = \frac{I_{ph} \cdot A_{pix} \cdot t_{INT}}{q} \qquad (3.41)$$

$$N_{ph} = \frac{P \cdot A_{pix} \cdot t_{INT}}{h\nu} \qquad (3.42)$$

where

I_{ph} is the photocurrent in [A/cm^2].
A_{pix} is the pixel size in [cm^2].
P is the optical input power in [W/cm^2].
t_{INT} is the integration time.
q is the electron charge.

Responsivity, $R(\lambda)$, is defined as the ratio of the photocurrent to the optical input power and is given by

$$R = \frac{I_{ph}[A/cm^2]}{P[W/cm^2]} = \frac{qN_{sig}}{h\nu N_{ph}} = QE \cdot \frac{q\lambda}{hc} \qquad (3.43)$$

Referencing Equation 3.43, spectral response can be represented two ways: using responsivity or quantum efficiency. An example is shown in Figure 3.18, in which

Basics of Image Sensors

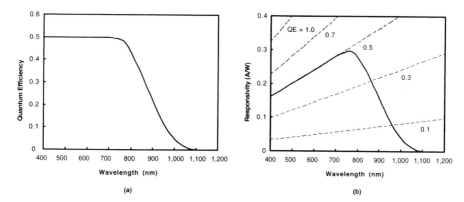

FIGURE 3.18 Spectral response: (a) spectral quantum efficiency; (b) spectral responsivity.

a virtual image sensor response with a constant QE value of 0.5 in the range of 400 to 700 nm, is assumed to highlight the differences between two representations. Also, the relative response, in which the response is normalized by its peak value, is often used. Overall color response is obtained by multiplying the color filter response shown in Figure 3.10 with the image sensor's response.

3.4.2 Mechanics of Photoconversion Characteristics

In this section, the photoconversion characteristics that demonstrate the relationship between the output voltage and exposure will be examined.

In DSC applications, exposure, using a standard light source, is most often expressed in lux-seconds. Because the procedure for estimating the number of incident photons coming from a standard light source is somewhat complicated, the photoconversion characteristics will be presented for monochrome light, for which the number of photons per incident light energy can be obtained easily and the mechanics of photoconversion characteristics analyzed. The method for estimating the number of incident photons from a standard light source is provided in Appendix A.

Figure 3.19 shows an example of photoconversion characteristics, illustrating signal, photon shot noise, and read noise (noise floor) as a function of incident photons.[30] To plot this figure, a virtual image sensor with a pixel size of 25 μm², C.G. = 40 μV/e⁻, a full-well capacity of 20,000 electrons, a noise floor of 12 electrons, and a detector's QE of 0.5 are assumed. Dark current shot noise is not included in the plot.

3.4.2.1 Dynamic Range and Signal-to-Noise Ratio

Dynamic range (DR) is defined as the ratio between the full-well capacity and the noise floor. Signal-to-noise ratio (SNR) is the ratio between the signal and the noise at a given input level. They are represented by

$$DR = 20 \log\left(\frac{N_{sat}}{n_{read}}\right) \text{ [dB]} \qquad (3.44)$$

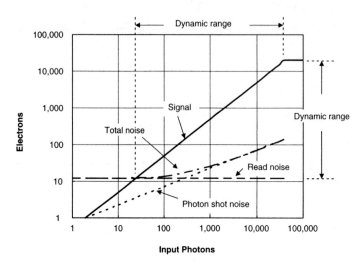

FIGURE 3.19 Example of photoconversion characteristics. $A_{pix} = 25$ μm²; C.G. = 40 μV/e⁻, $N_{sat} = 20{,}000$ e⁻, $n_{read} = 12$ e⁻.

$$SNR = 20\log\left(\frac{N_{sig}}{n}\right) \text{ [dB]} \qquad (3.45)$$

In the example of Figure 3.19, DR is calculated as $20\cdot\log(20{,}000/12) = 64.4$ dB. For SNR, the noise, n, is the total temporal noise at the signal level N_{sig}. When the read noise is dominant in the total noise, SNR is given by

$$SNR = 20\log\left(\frac{N_{sig}}{n_{read}}\right) \qquad (3.46)$$

and in cases in which the photon shot noise is dominant, it is represented by

$$SNR = 20\log\left(\frac{N_{sig}}{n_{photon}}\right) = 20\log\left(\frac{N_{sig}}{\sqrt{N_{sig}}}\right) = 20\log\sqrt{N_{sig}} \qquad (3.47)$$

Figure 3.20 shows the SNR as a function of the number of incident photons. From Equation 3.47, it is understood that the maximum SNR is determined by the full-well capacity only and is given by

$$SNR_{max} = 20\log\sqrt{N_{sat}} = 10\log(N_{sat}) \qquad (3.48)$$

Basics of Image Sensors

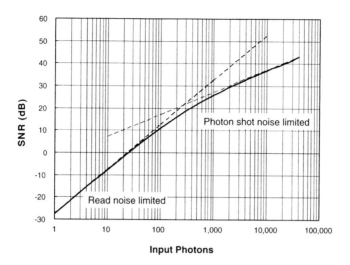

FIGURE 3.20 SNR as a function of incident photons.

3.4.2.2 Estimation of Quantum Efficiency

QE is readily obtained from Figure 3.19, which is 0.5 in this example. Also, from Equation 3.47, QE can be estimated using the shot noise dominant portion of the SNR plot shown in Figure 3.20:

$$QE = \frac{N_{sig}}{N_{photon}} = \frac{(S/N)^2}{N_{photon}} \qquad (3.49)$$

where (S/N) is N_{sig}/n_{photon}.

Next, the number of incident photons (the horizontal axis of Figure 3.19) must be converted to the exposure value, and the number of signal electrons (the vertical axis of Figure 3.19) to the output voltage (in this case, the sense-node voltage or the input referred voltage). For monochrome light, the number of incident photons is given by

$$N_{photon} = \frac{\lambda}{hc} \cdot P \cdot A_{pix} \cdot t_{INT} \qquad (3.50)$$

where P is the face-plate irradiance in W/cm², A_{pix} the pixel area in cm², and t_{INT} the integration time in seconds.

3.4.2.3 Estimation of Conversion Gain

Converting an image sensor's signal charge to signal voltage is accomplished using the following relationship:

$$V_{sig} = C.G. \cdot N_{sig} \qquad (3.51)$$

where C.G. is the conversion gain (see Equation 3.6). To estimate the conversion gain, use the photon shot noise, expressed in Equation 3.35 as

$$v_{photon} = C.G. \cdot \sqrt{N_{sig}} \qquad (3.52)$$

Equation 3.51 and 3.52 provide the following relationship:

$$v_{photon}^2 = (C.G.) \cdot V_{sig} \qquad (3.53)$$

Thus, the conversion gain can be obtained as a slope of $V_{sig} - v_{photon}^2$ plot. In this technique, the photon shot noise is used as a "signal" that provides useful information, allowing the relationship between the exposure value and the output voltage, as shown in Figure 3.21, to be obtained.

3.4.2.4 Estimation of Full-Well Capacity

Equation 3.48 implies that the full-well capacity can be obtained experimentally by measuring the maximum SNR as

$$N_{sat} = 10^{SNR_{max}/10} \qquad (3.54)$$

Also, because the intersection of the signal line and the photon shot noise line in Figure 3.21 occurs at $N_{sig} = 1$ (where the signal voltage corresponds to the conversion

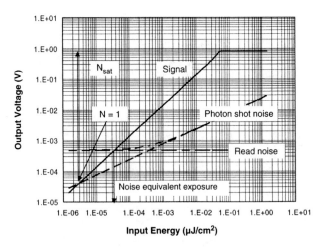

FIGURE 3.21 Photo-conversion characteristics. Exposure vs. signal voltage at the charge detection node for monochrome light with wavelength 550 nm; $QE = 0.5$; $A_{pix} = 25$ µm²; C.G. = 40 µV/e⁻; N_{sat} = 20,000 e⁻; and n_{read} = 12 e⁻.

gain), the full-well capacity or saturation charge, N_{sat}, can be estimated from the figure.

3.4.2.5 Noise Equivalent Exposure

Noise equivalent exposure can be defined as the exposure at which the signal level is equal to the read noise level, which corresponds to SNR = 1.

In reality, finding the relationship between the incident photons and signal charge requires a reverse path, starting with a measured relationship between the exposure and the output signal of an image sensor. This method assumes that the photoconversion characteristic is linear with no offset voltage. The nonlinear conversion characteristic and offset, such as that caused by dark current, if any, should be corrected before applying the preceding method to obtain the conversion gain.

Also, in actual devices, the SNR, including FPN, actually limits the real image SNR because a snapshot includes both sources. In addition, PRNU limits the total maximum SNR because it grows linearly, while the shot noise grows proportionally to the square root of the total signal electrons. In cases in which PRNU is linear at 1%, the SNR maximum, including PRNU, can never exceed 40 dB, no matter how large the full well becomes.

3.4.2.6 Linearity

The photon-to-electron conversion is inherently a linear process. However, the electron-to-signal charge conversion (i.e., the charge collection efficiency) and the signal charge-to-output voltage conversion could be nonlinear processes.

In a CCD image sensor, the nonlinearity may originate from the voltage-dependent floating diffusion capacitance and the nonlinearity of the output amplifier. However, these contributions are typically very small because the operating range is relatively limited (<1 V) compared to the usable range of the high-bias voltage (~15 V) for the output amplifier. The most noticeable nonlinearity in an IT CCD is seen near the saturation level in cases in which a vertical overflow drain structure is implemented in a pixel. The nonlinearity comes from the emission of electrons from the photodiode charge storage region to the vertical overflow drain.[31,32]

3.4.2.7 Crosstalk

Two components contribute to crosstalk: optical and electrical. The optical component originates in stray light or the angular dependence of an on-chip microlens array, as previously shown in Figure 3.11. Regarding the electric component, the signal charge generated deep in the photoconversion area by long wavelength light may diffuse into neighboring pixels. Methods of reducing this type of charge diffusion include:

- Make the effective photoconversion depth shallow while maintaining sufficient responsivity for the red region of the spectrum. For example, use a positively biased n-type substrate (see Figure 3.13b).
- Create an isolation region between pixels. An example for CMOS image sensors would be a pixel structure with a higher concentration of p-type isolation that surrounds a photodiode area.[33]

For small pixels, an on-chip microlens array is typically formed to increase sensitivity, as described in Section 3.2.3. The microlens also helps reduce crosstalk by focusing incident light rays on a spot at the center of the photodiode surface.[11]

3.4.3 SENSITIVITY AND SNR

The most important performance measure in image sensors is sensitivity, which is typically defined as the ratio of the output change to the input light change, expressed in volts per lux-second, electrons per lux-second, bits per lux-second, etc. However, this definition does not answer an essential question: "What is the lowest light level scene an image sensor can capture and still produce an output image?" To answer, it is necessary to know the image sensor's "sensitivity" (the ratio of the output/input changes) and its noise level. The noise equivalent exposure takes these two factors into account at very low light levels. SNR is considered a measure for true "sensitivity" of the image sensor when the entire illumination range from dark to light is considered.

3.4.4 HOW TO INCREASE SIGNAL-TO-NOISE RATIO

It is obvious that SNR can be improved by increasing the signal and reducing the noise. To increase the signal, it is necessary to increase the QE, as described by Equation 3.39, which is broken down further in Equation 3.40. Therefore, the following need to be increased:

- Transmittance of light [$T(\lambda)$] by
  - Reducing absorption by the color filters
  - Reducing reflection at the SiO_2/Si interface
- Fill factor [FF] by
  - Reducing the nondetector area inside a pixel
  - Optimizing the microlens structure
- Charge collection efficiency [$\eta(\lambda)$] by
  - Optimizing the detector structure while avoiding pixel-to-pixel crosstalk

All possible efforts should also be taken to reduce noise further. One possible noise reduction technique involves the selection of the charge-to-voltage conversion factor. As described in Equation 3.38, a higher conversion factor ($A_V \cdot C.G.$) provides lower input referred noise. This effectively provides higher signal gain before the signal enters noise-producing readout circuits. However, this technique may conflict with the camera's dynamic range requirement, especially in CMOS image sensors,

Basics of Image Sensors

because the higher conversion factor effectively reduces the full-well capacity when a limited power supply voltage is available. Choosing optimal values for sensitivity and dynamic range is an important design issue, especially with small pixel image sensors.

3.5 ARRAY PERFORMANCE

3.5.1 MODULATION TRANSFER FUNCTION (MTF)

Measuring the MTF is a technique used to characterize a system's frequency response or resolving capability. In a linear imaging system, the input $i(x, y)$ is a two-dimensional optical input, and the output $o(x, y)$ is a final reproduced image observed on a TV monitor or as a printed image. The relationship between the input and the output is represented by

$$o(x,y) = \iint h(x-x_0, y-y_0) \cdot i(x_0, y_0) \cdot dx_0 \cdot dy_0 \qquad (3.55)$$

where $h(x, y)$ is a system impulse response. The equivalent relationship in the frequency domain is obtained by performing the Fourier transform of Equation 3.55, as follows:

$$O(f_x, f_y) = H(f_x, f_y) \cdot I(f_x, f_y) \qquad (3.56)$$

where $H(f_x, f_y)$ is the transfer function; and $I(f_x, f_y)$ and $O(f_x, f_y)$ are the Fourier transform of $i(x, y)$ and $o(x, y)$, respectively. The MTF is the magnitude of $H(f_x, f_y)$, so that

$$H(f_x, f_y) = MTF(f_x, f_y) \cdot \exp\{-\phi(f_x, f_y)\} \qquad (3.57)$$

where $\phi(f_x, f_y)$ is the phase modulation function.

In a DSC system, the lens system, optical components (such as an IR cut filter and an optical low-pass filter), the image sensor, and the image-processing block are cascaded to form the total MTF. The total system MTF is simply the product of the individual MTFs of each component:

$$MTF_{System} = MTF_{Lens} \cdot MTF_{Optical\_Filter} \cdot MTF_{Imager} \cdot MTF_{Signal\_Processing} \qquad (3.58)$$

The imager and the signal-processing block involve time domain processing. An image sensor transforms a two-dimensional optical input into a time sequential signal, and the image-processing block transfers this signal from the image sensor to a final two-dimensional output image. Also, nonlinear processing is often applied

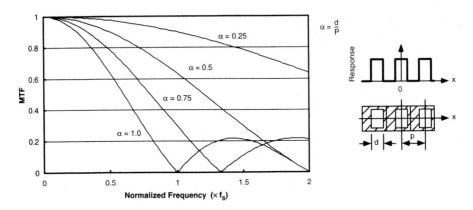

FIGURE 3.22 Example of MTF with uniform detector sensitivity.

in the image-processing block. Therefore, care should be taken when considering MTF_{Imager} and $MTF_{Signal\_Processing}$ in Equation 3.58.

3.5.2 MTF of Image Sensors, MTF_{IMAGER}

An image sensor is composed of discrete pixels that are arranged in a periodic row-and-column structure. Thus, the Nyquist sampling theorem, which describes the response of sampled-data systems, applies to an image sensor. The MTF of an image sensor is the magnitude of the Fourier transform of the sensitivity distribution inside a pixel and is given by

$$MTF(f_x, f_y) = \left| \iint S(x,y) \cdot \exp\{-j2\pi(f_x \cdot x + f_y \cdot y)\} \cdot dx \cdot dy \right| \quad (3.59)$$

where $S(x, y)$ is the sensitivity distribution inside a pixel. Usually $S(x, y)$ corresponds to the shape of the aperture (drawn photosensitive area), but it can also include the effects of a microlens and/or charge diffusion in the silicon bulk, as well as the effect of aperture shape.

Consider a one-dimensional example with a pixel pitch of p and a uniform sensitivity distribution across the photosensitive region d, as shown in Figure 3.22. In this case, the sampling frequency, f_S, and the Nyquist frequency, f_N, are given by

$$f_S = \frac{1}{p} \quad (3.60)$$

$$f_N = \frac{f_S}{2} = \frac{1}{2p} \quad (3.61)$$

and Equation 3.59 reduces to

$$MTF(f_x) = \left| \int_{-\infty}^{+\infty} Rect(x_0, d) \cdot \exp(-j2\pi f_x x) dx \right| = \frac{\sin\left(2\pi f_x \cdot \frac{d}{2}\right)}{2\pi f_x \cdot \frac{d}{2}} \quad (3.62)$$

As seen from the figure, higher MTF is obtained from a narrower sensitivity distribution. However, as the sampling theorem indicates, an original image whose highest frequency is lower than the Nyquist frequency f_N can only be reconstructed perfectly. If the input image has frequency components higher than the Nyquist frequency f_N, a false signal appears at $(f - f_S)$. This phenomenon is known as "aliasing," but it is often called "moiré" when applied to a two-dimensional image.

As the input frequency approaches the Nyquist frequency f_N, the phase between the input image and the pixels affects the output response. When the signal is in phase with the pixel period, the signal modulation has maximum amplitude and, conversely, an out-of-phase signal produces a minimum modulation signal. To avoid aliasing artifacts, an optical low-pass filter may be placed on an image sensor (see Section 2.2.3 in Chapter 2).

3.5.3 Optical Black Pixels and Dummy Pixels

Optical black (OB) pixels and dummy pixels are located at the periphery of an imaging array, as shown in Figure 3.23. The OB pixels play an essential role in determining a proper black level, which is a reference black level of a reproduced image. Dark current variation across the operating temperature range of the imager must be tracked by the OB pixels to ensure that a fixed-image black level can be determined. The number of columns used for the OB clamp pixels is established by the performance of the clamp circuit in the analog front-end.

Also, in IT CCDs, dark currents generated at V-CCD cause dark current shading in the vertical direction. Because the output from the OB columns contains the same dark currents, the dark current shading can be compensated. Dummy pixels are also

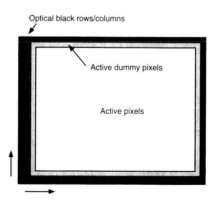

FIGURE 3.23 Optical black pixels and dummy pixels.

placed between the active array and the OB areas near the periphery of an imaging array. They allow color interpolation for all active pixels in the array because the color interpolation usually needs a mask of 5 × 5 pixels to interpolate the color value of a single pixel.

3.6 OPTICAL FORMAT AND PIXEL SIZE

3.6.1 Optical Format

Optical format, or optical image size, is represented by the diagonal measurement of an optical image projected on an image sensor by the imaging lens. Usually, the diagonal of a 1-in. optical image size is approximately 16 mm. This is not 25.4 mm (the metric equivalent of 1 in.) because the optical format was originally standardized to the diameter of vacuum tube imaging devices, which had larger diagonals than their actual optical image sizes. Approximate values of the diagonal of an image size can be estimated by assuming 1 in. equals 16 mm for solid-state image sensors with optical formats greater than 0.5 in. and 18 mm for image sensors with formats smaller than one third of an inch.[34] Until recently, optical format had been expressed in inches, a vestige of the days of imaging tubes. However, it is now expressed as "type," such as 1/1.8 type, instead of 1/1.8 in.

Imaging array sizes as a function of the optical format and standard array definitions for PC displays are summarized in Table 3.2 and Table 3.3, respectively.

TABLE 3.2
Optical Format and Effective Array Size

Format (type)	Diagonal (mm)	H (mm)	V (mm)	
1	16.0	12.80	9.60	16 mm/in.
2/3	11.0	8.80	6.60	
1/1.8	8.89	7.11	5.33	
1/2	8.00	6.40	4.80	
1/2.5	7.20	5.76	4.32	18 mm/in.
1/2.7	6.67	5.33	4.00	
1/3	6.00	4.80	3.60	
1/3.2	5.63	4.50	3.38	
1/4	4.50	3.60	2.70	
1/5	3.60	2.88	2.16	
1/6	3.00	2.40	1.80	

For DSLR

Format	Diagnal (mm)	H (mm)	V (mm)	Aspect ratio
35 mm	43.27	36.00	24.00	3:2
APS-DX	28.37	23.7	15.6	
APS-C	27.26	22.7	15.1	
APS-H	33.93	28.7	19.1	
Four-thirds	21.63	17.3	13.0	4:3

TABLE 3.3
Resolution of PC Displays

	Format	Resolution (pixels)
QCIF	Quarter common intermediate format	176 × 144
CIF	Common intermediate format	352 × 288
QVGA	Quarter video graphics array	320 × 240
VGA	Video graphics array	640 × 480
SVGA	Super video graphics array	800 × 600
XGA	Extended graphics array	1024 × 768
SXGA	Super extended graphics array	1280 × 1024
UXGA	Ultra extended graphics array	1600 × 1200
QXGA	Quad extended graphics array	2048 × 1536

(See Figure 4.31 in Chapter 4, which shows relationships between the number of pixels and pixel sizes with the optical format as a parameter.)

3.6.2 PIXEL SIZE CONSIDERATIONS

The state-of-the-art pixel size at the time of this writing is as small as 2.2 μm for CCD sensors and 2.25 μm for CMOS sensors.[35] However, it is obvious that sensitivity and full-well capacity decrease linearly with pixel size. As described in Section 3.4.2, the maximum SNR is determined by the full-well capacity only (assuming the SNR is photon shot noise limited). Also, as described in Section 2.2.4 in Chapter 2, the radius of the Airy disk is given by

$$r = 1.22\lambda F \tag{3.63}$$

Therefore, the resolving capability of small pixels suffers from the diffraction limit. Because the diffraction spot increases with the F-number and lower F-numbers produce large incident angles of light on pixels causing more shading and crosstalk, the usable range of F-settings is quite limited. In addition, as the objective depth becomes longer with smaller pixels (see Section 2.2.2), reproduced images taken by an image sensor with a small pixel tend to be pan-focus. Furthermore, a more complicated lens system design is needed to compensate for the several aberrations described in Section 2.1.3 in Chapter 2 because the pixel size is reduced. Maintaining image quality as pixel sizes decrease will require a larger role for back-end processors to compensate for the negative impacts caused by small pixels mentioned earlier.

On the other hand, high-resolution image sensors with a large pixel for digital SLR cameras can offer higher sensitivity, wider dynamic range, and a wider range of F-settings, which in turn permits a variety of photo shooting techniques. The expense/trade-off is a large image sensor chip, which requires a larger lens system, resulting in a larger, more expensive DSC. Also, as the physical silicon area grows, the number of imager die per wafer drops, driving the cost of each image sensor higher. The yield of good imager die per wafer may also be affected, driving sensor costs even higher.

Although image sensors now have 8 Mpixels for compact, point-and-shoot cameras and 17 Mpixels for DSLR cameras, resolutions of approximately 1.6 and 8.7 Mpixels are sufficient for reproducing 300 dots per inch (dpi), L-size (3.5 in. × 5.0 in.) prints and A4-size prints, respectively. Therefore, consumers should choose a DSC in the foreseeable future by considering the facts mentioned previously. For longer term considerations, see Chapter 11.

3.7 CCD IMAGE SENSOR VS. CMOS IMAGE SENSOR

Since the CMOS active-pixel sensor concept was introduced in the early 1990s,[36] CMOS image sensor technology has evolved to a level of performance that compares with CCD technology. Early CMOS image sensors suffered from large FPN caused by dark current nonuniformity. Many skeptics of CMOS imager technology stated that improving image quality was an issue for CMOS devices, even though they have many excellent features, such as low power and the capability of integrating drive/signal-processing circuits on-chip. However, with the recent introduction of pinned photodiode (PPD) technology, CMOS image quality issues are being resolved rapidly. The PPD active-pixel configuration, together with on-chip signal-processing circuits, offers even lower temporal noise performance than CCD image sensors.[37,38] High-resolution CMOS image sensors with large pixel sizes have actually installed in several DSLR cameras,[39–42] where their excellent image quality, higher-speed pixel rates, and lower power consumption have been demonstrated.

CMOS image sensor pixels were always larger than those in CCD image sensors. Consequently, except for the DSLR application, CMOS image sensors were used for low-end DSCs due to their lower prices. Again, however, the reduction of pixel sizes in CMOS image sensors has been remarkable in recent years, due to finer process technologies and the introduction of shared-pixel configurations (see Section 5.2.2). With these technology improvements, CMOS image sensors have become competitive alternatives in compact DSC and DSLR applications.

On the other hand, CCD image sensors have also made remarkable progress. In addition to inherently good image reproduction, recent CCD image sensors have several features that are especially useful for DSC applications (see Section 4.3 and Section 4.4 in Chapter 4).

REFERENCES

1. S.M. Sze, *Semiconductor Devices: Physics and Technology*, John Wiley & Sons, New York, 256, Chapter 7, 1985.
2. H.F. Wolf, *Silicon Semiconductor Data*, Pergamon Press, Oxford, 110, Chapter 2, 1969.
3. G.P. Weckler, Operation of p–n junction photodetectors in a photon flux integrating mode, *IEEE J. Solid-State Circuits*, SC-2(3), 65–73, September 1967.

4. W.F. Kosonocky and J.E. Carnes, Two-phase charge-coupled devices with overlapping polysilicon and aluminum gates, *RCA Rev.*, 34, 164–202, 1973.
5. M.H. White, D.R. Lampe, F.C. Blaha, and I.A. Mack, Characterization of surface channel CCD image arrays at low light levels, *IEEE J. Solid-State Circuits*, SC-9(1), 1–13, 1974.
6. B.E. Bayer, US patent 3,971,065, Color imaging array, July 20, 1976.
7. Y. Ishihara and K. Tanigaki, A high photosensitivity IL-CCD image sensor with monolithic resin lens array, *IEDM Tech. Dig.*, 497–500, December 1983.
8. M. Furumiya, K. Hatano, I. Murakami, T. Kawasaki, C. Ogawa, and Y. Nakashiba, A 1/3-in. 1.3-Mpixel, single-layer electrode CCD with a high-frame-rate skip mode, *IEEE Trans. Electron Devices*, 48(9), 1915–1921, September 2001.
9. H. Rhodes, G. Agranov, C. Hong, U. Boettiger, R. Mauritzson, J. Ladd, I. Karasev, J. McKee, E. Jenkins, W. Quinlin, I. Patrick, J. Li, X. Fan, R. Panicacci, S. Smith, C. Mouli, and J. Bruce, CMOS imager technology shrinks and image performance, *Proc. IEEE Workshop Microelectron. Electron Devices*, 7–18, April 2004.
10. M. Deguchi, T. Maruyama, F. Yamasaki, T. Hamamoto, and A. Izumi, Microlens design using simulation program for CCD image sensor, *IEEE Trans. Consumer Electron.*, 38(3), 583–588, August 1992.
11. G. Agranov, V. Berezin, and R.H. Tsai, Crosstalk and microlens study in a color CMOS image sensor, *IEEE Trans. Electron Devices*, 50(1), 4–11, January 2003.
12. J.T. Bosiers, A.C. Kleimann, H.C. Van Kuijk, L. Le Cam, H.L. Peek, J.P. Maas, and A.J.P. Theuwissen, Frame transfer CCDs for digital still cameras: concept, design, and evaluation, *IEEE Trans. Electron Devices*, 49(3), 377–386, March 2002.
13. H. Peek, D. Verbugt, J. Maas, and M. Beenhakkers, Technology and performance of VGA FT-imagers with double and single layer membrane poly-Si gates, *Program IEEE Workshop Charge-Coupled Devices Adv. Image Sensors*, R10, 1–4, June 1997.
14. H.C. van Kuijk, J.T. Bosiers, A.C. Kleimann, L.L. Cam, J.P. Maas, H.L. Peek, C.R. Peschel, Sensitivity improvement in progressive-scan FT-CCDs for digital still camera applications, *IEDM Tech. Dig.*, 689–692, 2000.
15. A. Tsukamoto, W. Kamisaka, H. Senda, N. Niisoe, H. Aoki, T. Otagaki, Y. Shigeta, M. Asaumi, Y. Miyata, Y. Sano, T. Kuriyama, and S. Terakawa, High-sensitivity pixel technology for a 1/4-inch PAL 430-kpixel IT-CCD, *IEEE Custom Integrated Circuit Conf.*, 39–42, 1996.
16. H. Mutoh, 3-D wave optical simulation of inner-layer lens structures, *Program IEEE Workshop Charge-Coupled Devices Adv. Image Sensors*, 106–109, June 1999.
17. M. Negishi, H. Yamada, K. Harada, M. Yamagishi, and K. Yonemoto, A low-smear structure for 2-Mpixel CCD image sensors, *IEEE Trans. Consumer Electron.*, 37(3), 494–500, August 1991.
18. F.A. Jenkins and H.E. White, *Fundamentals of Optics*, 4th ed., McGraw–Hill, New York, 526, Chapter 25, 1981.
19. I. Murakami, T. Nakano, K. Hatano, Y. Nakashiba, M. Furumiya, T. Nagata, T. Kawasaki, H. Utsumi, S. Uchiya, K. Arai, N. Mutoh, A. Kohno, N. Teranishi, and Y. Hokari, Technologies to improve photo-sensitivity and reduce VOD shutter voltage for CCD image sensors, *IEEE Trans. Electron Devices*, 47(8), 1566–1572, August 2000.
20. S.M. Sze, *Semiconductor Devices: Physics and Technology*, John Wiley & Sons, New York, 48–55, Chapter 2, 1985.
21. R.N. Hall, Electron–hole recombination in germanium, *Phys. Rev.*, 87, 387, July 1952.

22. W. Shockley and W.T. Read, Jr., Statistics of recombinations of holes and electrons, *Phys. Rev.*, 87, 835–842, September 1952.
23. R.F. Pierret, *Modular Series on Solid State Physics*, vol. VI, Addison–Wesley, Reading, MA, Chapter 5, 1987.
24. N. Teranishi, A. Kohno, Y. Ishihara, E. Oda, and K. Arai, No image lag photodiode structure in the interline CCD image sensor, *IEDM Tech. Dig.*, 324–327, December 1982.
25. W.C. McColgin, J.P. Lavine, J. Kyan, D.N. Nichols, and C.V. Stancampiano, Dark current quantization in CCD image sensors, *IEDM Tech. Dig.*, 113–116, December 1992.
26. N.S. Saks, A technique for suppressing dark current generated by interface states in buried channel CCD imagers, *IEEE Electron Device Lett.*, EDL-1, 131–133, July 1980.
27. C-C. Wang and C.G. Sodini, The effect of hot carriers on the operation of CMOS active pixel sensors, *IEDM Tech. Dig.*, 563–566, December 2001.
28. I. Takayanagi, J. Nakamura, E.-S. Eid, E. Fossum, K. Nagashima, T. Kunihiro, and H. Yurimoto, A low dark current stacked CMOS-APS for charged particle imaging, *IEDM Tech. Dig.*, 551–554, December 2001.
29. B. Razavi, *Design of Analog CMOS Integrated Circuits*, McGraw–Hill, New York, 209–218, Chapter 7, 2001.
30. J. Janesick, CCD characterization using the photon transfer technique, *Proc. SPIE, 570, Solid State Imaging Arrays*, 7–19, 1985.
31. E.G. Stevens, Photoresponse nonlinearity of solid-state image sensors with antiblooming protection, *IEEE Trans. Electron Devices*, 38(2), 299–302, February 1991.
32. S. Kawai, M. Morimoto, N. Mutoh, and N. Teranishi, Photo response analysis in CCD image sensors with a VOD structure, *IEEE Trans. Electron Devices*, 42(4), 652–655, April 1995.
33. M. Furumiya, H. Ohkubo, Y. Muramatsu, S. Kurosawa, F. Okamaoto, Y. Fujimoto, and Y. Nakashiba, High-sensitivity and no-cross-talk pixel technology for embedded CMOS image sensor, *IEEE Trans. Electron Devices*, 48(10), 2221–2227, October 2001.
34. N. Egami, Optical image size, *J. ITEJ*, 56(10), 1575–1576, October 2002 (in Japanese).
35. M. Mori, M. Katsuno, S. Kasuga, T. Murata, and T. Yamaguchi, A π-inch 2-Mpixel CMOS image sensor with 1.75 transistor/pixel, *ISSCC Dig. Tech. Papers*, 110–111, February 2004.
36. E.R. Fossum, Active pixel sensors: are CCDs dinosaurs? *Proc. SPIE, 1900, Charge-Coupled Devices and Solid-State Optical Sensors III*, 2–14, 1993.
37. L.J. Kozlowski, J. Luo, and A. Tomasini, Performance limits in visible and infrared image sensors, *IEDM Tech. Dig.*, 867–870, December 1999.
38. A. Krymski, N. Khaliullin, and H. Rhodes, A 2 e$^-$ noise, 1.3 megapixel CMOS sensor, *Program IEEE Workshop Charge-Coupled Devices Adv. Image Sensors*, May 2003.
39. S. Inoue, K. Sakurai, I. Ueno, T. Koizumi, H. Hiyama, T. Asaba, S. Sugawa, A. Maeda, K. Higashitani, H. Kato, K. Iizuka, and M. Yamawaki, A 3.25-Mpixel APS-C size CMOS image sensor, *Program IEEE Workshop Charge-Coupled Devices Adv. Image Sensors*, 16–19, June 2001.
40. A. Rush and P. Hubel, X3 sensor characteristics, *J. Soc. Photogr. Sci. Technol. Jpn.*, 66(1), 57–60, 2003.

41. G. Meynants, B. Dierickx, A. Alaerts, D. Uwaerts, S. Cos, and D. Scheffer, A 35-mm 13.89-million pixel CMOS active pixel image sensor, *Program IEEE Workshop Charge-Coupled Devices Adv. Image Sensors*, May 2003.
42. T. Isogai, T. Ishida, A. Kamashita, S. Suzuki, M. Juen, and T. Kazama, 4.1-Mpixel JFET imaging sensor LBCAST, *Proc. SPIE*, 5301, 258–263, 2004.

4 CCD Image Sensors

Tetsuo Yamada

CONTENTS

4.1 Basics of CCDs 95
    4.1.1 Concept of the Charge Coupled Device 95
    4.1.2 Charge Transfer Mechanism 97
    4.1.3 Surface Channel and Buried Channel 99
    4.1.4 Typical Structures and Operations (Two- and Four-Phase Clocking) 105
    4.1.5 Output Circuitries and Noise Reduction: Floating Diffusion Charge Detector and CDS Process 107
4.2 Structures and Characteristics of CCD Image Sensor 110
    4.2.1 Frame Transfer CCD and Interline Transfer CCD 110
    4.2.2 P-Substrate Structure and P-Well Structure 112
    4.2.3 Antiblooming and Low-Noise Pixels (Photodiode and VCCD) 114
    4.2.4 CCD Image Sensor Characteristics 117
        4.2.4.1 Photoelectric Conversion Characteristics 117
        4.2.4.2 Smear and Blooming 119
        4.2.4.3 Dark Current Noise 119
        4.2.4.4 White Blemishes and Black Defects 120
        4.2.4.5 Charge Transfer Efficiency 120
    4.2.5 Operation and Power Dissipation 121
4.3 DSC Applications 122
    4.3.1 Requirements from DSC Applications 122
    4.3.2 Interlace Scan and Progressive Scan 123
    4.3.3 Imaging Operation 126
    4.3.4 PIACCD (Super CCD) 129
    4.3.5 High-Resolution Still Picture and High-Frame-Rate Movie 132
    4.3.6 System Solution for Using CCD Image Sensors 137
4.4 Future Prospects 138
References 139

4.1 BASICS OF CCDS

4.1.1 CONCEPT OF THE CHARGE COUPLED DEVICE

The concept of a charge coupled device (CCD) encompasses storing and transferring electrons (occasionally holes) as charge packet signals in a semiconductor.[1] A build-

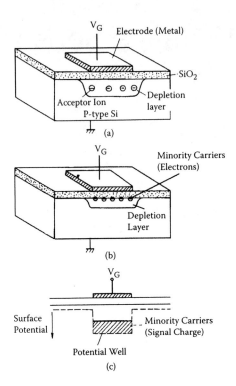

FIGURE 4.1 MOS capacitor: (a) thermal nonequilibrium deep depletion condition; (b) electron storage condition; (c) liquid model of charge storage in a potential well.

ing block of the CCD is a metal-oxide-semiconductor (MOS) capacitor as shown in Figure 4.1(a). When a positive voltage is applied to the metal electrode, majority carrier holes in the p-type silicon substrate are repelled from the silicon surface region, and a depletion layer is formed in the surface region. Then, electric field (flux) lines from the electrode are terminated on negative space charges of acceptor ions in the depletion layer. Under this thermal nonequilibrium condition, minority carrier electrons injected in this layer are attracted to the Si–SiO$_2$ interface beneath the electrode, as shown in Figure 4.1(b). This means that a potential well for electrons is formed at the Si–SiO$_2$ interface. Usually, a liquid model is used to represent charge packet storage and transfer situations, as shown in Figure 4.1(c).

Next, we consider an interaction between two MOS capacitors using Figure 4.2. Figure 4.2(a) shows a case in which two electrodes, G1 and G2, are placed with wide space, where positive high voltage is applied on both of the electrodes; the potential well formed beneath G1 stores an electron packet; and that beneath G2 is empty. In this case, no interaction occurs between the MOS capacitors. When G1–G2 space is narrowed extremely, the two potential wells are coupled together. Thus, the electron charge packet stored beneath G1 spreads over the coupled potential well beneath G1 and G2, as shown in Figure 4.2(b). Then, the electron packet is transferred into the potential well beneath G2 completely by lowering the voltage applied on G1, as shown in Figure 4.2(c). Therefore, a charge packet can be transferred from the G1 position to the G2 position.

CCD Image Sensors

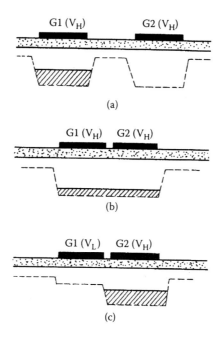

FIGURE 4.2 Principle of charge transfer: (a) G1 and G2 are biased high level and are separated. (b) G1 and G2 are closely spaced. (c) The bias of G1 becomes low level.

If a large number of closely spaced MOS capacitors are formed in a line, charge packets can be transferred from capacitor to capacitor freely by controlling potential wells. In this way, the CCD acts as an analog shift register in which signals are expressed with electron numbers in charge packets. A key point of the charge transfer in the CCD is on transferring all electrons stored in a potential well to the next well completely; this is called the complete charge transfer mode. The complete charge transfer mode makes the CCD the best architecture for an image sensor application because an electron number in a well is preserved from fluctuations of voltage and current. Therefore, the CCD has extremely high signal to noise ratio (S/N).

4.1.2 Charge Transfer Mechanism

The basic charge transfer is governed by three mechanisms: self-induced drift, thermal diffusion, and fringing field effect. Figure 4.3(a) and (b) shows a simple model of charge transferring from the electrode G2 to the electrode G3 without considering the fringing field. When a charge packet is large, at the beginning of charge transfer, for example, the self-induced drift caused by electrostatic repulsion of the carriers controls the transfer. The remaining charge under G2 after transfer time t is expressed as the next approximate equation[2] is:

$$Q(t)/Q_0 \approx t_0/(t_0 + t) \qquad (4.1)$$

$$t_0 = \pi L^3 W C_{\mathit{eff}}/2\mu Q_0 = \pi/2 \cdot L^2/\mu(V_1 - V_0), \qquad (4.2)$$

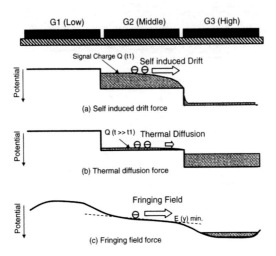

FIGURE 4.3 Charge transfer mechanism: (a) self-induced drift force; (b) thermal diffusion force; (c) fringing field force.

where L and W are the length and the width of the electrode G2, respectively; μ is the carrier (electron) mobility; and C_{eff} is the effective storage capacitance per unit area, which corresponds to the gate oxide capacitance for the previously mentioned prototype CCD.

$V_1 - V_0 = Q_0/LWC_{eff}$ is the initial voltage to move the carriers into the next electrode G3. Equation 4.1 and Equation 4.2 mean that the decay speed is proportional to the initial charge density, Q_0/LW. When the channel voltage of the remaining charge under G2 becomes as small as thermal voltage kT/q (26 mV at room temperature), as shown in Figure 4.3(b), the thermal diffusion occupies the transfer process, which decreases the remaining charge under G2 exponentially with time. The time constant, τ_{th}, of the thermal diffusion is expressed by next equation.

$$\tau_{th} = 4L^2/\pi^2 D, \qquad (4.3)$$

where D is the carrier diffusion constant. In case the fringing field is not considered, the thermal diffusion determines the charge transfer performance because the final remaining charge should be as extremely small as several electrons. Actually, the fringing field, E_y, is caused by a voltage difference between two electrodes, as shown in Figure 4.3(c), and accelerates the transfer for remaining charge at the final process. The fringing field intensity and profile depend on the gate oxide thickness, the impurity profile in Si, and the electrode voltage difference. The unit carrier transit time, t_{tr}, through an L length electrode is expressed as

$$t_{tr} = \frac{1}{\mu} \int_0^L (1/E_y) dy . \qquad (4.4)^2$$

CCD Image Sensors

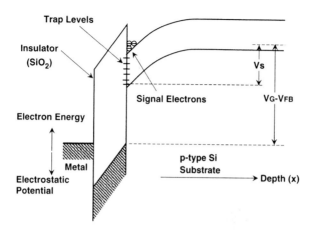

FIGURE 4.4 Energy band diagram in surface channel CCD with electron storage.

The fringing field is a most important force for high-speed operations such as over 10-MHz clocking. It is therefore important to consider how to intensify the fringing field in designing a CCD.

The transfer efficiency, η, is useful to evaluate a performance of CCD,[4] which is defined as

$$\eta = [Q_t/Q_i]^{1/N} \times 100 \ [\%], \quad (4.5)$$

where Q_i is the input impulse signal charge; Q_t is the transferred charge in the leading charge packet; and N is the total number of transfer stages.

4.1.3 SURFACE CHANNEL AND BURIED CHANNEL

In the case of a previously introduced CCD, a charge packet is stored in and transferred along the Si surface of an MOS structure, as shown in Figure 4.1 and Figure 4.2. This type is called the surface channel CCD. Figure 4.4 shows the one-dimensional energy bands bent by applying positive voltage, V_G, on the electrode with storing a charge packet, Qs. The surface potential, V_s, can be obtained by solving the Poisson equations with the depletion approximation as follows[5]:

$$Vs = qN_A \, x_d^2/2\varepsilon_s \quad (4.6)$$

$$V_G - V_{FB} = Vs + Vox = Vs + qN_A \, x_d/Cox + Qs/Cox, \quad (4.7)$$

where

N_A is the doping density of acceptor ions.
x_d is the depletion region width.
ε_s is the dielectric constant of Si.

Vox is the voltage drop across the oxide.
V_{FB} is the flat band voltage.
Qs is the stored charge per unit surface area.
Cox is the oxide capacitance (oxide dielectric constant ε_{ox}/oxide thickness t_{ox}).

By solving Equation 4.6 and Equation 4.7 for *Vs*, the next equation can be obtained:

$$Vs = V_G' + V_0 - (2 V_G' V_0 + V_0^2)^{1/2}, \qquad (4.8)$$

where

$$V_G' = V_G - V_{FB} - Qs/Cox \qquad (4.9)$$

$$V_0 = V_G - V_{FB} - qN_A\varepsilon_s/Cox^2. \qquad (4.10)$$

From Equation 4.8, the surface potential, *Vs*, is plotted for V_G in Figure 4.5 with N_A and t_{ox} as parameters. As N_A becomes lower and V_G rises higher, the curve approximates linear with a slope of one.

The surface channel has a serious disadvantage: the high density of carrier trap energy levels is introduced in the forbidden gap of the Si surface; this is called the surface state or the interface state, due to the drastic irregularity of crystal lattice at Si surface. Thus, signal electrons (rarely holes) transferred along the Si surface are

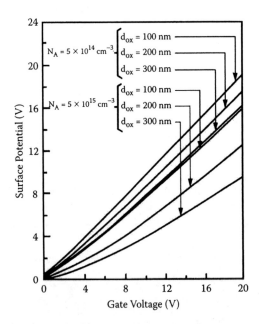

FIGURE 4.5 Surface channel potential vs. gate (electrode) voltage with three gate-oxide thicknesses and two substrate impurity densities.

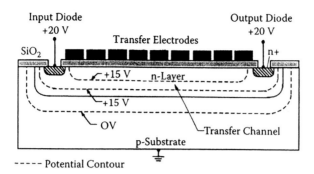

FIGURE 4.6 Cross-sectional view of a buried channel CCD. (After Walden, R.H. et al., *Bell Syst. Tech. J.*, 51, 1635–1640, 1972.)

trapped in the surface states, with the probability determined by the trap level distribution, and suffer significant losses of electrons through the overall transfer process.[6] In other word, the surface channel cannot transfer charge packets with high transfer efficiency, i.e., higher than 99.99% per unit transfer stage, and is not a good fit for a large-scale CCD.

To overcome this problem, the buried channel CCD (BCCD), which has essentially highly efficient transfer capability, was developed.[7] The cross-section is shown in Figure 4.6. The buried channel consists of an n-type impurity layer on a p-type Si substrate and is completely depleted by applying a reverse bias between the n-layer and the p-substrate at the initial condition. The channel potential of BCCD is controlled by applying a voltage on the electrodes formed over the n-layer in the same manner as that for the previously mentioned surface channel CCD.

The one-dimensional energy band diagram of BCCD is shown in Figure 4.7. Figure 4.7(a) and (b) show band diagrams for a zero reverse bias condition and that of a completely depleted condition by applying sufficient reverse bias voltage on the n-layer with a roughly zero electrode voltage, respectively. As shown, the energy band is pulled up toward the Si surface by the electrode potential. Figure 4.7(c) is a case in which a signal charge packet is in the n-layer. Electrons are stored in the minimum potential region as shown and are separated from the Si surface by the surface depletion layer.

Because no electrons interact with the interface states (trap levels), excellent charge transfer efficiency is achieved with the BCCD. The one-dimensional potential profile can also be derived analytically from Poisson equations with depletion approximation for the uniform impurity profiles in the n-layer and p-substrate as[8]:

$$d^2V_B/dx^2 = -qN_D/\varepsilon_s, \ (0 \leq x \leq x_j) \qquad 4.11$$

$$d^2V_B/dx^2 = qN_A/\varepsilon_s \ (x_j < x) \qquad 4.12$$

where V_B is channel potential of BCCD; N_D is the impurity concentration of the donor ion in the n-layer; and x_j is the p-n junction depth. The simple model for the analysis is given in Figure 4.8. By solving Equation 4.11 and Equation 4.12 for the

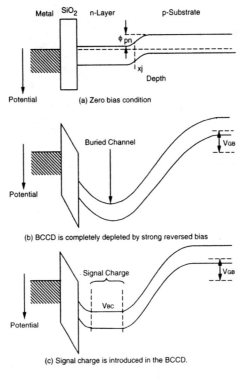

FIGURE 4.7 One-dimensional energy band diagram of BCCD: (a) zero bias condition. (b) BCCD is completely depleted by strong reversed bias. (c) Signal charge is introduced in the BCCD.

boundary condition that the electrode is biased at V_{GB} (the potential and the dielectric displacement should be continuous at $x = x_j$ and $x = 0$, respectively), the maximum (minimum for electrons) channel potential, V_{MB}, can be expressed as a function of the gate electrode potential, $V_{GB} - V_{FB}$.

$$\sqrt{V_{MB}} = \sqrt{V_K} - \left[V_K + (V_{GB} - V_{FB} - V_I) \cdot (N_A + N_D)/N_D\right]^{1/2}, \quad (4.13)$$

where

$$V_K = qN_A(N_A + N_D)(t_{OX}\varepsilon_S/\varepsilon_{OX} + x_j)^2/N_D \quad (4.14)$$

$$V_I = qN_D x_j (2t_{OX}\varepsilon_S/\varepsilon_{OX} + x_j)/2\varepsilon_S. \quad (4.15)$$

Figure 4.9 shows calculated $V_{MB} - V_{GB}$ curves with the experimental values for three doping densities of the n-layer as device parameters. In this figure, each curve has a knee point at the negative voltage of V_{GB}. These knee curves are caused by

CCD Image Sensors

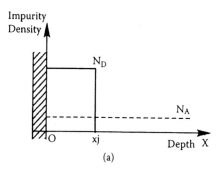

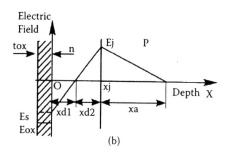

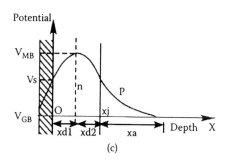

FIGURE 4.8 One-dimensional analysis model of BCCD: (a) impurity profiles with box approximation; (b) electric field; and (c) electrostatic potential in BCCD without signal charge and with depletion approximation.

holes injected into the surface of the n-layer from a p-type region surrounding the BCCD such as p-type channel stops. The injected holes terminate electric field (flux) lines and pin the surface potential to the substrate potential. This phenomenon, illustrated in Figure 4.10, is called the valence band pinning; it improves the BCCD characteristics drastically.[9] Actually, the holes accumulated in the surface suppress the generation current generated thermally through the surface states distributed near the midband in the forbidden gap. The thermally generated current is called the dark current, and appears as a temporal noise or a fixed pattern noise (FPN) for image sensor applications (explained in Section 4.2). In the valence band pinning condition, the dark current generated from the surface is expressed as[10]:

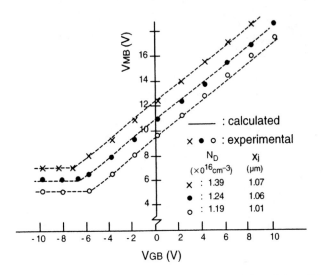

FIGURE 4.9 Calculated and experimental V_{MB}–V_{GB} characteristics of three BCCD samples under zero-charge condition.

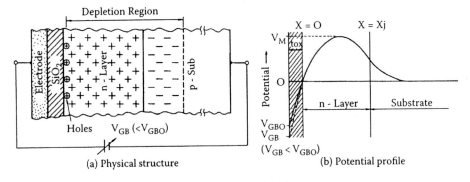

FIGURE 4.10 Buried channel, under valence band-pinning condition: (a) physical structure; (b) potential profile.

$$|Is| = -e\, U_s = eS_{on}\, n_i^2/p_s$$

$$= (eS_{on}\, n_i/2) \times (2n_i/p_s), \qquad (4.16)$$

Dark current in the depleted surface   Suppression factor

where U_s is the surface generation rate; S_{on} is the surface generation velocity; p_s is the hole density in the surface region; and $n_s p_s \ll n_i^2$ is presumed for the thermal nonequilibrium condition. For example, p_s is more than 10^{17} for over 1 V from the critical pinning voltage and $2\, n_i/p_s$ becomes about 10^{-7} at room temperature. This means that the dark current becomes negligibly small.

Substantially, the dark current decreases drastically when the pinning occurs, as shown in Figure 4.11, which is a measurement curve of the dark current in a BCCD.[11,12]

CCD Image Sensors

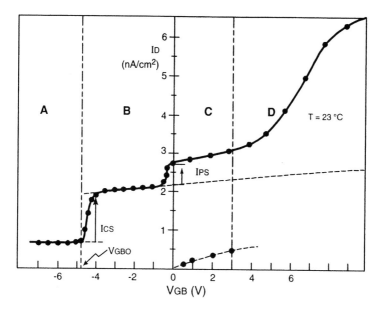

FIGURE 4.11 Measured curve of dark current vs. the electrode voltage.

In this figure, V_{GB0} is the electrode voltage critical to causing the valence band pinning. Therefore, applying a negative voltage as the low level of clock pulses can suppress the dark current noise of the vertical transfer CCD (VCCD) by utilizing the valence band pinning. Another advantage of BCCD is that an enhanced fringing field in the charge transfer channel can be obtained because the distance between the electrode and the maximum channel potential position is longer than that of the SCCD.[7] Moreover, the electron mobility in the bulk is two or three times higher than that at the surface. The enhanced fringing field and the high mobility accelerate the charge transfer velocity of the BCCD. With these advantages, the BCCD has become the standard of the CCD, which can realize charge transfer efficiency higher than 99.9999%.

4.1.4 TYPICAL STRUCTURES AND OPERATIONS (TWO- AND FOUR-PHASE CLOCKING)

CCD requires that the space between electrodes be as narrow as possible, for example less than 0.1 μm. Therefore, the double-layer poly-Si overlapping electrodes are used in general.[13] Figure 4.12(a) and Figure 4.13(a) show cross-sectional views of two typical CCD electrodes, i.e., Figure 4.12(a) is a two-phase driving CCD and Figure 4.13(a) is a four-phase driving CCD, both of which are the BCCD. In the case of the two-phase CCD, first- and second-layer poly-Si electrodes are connected electrically in pairs; the doping density of n-layer under the second-layer electrode is made lower than that under the first-layer electrode by a counter implantation of p-type ion, i.e., boron, generally.[14]

Thus, the potential step is formed in the channel under the same phase electrodes as shown in Figure 4.12(b), which can transfer charge packets by applying two-phase clock pulses. The potential step prevents the charge transfer from running

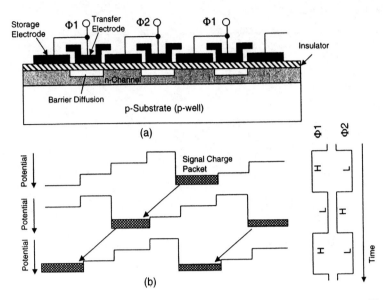

FIGURE 4.12 Two-phase CCD: (a) cross-sectional view; (b) channel potential profiles under applying the two-phase pulse set (c) of $\Phi 1$ and $\Phi 2$.

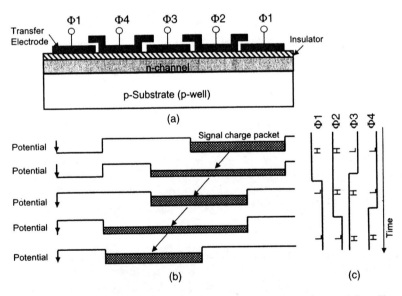

FIGURE 4.13 Four-phase CCD: (a) cross-sectional view; (b) channel potential profiles under applying the four-phase pulse set (c) of $\Phi 1$, $\Phi 2$, $\Phi 3$, and $\Phi 4$.

back, determines the charge transfer direction, and allows charge packets to transfer with even slightly degraded clock pulses. The charge transfer capability of the two-phase CCD is determined by the potential barrier under the second-layer electrode (transfer electrode) and the area of the first-layer electrode (storage electrode). The two-phase CCD is suitable for a high-speed transfer even though the charge storage capability is smaller than that of a four-phase CCD.

In the case of the four-phase CCD, the channel impurity density is basically uniform, and neighboring two electrodes act as storage electrodes; the other two electrodes act as barrier electrodes, as shown in Figure 4.13.[4] A unit transfer stage consists of four electrodes, to which four-phase clock pulses are applied. The duty ratio of each pulse should be more than 0.5 because the forward electrode should become a high level at the beginning of transfer process; after that, the rear electrode becomes its low level to secure the proper charge transfer action, as shown in Figure 4.13(b). The four-phase CCD usually has several times higher charge transfer capability than the two-phase CCD, and is well suited to be applied in highly integrated interline transfer CCD image sensors as mentioned in a later section. Generally, the maximum transferable charge, Q_m, of a multiphase (three phases or more) CCD is given as the next equation for same unit transfer stage length, L_s:

$$Q_m = q_i \, W(M - 2) \, L_s/M, \tag{4.17}$$

where q_i is the transferable charge per unit area; W is the channel width; and M is the phase number of unit transfer stage. Equation 4.17 means that Q_m becomes larger as M increases. Moreover, in case the length of all electrodes is fixed at L_e, Q_m is given by

$$Q_m = q_i \, W(M - 2) \, L_e. \tag{4.18}$$

Thus, increasing the phase number M is useful to enlarge the charge transfer capability. For example, Figure 4.14 shows a ten-phase CCD, which can use about 80% of the unit transfer stage area as active storage area.[15]

4.1.5 OUTPUT CIRCUITRIES AND NOISE REDUCTION: FLOATING DIFFUSION CHARGE DETECTOR AND CDS PROCESS

Output circuits act to convert a signal charge packet to the signal voltage and amplify the signal current. The most popular output circuit consists of the floating diffusion (FD) charge detector followed by two- or three-stage source follower

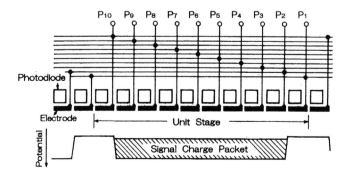

FIGURE 4.14 Electrode arrangement and potential profile in a ten-phase CCD.

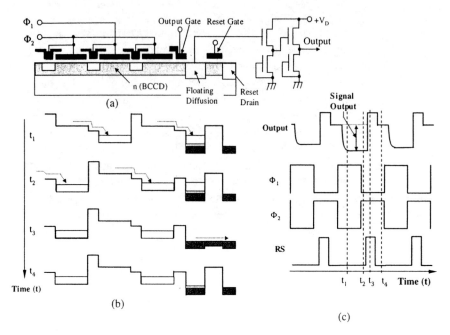

FIGURE 4.15 Principle of signal charge detection: (a) cross-sectional view of HCCD and floating diffusion type charge detector; (b) potential profiles and charge transfer packets at t_1, t_2, t_3, and t_4, shown in (c) timing chart.

circuits. Figure 4.15 shows a cross-section of the FD charge detector and the final stage of two-phase CCD with the two-stage source follower circuits.[13]

In this figure, the timing chart is also presented with an output signal wave form. At time t_1, FD is reset to the reference voltage level by applying high voltage of RS pulse on the reset gate (RS). FD becomes floating during a transition of the RS pulse from high to low, and its electrostatic level is somewhat lowered; this is induced by capacitance coupling of the RS pulse. The output level becomes as shown at time t_2, which is called the feed-through level. At $t = t_3$, the final CCD electrode, $\Phi 2$, becomes its low level; the charge packet is transferred to the FD detector over the output gate potential; and the FD potential changes to a low level according to the electron number in the charge packet. In this case, a thermal noise generated in the reset gate channel is held on FD after the reset gate is closed. Because the thermal noises in the feed-through level and the signal level are identical, the thermal noise can be canceled by subtracting the feed-through level from the signal level. This noise reduction technique is called the correlated double sampling (CDS).[16]

The most important performance of the charge detector is the charge detection sensitivity or conversion gain (C.G.). This is determined by the electrostatic capacitance, C_{FD}, of the FD node and the source follower circuitry gain as:

$$C.G. = A_G \cdot \frac{q}{C_{FD}}, \quad (4.19)$$

where A_G is the voltage gain of the source follower circuitry.

Thus, C_{FD} should be as small as possible for high-sensitivity charge detection, which consists of the p–n junction capacitance; coupling capacitances to the reset gate and the output gate; and parasitic capacitance of the wiring to the source follower circuitry and the first source follower gate capacitance, which contains the drain coupling and the channel modulation capacitance as shown in Figure 4.16. An example of the reducing C_{FD} technique is shown in Figure 4.17, where the reset gate and the output gate are separated from the primary charge storage area (high-density impurity region) of FD, which also reduces the coupling reset noise.[17] A high conversion gain of 80 µV/electron has been realized.

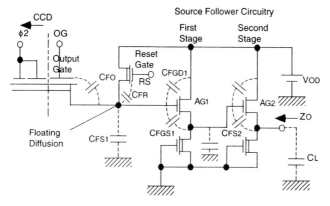

FIGURE 4.16 Equivalent circuitry for signal charge detection.

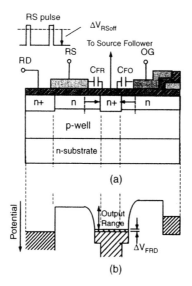

FIGURE 4.17 Key technology to enhance FD charge detector sensitivity.

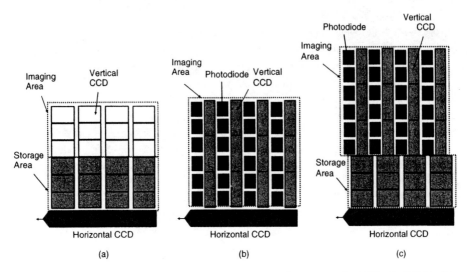

FIGURE 4.18 Three types of CCD image sensor: (a) frame transfer CCD (FTCCD); (b) interline transfer CCD (ITCCD); (c) frame-interline transfer (FIT) CCD.

4.2 STRUCTURES AND CHARACTERISTICS OF CCD IMAGE SENSOR

4.2.1 FRAME TRANSFER CCD AND INTERLINE TRANSFER CCD

Three types of CCD image sensor are shown in Figure 4.18. Figure 4.18(a) is a block diagram of a frame transfer CCD (FTCCD), which consists of an imaging area, a charge storage area, a horizontal charge transfer CCD (HCCD), and an output circuitry.[18] The imaging area and the storage area are made up of a multichannel vertical transfer CCD, which transfers charge packets in parallel vertically. The storage area and HCCD are covered by a light shield metal. Incident light rays through poly-Si electrodes in the imaging area are absorbed in Si bulk and generate electron–hole pairs. The electrons generated in or near potential wells formed in the imaging area are gathered and integrated in the potential wells as imaging signal charge packets. The signal charge packets integrated for a predetermined period are transferred in parallel toward the storage area, which acts as an analog frame memory.

After this operation, the charge packets in a horizontal line are transferred into the HCCD during a horizontal blanking period, referred to as the line shift, and are transferred to the output circuitry serially and output as voltage signals, one after another. Because FTCCD has a simple pixel structure, it is relatively easy to make a pixel size small. However, FTCCD has a severe problem of the spurious signal called smear that is caused by superimposed light-generated carriers during the signal charge transfer through the light incident section of the imaging area. To reduce the smear, the frame transfer speed from the imaging area to the storage area should be high enough with high-frequency clocking. The smear ratio SMR to signal is inversely proportional to the frame transfer frequency, f_F, as:

CCD Image Sensors

$$SMR \propto \frac{t_F}{T_{INT}} = \frac{1}{f_F \cdot T_{INT}}, \qquad (4.20)$$

where T_{INT} is the signal integration time; and t_F is a period of the frame transfer clock.

Moreover, the poly-Si electrodes mainly absorb short wave length light such as blue. Thus, a large part of blue light cannot arrive at the photo diodes in the CCD channel, resulting in low sensitivity for blue light. Therefore, a transparent electrode, such as ITO, is needed to avoid the loss in the blue light region of the spectrum, or the virtual phase CCD may be a solution for this problem.[19] Though FTCCD has a simple structure, it has the disadvantage of requiring extra charge storage area. If the application is limited only in shooting still images with using a mechanical shutter, the storage area can be removed from FTCCD. This type of image sensor is called the full-frame transfer CCD.[20]

Figure 4.18(b) is a block diagram of an interline transfer CCD (ITCCD), which is the most popular image sensor for applying camcorders and digital still cameras.[21] Photodiodes are arranged at rectangle lattice positions and vertical transfer CCDs (VCCDs) are placed on the interlined position of the photodiode rows. In this case, the interlined VCCDs are light shielded and act as an analog frame memory. Signal charge packets integrated in the photodiodes are transferred into the VCCDs momentarily during microsecond order by opening the transfer gate. The charge packets in the VCCDs are transferred to the HCCD and output from the output circuitry in the same manner for FTCCD.

A cross-section of an ITCCD pixel is shown in Figure 4.19(a). As shown, the transfer gate is made with a part of the VCCD electrode, which forms a surface channel varied from the buried channel of VCCD. The VCCD is usually operated with applying negative voltage pulses, so the surface channel beneath the transfer gate becomes hole-accumulation condition and forms the potential barrier for electrons; the buried channel region forms potential well for charge transfer as shown in Figure 4.19(b). A positive high-voltage pulse is applied on the VCCD electrodes to transfer the signal charge packets stored in the photodiodes into the buried channel in the VCCDs, as shown in Figure 4.19(c). Therefore, a three-level pulse is needed for the VCCD operation because the VCCD electrodes are also used for the transfer gates for simplifying the pixel structure as mentioned earlier.

In the case of ITCCD, the smear problem can essentially be avoided because VCCD is almost completely light shielded. Moreover, because the photodiode is separated from VCCD, it can be designed as an individually optimized structure to have the best performance, such as high sensitivity, low noise, and wide dynamic range. ITCCD has become a standard image sensor for VTR camera and DSC use. Therefore, the ITCCD is mainly described as a CCD image sensor in this text.

The other well-known type of CCD image sensor is the frame-interline transfer (FIT) CCD, which is shown in Figure 4.18(c).[22] FIT consists of the imaging area and the storage area like FTCCD and has the same pixel structure as ITCCD. The merit of FIT is in high smear protection. Although the VCCD used in ITCCD is light shielded, leakage light rays into VCCD generate smear electrons and a part of photogenerated electrons in Si bulk also diffuse into VCCD as smear electrons. FIT

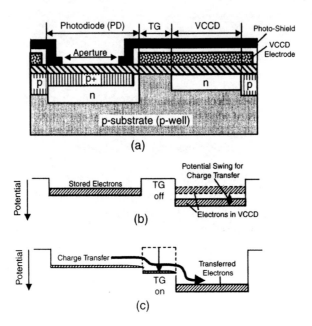

FIGURE 4.19 An ITCCD pixel: (a) a pixel cross-section of ITCCD; (b) potential profiles under the signal charge transferring vertically in VCCD; (c) Potential profile under the signal charge transferring from photodiode to VCCD.

reduces the smear by f_H/f_F, where f_H is the line shift frequency determined by a required time to output total charge packets in a horizontal video line. However, FIT loses the simple pixel structure of FTCCD and has some of the disadvantages of FTCCD, such as requiring the storage area. This results in a large chip size and the large load clocking for the frame transfer, which results in high power dissipation. For these reasons, FIT has been used in relatively expensive camera systems such as broadcasting cameras.

4.2.2 P-Substrate Structure and P-Well Structure

In the early stage of CCD developments, ITCCD image sensors were constructed on a p-type Si substrate. With this substrate, photogenerated electrons in deep Si bulk, which is neutral electrically, diffuse homogeneously, causing a signal cross talk between neighboring photodiodes and smear, as shown in Figure 4.20. Thermal diffusion current of minority carrier (electrons) from the Si bulk also flows into the photodiodes and VCCDs and becomes a dark current noise. These diffusion currents are given by solving the one-dimensional diffusion equation along depth x from the edge of depletion region of p–n junction between the n-type of photodiode or the buried channel and the p-substrate with the appropriate boundary condition as follows[10]:

$$D_n \, d^2 n_p/dx^2 + G_L - (n_p - n_{p0})/\tau_n = 0, \qquad (4.21)$$

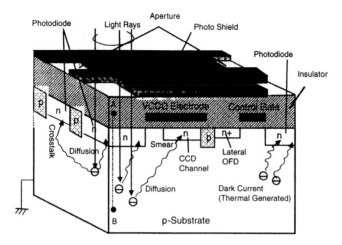

FIGURE 4.20 Problems caused by electrons generated in p-type substrate.

where

D_n is the diffusion constant of electron.
G_L is the photogeneration rate.
τ_n is the electron lifetime.
n_p is the electron density.
n_{p0} is the electron density under thermal equilibrium condition in the p-type Si substrate.

Equation 4.21 is solved for n_p as next with the boundary condition that (1) n_p becomes zero at the edge of depletion region, namely, $n_p(0) = 0$; and (2) n_p becomes the constant value at deep region in Si substrate as $n(\infty) = n_{p0} + \tau_n G_L$.

$$n_p(x) = (n_{p0} + \tau_n G_L)[1 - \exp(-x/L_n)], \quad (4.22)$$

where L_n is the diffusion length defined as $L_n = \sqrt{D_n \tau_n}$. In this case, G_L is constant to simplify the discussion. The diffusion current density, I_{DF}, flowing into the photodiode or VCCD is expressed as

$$I_{DF} = eD_n \left. (dn_p/dx) \right|_{x=0} = eD_n^{1/2} (\tau_n^{-1/2} n_{p0} + \tau_n^{1/2} G_L). \quad (4.23)$$
$$\text{Dark current} \quad \text{Photo current}$$

The equation means that, although a long lifetime of τ_n can suppress the dark current noise, it increases the photogenerated diffusion current, which results in signal cross talk and smear. In other words, there is no solution to suppress the diffusion current and the undesirable photogenerated current insofar as using a p-type substrate.

Moreover, the excess electrons generated by excessive strong light rays overflow from the photodiodes into Si bulk and diffuse into the VCCD and the photodiode. The overflowed electrons spread from a position where the excessive light is irra-

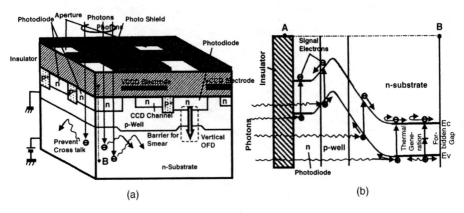

FIGURE 4.21 Advanced ITCCD constructed in p-well: (a) cross-sectional view and effects brought by reverse biased p-well; (b) potential profile from photodiode toward n-substrate.

diated to the surrounding image region as they paint the image area white as a spurious image, which is called blooming. Therefore, an antiblooming drain must be constructed on a bordered area of the photodiode to drain out the excess electrons, as shown in Figure 4.20. This is called the lateral overflow drain.[23] However, it occupies a rather large area in a pixel and prevents reduction in the pixel size.

To overcome these problems, a p-well structure was introduced in the almost CCD image sensors.[24] Figure 4.21 shows a cross-section of a typical CCD image sensor constructed in the p-well on an n-type substrate. The n-substrate is reverse biased to the p-well that is usually grounded. With this structure, a potential barrier is formed to prevent the diffusion currents from flowing into the photodiode and VCCD. In this way, the p-well structure improves sensor performance, but changes the spectral response and, especially, lowers the response in the long light wavelengths such as red and infrared bands. The spectral response can be analytically obtained by solving the diffusion equation. The theoretical responses are shown in Figure 4.22 for three junction depths together with that of a p-substrate. These spectrum response curves mean that the p-well structure can make the spectral response similar to that of human eyes. The p-well structure has been confirmed as suitable for visible image sensors.

4.2.3 ANTIBLOOMING AND LOW-NOISE PIXELS (PHOTODIODE AND VCCD)

The use of a p-substrate requires a lateral overflow drain comprising a high-density n-type drain and an overflow control electrode for antiblooming, which occupies a large area in the pixel surface as previously mentioned. In the case of the p-well structure, the n-substrate can be used as an antiblooming drain. This overflow drain, called vertical overflow drain (VOD), prevents blooming, without sacrificing an effective light-sensing area inside a pixel, because it has a three-dimensional structure.[25,26] The principle of VOD is shown in Figure 4.23.

By applying an appropriate reverse bias on the n-substrate, the p-well beneath the photodiodes is depleted where a punch-through condition from the n-substrate

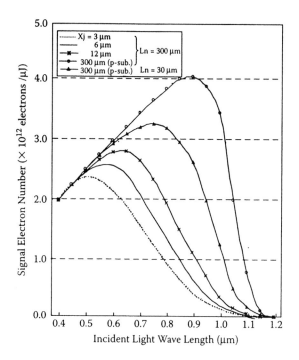

FIGURE 4.22 Calculated spectral responses for three p–well depths and two p-substrate diffusion lengths.

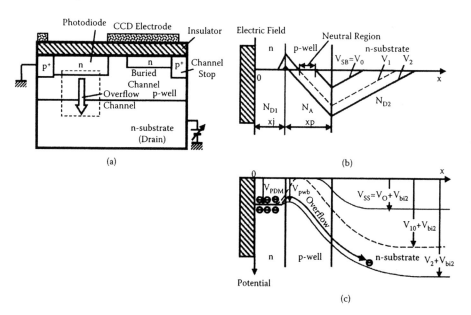

FIGURE 4.23 Principle of vertical overflow drain: (a) cross-sectional view; (b) electric fields for three reverse biases; (c) the potential profiles for three reverse biases and the overflow action for excess electrons.

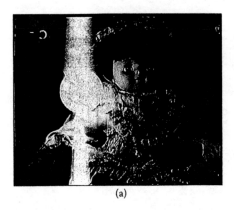 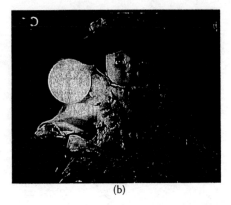

(a)   (b)

FIGURE 4.24 (See Color Figure 4.24 following page 178.) A blooming image and an improved image using VOD: (a) a spurious image caused by the blooming; (b) right image brought by VOFD.

to the n-layer of the photodiode is realized, and the electrostatic potential of the p-well is controlled by the reverse bias so as to ensure that the excess electrons are absorbed into the n-substrate. Figure 4.24 shows a blooming image and an improved image with using VOD. The antiblooming ability withstands brightness over several 10^5 lx. This technology remarkably contributes to the progress of pixel size reduction.

The other important pixel technology is the buried photodiode whose cross-section is shown in Figure 4.25 with its potential profile.[27] The high-density p-type (p^+) layer is formed in the surface region over the n-type charge storage layer of the photodiode, which is fixed at ground level and is electrically neutral. The p^+-layer suppresses the dark current generated thermally from the surface similar to the effect brought by the valence band pinning described in Section 4.1.3. In this structure, the dark current, Is, flowing into the n-type photodiode layer from the surface through the p^+ layer is given by Equation 4.16, while p_s is being replaced by the surface impurity density, N_{SA}, and is thus given by

$$|Is| = -e\, U_s = eS_{on}\, n_i^2/N_{SA} = (eS_{on}\, n_i/2) \times (2n_i/N_{SA}). \qquad (4.24)$$

Dark current in the   Suppression
depleted surface   factor

When N_{SA} is in the order of $10^{17}/cm^3$ or more, the suppression factor is less than $1/10^7$, and Is becomes negligibly small.

The charge transfer from a photodiode into the VCCD should be performed also in the complete transfer mode, as discussed in Section 4.1.1. If the electrons in the photodiode are transferred to VCCD in an incomplete transfer mode, the number of transferred signal electrons fluctuates by a thermal noise caused in the transfer channel resistance. Moreover, a number of the residual electrons with high thermal energy are emitted to VCCD during the transfer period in the next field according to Boltzmann's distribution function, even if the scene has changed to dark. This phenomenon appears as an image lag in the reproduced image.

In the case of a buried photodiode, its empty potential is lowered to 4 or 5 V, because the p^+ surface layer biased at ground level pulls down the photodiode

CCD Image Sensors

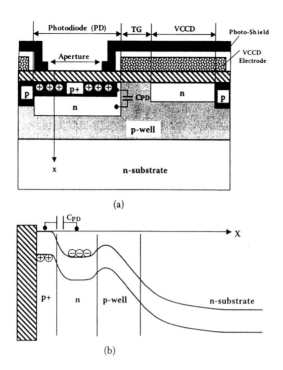

FIGURE 4.25 Buries photodiode: (a) cross-section of buried photodiode in ITCCD pixel; (b) potential profile in buried photodiode with VOFD.

potential. Therefore, a complete depletion of the photodiode is easily accomplished and the complete charge transfer from the photodiode to VCCD is achieved.[27] Moreover, the p–n junction capacity between the surface p+-layer and the n-layer of the photodiode increases the charge storage capability of the photodiode.[28]

Using these technologies, the standard pixel structure has been established as shown in Figure 4.26. In this figure, a color filter covers the pixel to sense color images, and a microlens is formed over the color filter to gather incident light rays effectively into the photodiode through the window of photo-shield metal.

4.2.4 CCD Image Sensor Characteristics

4.2.4.1 Photoelectric Conversion Characteristics

Photoelectric conversion characteristics are shown in Figure 4.27. The signal output is substantially proportional to the light intensity and the integration time. The photosensitivity is roughly proportional to the pixel size, with an assumption that the on-chip micro lenses are completely covering the pixel area. Generally, the saturation voltage (V_{sat}) is determined by the maximum storage electrons or full well capacity of the photodiode, and the dynamic range is defined as the ratio of V_{sat} to the dark noise level. S/N (signal-to-noise) ratio under illumination is determined by the number of generated signal electrons, N, to the optical shot noise given by $\sqrt{N}$, which is the fluctuation of the number of photogenerated carriers. This is given as the standard deviation of the normal distribution. S/N is given by:

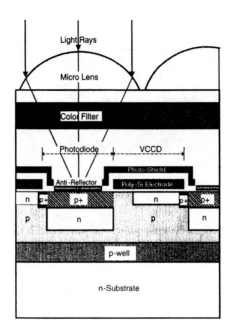

FIGURE 4.26 Cross-sectional view of an advanced ITCCD pixel.

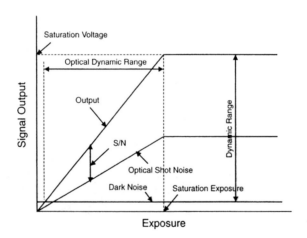

FIGURE 4.27 Photoelectric conversion characteristics.

$$S/N = \frac{N}{\sqrt{N}} = \sqrt{N} \ . \tag{4.25}$$

Thus, S/N of an illuminated image only depends on how many electrons are contained in the signal charge packets. The number of signal electrons can be calculated with dividing the output voltage by the conversion gain of the output circuitry given by Equation 4.19. On the other hand, S/N for a gloomy image is determined by dark noise, such as thermal noise and 1/f noise of the MOS source

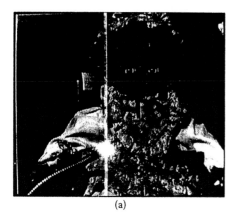
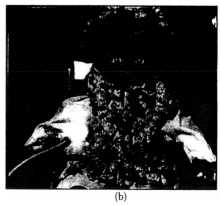

(a) (b)

FIGURE 4.28 A smear image and an improved image: (a) spurious image caused by the smear; (b) improved image brought by p-well structure.

follower circuitry.[29] The number of equivalent noise electrons under dark conditions is as small as several electrons; therefore, CCD image sensors can usually provide high S/N signal.

4.2.4.2 Smear and Blooming

Smear is a peculiar phenomenon for CCD image sensors. It is caused by unfavorable electrons such as photogenerated electrons diffusing into VCCD and electrons generated in VCCD by stray light rays. The former are drastically decreased by a use of a p-well or VOD structure, and the latter can be diminished by optimizing the photo-shield metal structure that covers VCCD. The optimization includes minimizing the space between the metal edge and the photodiode surface, as shown in Figure 4.26. The smear appears as a white vertical stripe image (Figure 4.28) and is defined as the smear-to-signal ratio under 10% vertical height illumination (see Section 6.3.6). Low smear-to-signal ratio of less than −100 dB has been accomplished with advanced CCD sensors. Blooming is sufficiently suppressed for imaging natural scenery by using VOD. Its antiblooming ability withstands brightness over several 10^5 lx, as described in Section 4.2.3.

4.2.4.3 Dark Current Noise

Almost all of the dark current, I_D, is a generation–recombination current generated from the depleted surface in VCCDs.[10] As the electrons are thermally generated from the valence band to the conduction band through the level around the midband in the forbidden energy gap, E_g, of 1.1 eV of Si semiconductor, the activation energy of the dark current becomes about $E_g/2$:

$$I_D \propto \exp(-E_g/2kT). \qquad (4.26)$$

An example of measured dark current is shown in Figure 4.29. In this case, the activation energy is about 0.6 eV, and the dark current is doubled with every 8°C

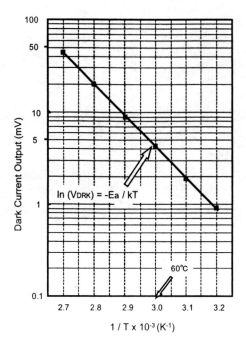

FIGURE 4.29 Dark current characteristics with temperature.

rise. Deviation in dark currents appears as FPN. Also, dark current fluctuates in time, which is observed as dark current shot noise with its standard deviation proportional to $\sqrt{N_D}$, where N_D is the average number of dark electrons.

4.2.4.4 White Blemishes and Black Defects

The white blemish appears as a white spot on the reproduced image. It is mostly brought by enormously large dark current generated in a local pixel. The generation center of such a large dark current is induced mainly by heavy metal ions such as the Fe ion, which generates energy levels near midband in Si band gap. Thus, the white blemish has thermal characteristics similar to that of the previously mentioned dark current and can be suppressed by cooling the sensor chip. On the other hand, the black defect is caused by various factors, such as a particle, dust, or residue on the pixel. In many cases, the output of the black defect is proportional to exposure, but its response is lower than that of the normal pixel. Therefore, its degree can be expressed as a percentage to the normal response.

4.2.4.5 Charge Transfer Efficiency

Recently, the charge transfer efficiency has become higher than 99.9999%; this can realize over 10,000 transfer stages with the buried channel and reduced electrode-to-electrode spacing. However, if any kind of defect exists in VCCD, it causes a local inefficiency, which appears as a vertically striped scratch.

CCD Image Sensors

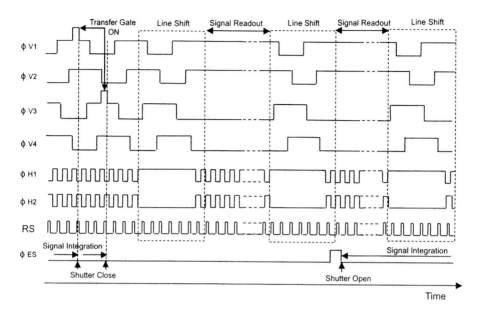

FIGURE 4.30 A timing chart to drive ITCCD image sensor.

4.2.5 OPERATION AND POWER DISSIPATION

Generally speaking, ITCCD image sensors operate with three power supplies, e.g., 15, 3, and −8 V. A typical driving pulse set is shown in Figure 4.30. VCCD is driven by three levels of pulses. The high-level pulse applied on VCCD electrodes transfers signal charge packets from photodiodes into VCCDs and -8V-to-0V swing transfers the charge packets in a vertical direction in VCCDs. When −8 V is applied, the buried channel in VCCD is forced to the valence band pinning condition, which suppresses the dark current generation in VCCD. HCCD can be driven by about 0- to 3-V swing clocks because HCCD channel width can be made wide as several tens of micrometers.

The reason that HCCD drive pulses are positive is found in the smooth transfer from the VCCD into the HCCD by the strong fringing field caused by voltage difference between a VCCD electrode of −8 V and HCCD of 3 V. The buried channel potential of HCCD is usually 7 to 9 V for the electrode voltage of 0 V, and output gate potential is 1 to 3 V higher than that of HCCD. Therefore, the FD potential should be higher than about 10 V, and the reset drain voltage is set to 12 to 15 V with operational margin in many DSC applications.

When a sufficiently high-voltage pulse (electronic shutter: ES pulse) is applied on the n-substrate, the electrons stored in the photodiode are drained into the VOD, and the charge integration is restarted by turning the ES pulse off. The electronic shutter speed is determined by the period between the turn-off timings of ES pulse and TG pulse, as shown in Figure 4.30.[30] The ES pulse voltage can be made by coupling −8 with 15 V, which becomes 23 V.

Power consumption, P_C, of driving the CCD electrodes is given by:

$$P_C = f_C C_{CL} V_C^2, \qquad (4.27)$$

where f_C is the transfer pulse frequency; C_{CL} is the load capacitance of CCD; and V_C is the transfer pulse voltage.

Because TG and ES pulses are applied once a frame period, those power dissipations are negligibly small. Generally, the load capacitance of VCCD is larger than that of HCCD by one or two orders, but the line shift frequency is lower than that of HCCD by three orders or more in ITCCD for megapixel DSC applications. Therefore, the power is dissipated mainly in driving HCCD. Another dominant factor of power dissipation is in the source follower current buffer. For example, the total power consumption in a 3-Mpixel image sensor becomes about 120 mW with 24-MHz data rate, 3.3-V HCCD clocking, and a drain voltage of 15 V. The important design point for saving power dissipation is in how to reduce the HCCD clocking voltage and the source follower drain voltage.

4.3 DSC APPLICATIONS

4.3.1 Requirements from DSC Applications

Required items for image sensors from digital still camera (DSC) applications are summarized as follows:

1. High resolution to take high-quality pictures comparable with those taken by conventional film cameras, which requires the number of pixels to be at least 2 million or more
2. High sensitivity to make a high S/N picture and to allow a high-speed shot to avoid fussy pictures caused by camera shake
3. Wide dynamic range to cover a scene containing objects from dark to high light
4. Synchronous progressive scan to use the complete electronic shutter displacing mechanical shutters
5. Low smear and antiblooming, especially for the mechanical shutterless cameras
6. Low dark current to achieve long integration time needed for night view photography, etc.
7. Reproduction of real color images
8. Compatibility of a high-resolution still picture and a high-frame-rate movie
9. Low power consumption, compact size, low cost, etc.

Generally, the pixel number should be increased without enlarging the imaging size, i.e., the chip size, to keep the sensor cost low and the optical lens system compatible. Figure 4.31 shows pixel size vs. pixel number for widely used imaging sizes of consumer DSCs. As shown, pixel size is reduced in inverse proportion to the increase in pixel number. The pixel size reduction lowers the sensitivity and narrows the dynamic range by the same reduction factor because they are roughly proportional to pixel size. The sensitivity and the dynamic range are main factors

CCD Image Sensors

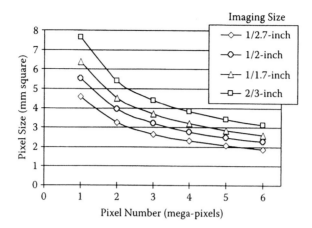

FIGURE 4.31 Pixel size vs. pixel number.

to determine the image sensor performance as well as the resolution. High resolution (1) is thus incompatible with high sensitivity (2) and wide dynamic range (3). The design challenge in making the pixel small is in how to reduce a loss of the incident light rays by optimizing the on-chip microlenses and introducing an antireflection layer on the Si–SiO$_2$ interface. How to enlarge the charge storage and transfer capability by optimizing the pixel pattern layout and the impurity profiles in Si is also a challenge. The cross-section of an advanced pixel is shown in Figure 4.26.

ITCCD can realize the synchronous progressive scan (4) by using the three-layer poly-silicon electrode technology.[31] However, this technology requires an extra and sophisticated fabrication process and reduces the active area in a pixel, as discussed in Section 4.3.2. Therefore, a lot of DSCs unwillingly adopt the interlace scan type with use of a mechanical shutter. Low-smear, antiblooming (5), and low dark current (6) are basic requirements for image sensors. The reproduction of a real color image (7) is especially emphasized in DSC applications. This requirement leads to the use of primary color filters of red, green, and blue, from the use of complementary color filters used in almost all image sensors for video camcorder applications. The most popular primary color filter pattern is the Bayer arrangement shown in Figure 4.32. Low power consumption, compact size, and low cost (9) are also common requests. The compatibility of a high-resolution still picture and a high-frame rate-movie (8) is an individual requirement from DSC applications. This subject will be discussed in Section 4.3.5.

4.3.2 INTERLACE SCAN AND PROGRESSIVE SCAN

In DSC applications, two types of ITCCD image sensor have been mainly used: an interlace scan type and a progressive scan type. A signal readout sequence of the interlace scan type is shown in Figure 4.33. At first, signal charge packets in the odd number of horizontal lines are transferred from the photodiodes into VCCDs. Next, all the signal packets in VCCDs are transferred into HCCD line by line. After all the charge packets in the odd lines (called an odd field) are read out, the signal charge packets in the even number of horizontal lines are transferred from the

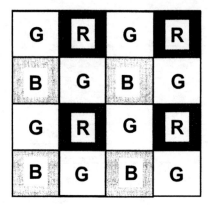

FIGURE 4.32 Bayer color filter arrangement.

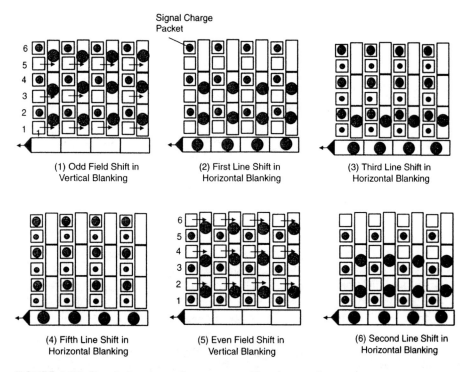

FIGURE 4.33 Signal charge transfer sequence of interlace readout.

photodiodes into VCCDs and then to HCCD, forming an even field. This scan type fits an interlace scan format movie such as NTSC or PAL. The image shooting time of the even field differs from that of the odd field in this mode.

For DSC application, signal charge integration starts at the same time for the even and odd fields, but the fields are read out sequentially after the charge integration is completed. Therefore, a precise mechanical shutter is required for using the interlace scan type image sensor to block light during the charge readout period.

On the other hand, the progressive scan type CCD transfers all the signal charge packets from photodiodes into VCCD simultaneously, as shown in Figure 4.34. Thus,

CCD Image Sensors

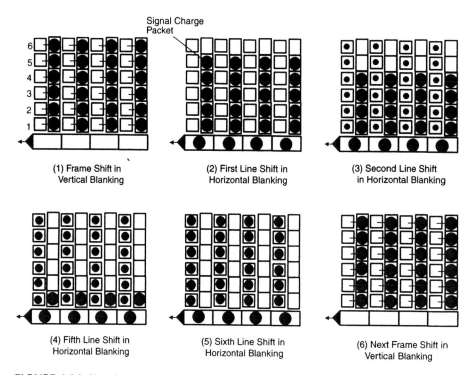

FIGURE 4.34 Signal charge transfer sequence of progressive readout.

each signal charge packet in a frame has the information from the same shooting time. As a result, this type of CCD can provide nonfailed still pictures without a mechanical shutter, even though the object is moving.

Images taken by an interlaced scan CCD and a progressive scan CCD, both without a mechanical shutter, are shown in Figure 4.35. In this figure, an image taken by a line-address type imager, such as a CMOS image sensor, is also presented.

From the fabrication point of view, the interlace scan type can be easily made with a standard double-layer poly-silicon technology. However, the progressive scan type needs the triple-layer poly-silicon electrodes and thus requires an extra sophisticated fabrication process. The two types of pixel structures are shown in Figure 4.36. As shown, a large area in the imaging region is occupied by extra electrode wiring between VCCDs in the progressive scan type; this region acts as dead space for the silicon bulk. These structural complexities and active space loss prevent reducing the pixel size and increasing the pixel number. Therefore, the most high-resolution DSCs have adopted the interlace scan type using a precise mechanical shutter. When a mechanical shutter is used, storage electrons in a photodiode are thermally emitted to antiblooming drain after closing of the mechanical shutter, as shown in Figure 4.37. This results in a severe loss of the maximum signal voltage (the saturation voltage). To suppress this phenomenon, bias level for the overflow control should be modulated to a lower level at a timing before or after closing the mechanical shutter.

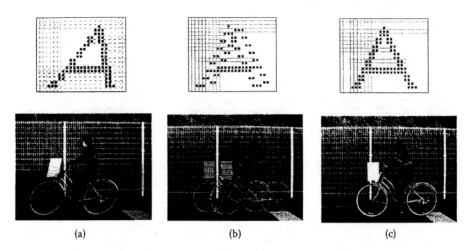

FIGURE 4.35 (See Color **Figure 4.35** following page 178.) Shot images: (a) a shot image taken by raster (or line address) scan CMOS imager; (b) a shot image taken by interlace scan CCD imager; (c) a shot image taken by progressive scan CCD imager.

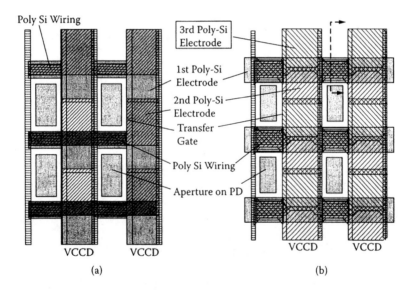

FIGURE 4.36 A pattern layouts of ITCCD: (a) interlace scan type; (b) progressive scan-type.

4.3.3 Imaging Operation

As previously mentioned, a DSC can realize a complete electronic shutter by using the progressive scan CCD image sensor. Figure 4.38 shows the principle of the complete electronic shutter operation. At first, the integration pulse, Φ_{ES}, is superimposed upon the overflow control bias applied to the n-type substrate to drain electrons integrated in a photodiode. It starts the integration of photogenerated electrons, which corresponds to opening the electronic shutter. After the exposing period for imaging an object, the transfer gate pulse, Φ_{TG}, is applied on the VCCD

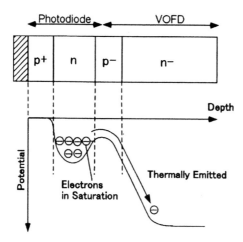

FIGURE 4.37 Thermal emission from filled photodiode into n-substrate.

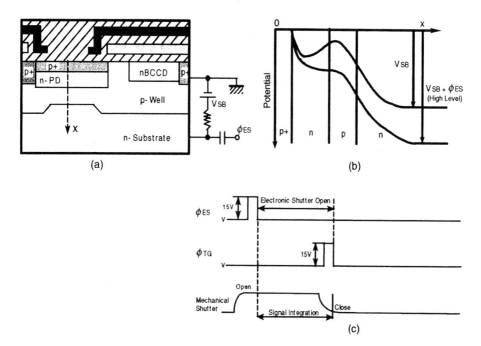

FIGURE 4.38 Principle of electronic shutter: (a) cross-sectional view of ITCCD pixel; (b) potential profiles for signal charge integration and integration clear condition; (c) a timing chart for electronic shutter operation.

electrode, and the integrated electron packet is transferred from the PD into the VCCD. This operation corresponds to the action of closing the electronic shutter. The period between the falling edges of Φ_{ES} and Φ_{TG} becomes the shutter speed. It is easy to set the period shorter than 100 μsec; thus, a super high-speed shutter action over 1/10,000 sec can be realized. An example picture shot with 1/10,000 sec is shown in Figure 4.39.

FIGURE 4.39 (See **Color Figure 4.39 following page 178.**) A picture shot with 1/10,000 sec electronic shutter.

In the case of using a mechanical shutter for a consumer use, the shutter-opening action is done by applying Φ_{ES} on the substrate in the same manner as for the electronic shutter and the shooting is finished by closing the mechanical shutter. This sequence is also shown in Figure 4.38. Generally, the time response of the mechanical shutter is not so fast, the super high-speed shutter is impossible, and the maximum speed is an inaccurate 1/2000 sec at the best. On the other hand, in the case of single-lens reflex (SLR) DSCs, all shutter actions are generally done by the focal plane type mechanical shutter, of which shutter speed is 1/4000 sec (max.). Recently, a kind of digital SLR was introduced that offers a super high-speed electronic shutter — for example, 1/8000 sec — by using a progressive scan CCD.

The use of a mechanical shutter has two advantages that provide completely smear-less imaging and multifield readout. After closing the mechanical shutter, smear electrons accumulated in VCCD can be swept out by clocking the VCCDs before transferring signal electrons from photodiodes into the VCCDs. Then, the signal electrons can be transferred and read out through the clean VCCDs. The multifield readout is a method expanded from the previously mentioned interlace scan: namely, all signal packets are read out separately within three fields, four fields, or more. In a case in which an ITCCD, which has interlace scan type electrodes as shown in Figure 4.36(a), is operated with the four-field readout, eight electrodes can be used for a transfer stage of eight-phase VCCD.[32] Because the eight-phase VCCD has a transfer capability three times larger than that of the four-phase VCCD from Equation 4.18, the channel width of VCCD can be practically narrowed to the utmost. This contributes to reduce the pixel size.

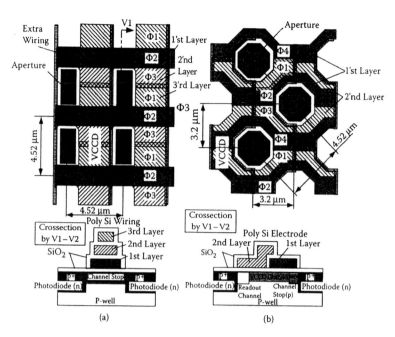

FIGURE 4.40 Pixel pattern layout of (a) ITCCD and (b) PIACCD.

4.3.4 PIACCD (Super CCD)

As previously mentioned, ITCCD fits to the interlace scan, but it is not basically suited for the progressive scan because it has been developed for movie applications with interlace scan formats such as NTSC or PAL. Moreover, it suffers from an essential loss of an active area in the pixel because of the extra wiring to supply transfer pulses to every VCCD electrode.

The pixel interleaved array CCD (PIACCD) has been developed as an image sensor suitable for the progressive scan, which can be fabricated with a standard double poly-silicon technology.[33] The pixel pattern layout of PIACCD compared with ITCCDs is shown in Figure 4.40. As shown, the four-phase VCCDs are composed of standard double-layer poly-Si electrodes of $\Phi 1$, $\Phi 2$, $\Phi 3$, and $\Phi 4$. Each VCCD channel meanders along the photodiodes and borders on the next VCCD channel without extra wiring, which has 1.5 times larger charge handling capability than a three-phase CCD used well for ITCCD in accordance with Equation 4.17. Curves in Figure 4.41 are the calculated relative active areas in ITCCD and PIACCD for pixel sizes using the design rule shown in the figure.

As shown, PIACCD enlarges the relative active area in the pixel by 1.3 times against ITCCDs. The saturation voltage, V_{SAT}, of PIACCD is also about 1.3 times larger than that of ITCCD because V_{SAT} is roughly proportional to this active area. In Figure 4.40, the octagonal regions are photodiodes, and the gray area is an aperture opened above each photodiode. The aperture size on the 1.3 times enlarged photodiode is expanded by 1.4 times with constant photo-shield brim length. This enlarged equilateral aperture has the advantage of gathering more rays of light passing through

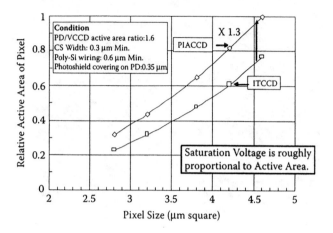

FIGURE 4.41 Relative active area for pixel size.

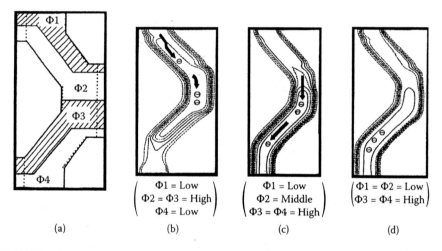

FIGURE 4.42 Channel potential profiles in PIACCD. (a) Electrode pattern layout; (b–d) potential contours formed in the channel.

an on-chip microlens than the oblong aperture of ITCCD.[34] The high spatial efficiency also means that PIACCD is suitable for finer pixel integration, especially for progressive scan applications.

PIACCD has the uniquely shaped four-phase electrodes and two times longer transfer channel in the pixel. The charge transfer process in this channel is presented with the potential contour shown in Figure 4.42; these are calculated with a three-dimensional device simulator.[35] The widened Φ1–Φ2 interface shortens the effective transfer length, L_{eff}, and shortens the charge transfer time τ because τ is roughly given by the relation of

$$\tau \propto L_{eff}^2 .$$
(4.28)[3]

CCD Image Sensors

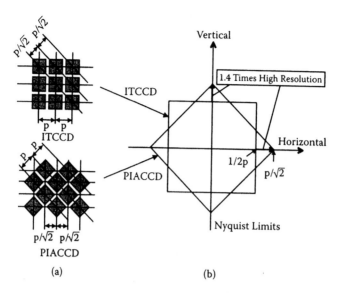

FIGURE 4.43 (a) Pixel arrangement and (b) Nyquist limits of ITCCD and PIACCD.

Moreover, the electric field from the narrow electrode portion toward the wide electrode portion beneath $\Phi 2$ and $\Phi 4$, which is generated by the narrow channel effect, accelerates the charge transfer in this VCCD channel.[36] The potential profile behavior in Figure 4.42 means that PIACCD can transfer the large number of signal electrons smoothly without sacrificing the transfer speed.

The resolution characteristics of PIACCD differ from that of ITCCD. Figure 4.43(a) shows the pixel arrangement of ITCCD and PIACCD. The horizontal and vertical pixel pitch in ITCCD is p, and that of PIACCD is $p/2^{1/2}$. Because these pixel pitches are the same as the spatial sampling pitches, the Nyquist limits of ITCCD and PIACCD are given as $1/(2p)$ and $1/(2^{1/2}p)$, respectively, in the horizontal and the vertical axes, as shown in Figure 4.43(b). This means that PIACCD widens the spatial frequency response — namely, heightens the resolution by $2^{1/2}$ in the horizontal and the vertical directions. However, in the 45° tilted directions, this resolution supremacy moves to ITCCD.

On the other hand, it is reported that human eyes are most sensitive for vertical and horizontal fine patterns, and the high-frequency spatial power spectrum of nature scenes concentrates in the vertical and horizontal directions, as shown in Figure 4.44 and Figure 4.45.[37,38] Therefore, the resolution characteristics of the spatially interleaved pixel arrangement of PIACCD fit the properties of the human eyes and natural scenery.

The spatially interleaved pixels can make the same number of virtual pixels in between the real pixels by a signal processing of pixel interpolation, as shown Figure 4.46; the effective pixel number is increased by twofold. A block diagram of PIACCD with color filter arrangement is shown in Figure 4.47.

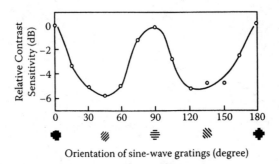

FIGURE 4.44 Spatial response of the human visual system. (Data from Watanabe, A. et al., *Vision Res.* 8, 1245–1263, 1968.)

FIGURE 4.45 Average spatial power spectrum distribution of about 500 scenes.

4.3.5 HIGH-RESOLUTION STILL PICTURE AND HIGH-FRAME-RATE MOVIE

In the DSC system, it is no wonder that a real-time movie displayed on an electronic viewer such as a liquid crystal display is used to confirm the objects and take the favorite snapshot. The monitor movie can be easily produced from the image sensors, which have less than about 1.3 Mpixels (SXGA). Moreover, it has become a sales point that the high-grade movie can be played back on a TV screen.

In the Bayer filter arrangement of ITCCD, each video line has merely two color signals: G and R or G and B, as shown in Figure 4.48(a). Therefore, joining two video lines is required to make one scanning line by using a line memory for a real-

CCD Image Sensors

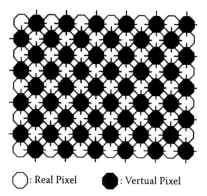

FIGURE 4.46 A concept of pixel interpolation for reproducing images.

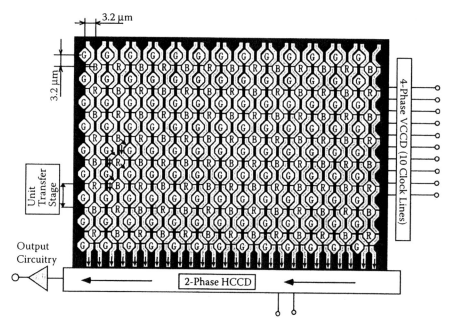

FIGURE 4.47 Block diagram of the fabricated PIACCD.

time video monitor, which needs to contain R, G, and B signals in every line, as shown in Figure 4.48(c). On the other hand, because each video line of PIACCD has three color signals of R, G, and B, a monitor scanning line can easily be made from any selected line without using a line memory, as shown in Figure 4.48(b) and (d). However, as the number of pixels increases over 2 or 3 millions, it becomes difficult to make high-grade movies such as high-frame-rate VGA only by extracting the video lines.

The line extraction can be easily done by supplying the transfer pulses to the selected VCCD electrodes, which decreases the number of vertical pixels. However, this does not decrease the number of horizontal pixels in each video line. In case of 3-Mpixel PIACCD (1408 × 2 pixels in a horizontal line by 1060 video lines), the

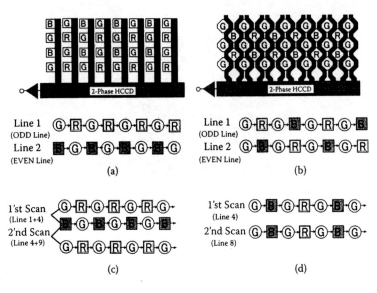

FIGURE 4.48 Full frame signal sequences of (a) ITCCD(Bayer) and (b) PIACCD, and the every fourth line selected signal sequences of (c) ITCCD (Bayer) and (d) PIACCD.

number of movie lines is decreased by half (530 lines) with every second line addressing. However, the number of horizontal pixels is too large to make a VGA movie and requires about 72-MHz data rate for 30 fps, resulting in useless high-power dissipation. Thus, the technology to decrease the number of horizontal pixels is the key for making a high-quality movie compatible with a high-resolution still picture.

Architecture that can mix the same color signal charge packets in a horizontal line for decreasing the number of horizontal pixels has been developed. It is composed of a CCD line memory (LM) bounded on the final VCCD stage and multiphase HCCD, e.g., four, six, or eight phases.[39] A block diagram of the PIACCD with the charge mixing circuitry is shown in Figure 4.49. The number of transfer stages of HCCD is reduced by half to transfer the mixed charge packets without increasing the HCCD driving frequency twofold. Signal charge packets in a horizontal line are transferred from VCCD into the line memory in parallel, and the charge packets selected from the horizontal line stored there are transferred into HCCD by applying a high-level clock on the linked HCCD electrodes selectively.

This selective charge transfer mechanism is shown in Figure 4.50. The charge packet is only transferred in the condition in which the line memory is biased at a low level and the HCCD electrode is biased at a high level. The charge packet mixing process is shown in Figure 4.51. By clocking HCCD with multiphase pulses, each charge packet selected first and transferred into HCCD can be transferred toward left in the figure as far as the electrode-bounded LM site in which the charge packet to be mixed is stored. Next, the charge packets kept in the LM are transferred into the HCCD to be mixed, respectively, with the partner charge packets by the previously mentioned selective transfer process. As a result, each color signal charge

CCD Image Sensors

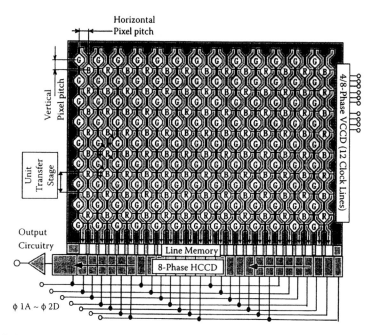

FIGURE 4.49 Block diagram of the PIACCD with the charge mixing circuitry.

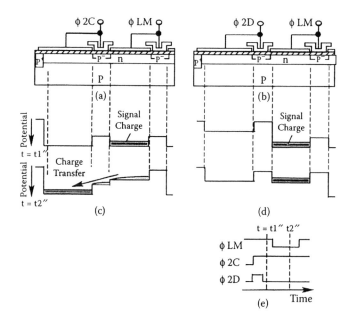

FIGURE 4.50 Selective charge transfer mechanism: (a) cross-section of a selected channel; (b) cross-section of an unselected channel; (c) selected channel potential profiles; (d) unselected channel potential profiles; (e) timing diagram.

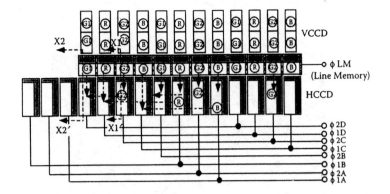

FIGURE 4.51 Charge packet mixing process.

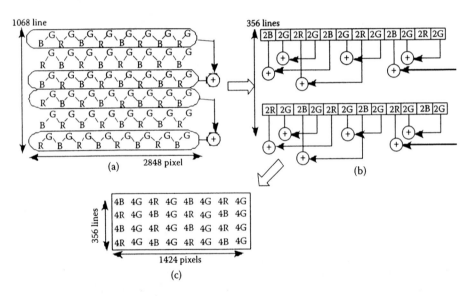

FIGURE 4.52 Combination of the charge mixing of vertical and horizontal: (a) horizontal line selection; (b) signal mixing in horizontal; (c) output signal arrangement for movie.

packet of G, R, and B is mixed, respectively, with the neighboring same color signal, and the signal sequence in the horizontal line becomes 2G-2R-2G-2B-2G-2R-2G-2B.

With introducing the horizontal charge mixing, a real-time movie with a pixel number of 1408 (H) × 530 (V) can be output at 30 fps with 36-MHz data rate from the 3-Mpixel image sensor. This operation provides a high-quality VGA picture because 708 green pixels are contained in the each video line. The charge packet mixing is also applicable to the mixing between video lines, which becomes a vertical charge packet mixing. Figure 4.52 shows an example of a combination of vertical and horizontal charge mixing. In this case, each color signal output is enlarged

CCD Image Sensors

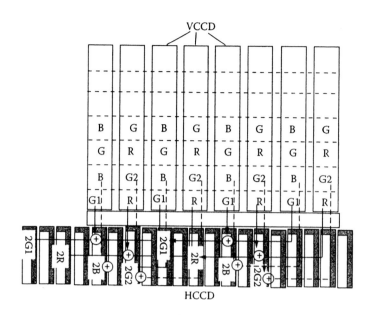

FIGURE 4.53 Another approach for horizontal charge packet mixing.

fourfold, which means that the mixed signals have four times higher sensitivity. The high sensitivity is especially important to monitor and shoot dark scenes.

Another approach for horizontal charge packet mixing is shown in Figure 4.53. In this case, the number of transfer stages in HCCD is not reduced by half, and the mixed charge packets of a horizontal line are distributed in every second stage. The empty transfer stages are used to mix the charge packet pair in the following horizontal video line. This means that the two movie lines are multiplexed in HCCD and output, one after the other.

4.3.6 SYSTEM SOLUTION FOR USING CCD IMAGE SENSORS

The fabrication technology to make a CCD image sensor differs from that of general CMOS logic, and the Si chip size is not so small, especially over a megapixel case. Therefore, it is not efficient to integrate peripheral circuitries, i.e., timing generator (TG); clock drivers; CDS circuitry; analog to digital converter (ADC); and image signal processor (ISP), in one chip. A CCD camera system uses three or four ICs as a chip set. An example of a solution for a CCD imaging system is represented in Figure 4.54. The VCCD driver chip contains the power supply circuitry and supplies VCCD driving pulses of a 0- to −7- or −8-V swing; TG pulse of a 12- to 15-V swing; DC power of 12 to 15 V for operating the output circuitry in a CCD chip and so on. The analog front-end (AFE) chip is composed of the CDS circuitry, ADC, and TG with HCCD driver. The various functions of ISP are described in Chapter 8 and Chapter 9.

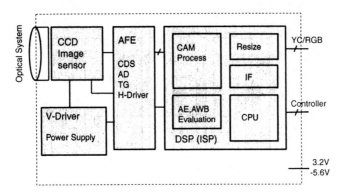

FIGURE 4.54 Block diagram of a solution for CCD imaging system.

4.4 FUTURE PROSPECTS

The development of CCD image sensors for DSC application has been focused mainly on how to reduce the pixel size and how to increase the number of pixels. The pixel size has already been reduced to about 2 μm square. However, the pixel reduction degrades image quality due to lowering sensitivity and narrowing dynamic range. The resolution made from a DSC with over 6 million pixels will be too high for consumer use because the resolution is already comparable to that of the film camera. The dynamic range of the pictures taken by a DSC is extremely narrow in comparison with that taken by a film camera, which has very wide latitude.

One of the key technologies on the next development stage for CCD image sensors will be in widening the dynamic range to be compatible to the photo films. The wide dynamic range for DSC means large saturation exposure, which is provided with low sensitivity as shown in Figure 4.55. However, high sensitivity is the most important performance measure for image sensors, especially when shooting dark scenes. The solution to causing wide dynamic range (wide latitude) and high sensitivity to go together is in combining the high- and low-sensitivity characteristics.

As an approach to realize this concept, a pixel with two photodiodes has been developed; here, a pixel is divided two areas: one for a photodiode with a high-sensitivity, large area and the other for a low-sensitivity, small area photodiode, as shown in Figure 4.56.[40] The imaging signals picked up at the high-sensitivity photodiodes and that of the low-sensitivity photodiodes are readout individually, and both of the images are compounded in the ISP with appropriate signal processing. The dynamic range is determined by the saturation exposure of the low-sensitivity photodiodes. The reason to read out both images individually is to prevent FPN appearing on pictures caused by the variation of saturation voltage. If the saturation voltages of every pixel could be flat, the high- and low-sensitivity signals can be mixed in CCD. This will simplify the signal processing drastically and determine an epoch in wide dynamic technology with utilizing the saturation property.

The compatibility between a high-resolution still and a high-quality movie on the grade higher than VGA, such as the high-definition (HD) movie, will be important and requires a new design to increase data rate. The solution for this requirement

CCD Image Sensors

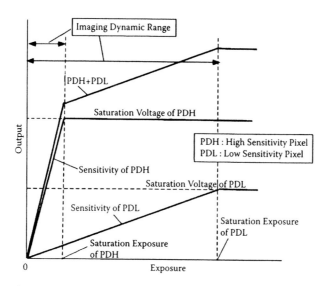

FIGURE 4.55 Principle of widening dynamic range without sacrificing high sensitivity.

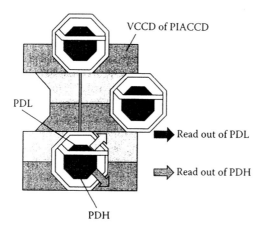

FIGURE 4.56 Double photodiode pixels for wide dynamic range imaging.

may be in the parallel signal readout technologies. Obviously, lowering the power dissipation is also very important for CCD image sensors. The key point to lowering the power dissipation will be in the low voltage driving of CCD and the output circuitry. Thus, CCD image sensor technologies hereafter will be expanded into various directions for enhancing performances and creating new functions to meet the requirements for future DSC systems.

REFERENCES

1. W.S. Boyle and G.E. Smith, Charge-coupled semiconductor devices, *Bell Syst. Tech. J.*, 49, 587–593, 1970.

2. C.H. Sequin and M.F. Tompsett, *Charge Transfer Devices*, Academic Press, New York, 1975.
3. S.M. Sze, *Physics of Semiconductor Devices*, 2nd ed., A Wiley-Interscience Publication, John Wiley & Sons, New York, 415, 1981.
4. D.F. Barbe, Imaging devices using the charge-coupled concept, *Proc. IEEE*, 63(1), 38–67, 1975.
5. G.F. Amelio, W.J. Bertram, Jr.,and M.F. Tompsett, Charge-coupled imaging devices: design considerations, *IEEE Trans. Electron Devices*, ED-18(1)1, 986–992, 1971.
6. J.E. Carnes and W.F. Kosonocky, First interface-state losses in charge-coupled devices, *Appl. Phys. Lett.*, 20, 261–263, 1972.
7. R.H. Walden, R.H. Krambeck, R.J. Strain, J. McKenna, N.L. Schryer, and G.E. Smith, The buried channel charge coupled devices, *Bell Syst. Tech. J.*, 51, 1635–1640, 1972.
8. A.W. Lees and W.D. Ryan, A simple model of a buried-channel charge-coupled device, *Solid-State Electron.*, 17, 1163–1169, 1974.
9. T. Yamada, H. Okano, and N. Suzuki, The evaluation of buried channel layer in BCCDs, *IEEE Trans. Electron Devices*, ED-25(5), 544–546, 1978.
10. A.S. Grove, *Physics and Technology of Semiconductor Devices*, John Wiley & Sons, Inc., New York 136–140, 267, 1967.
11. T. Yamada, H. Okano, K. Sekine, and N. Suzuki, Dark current characteristics in a buried channel CCD, Extended Abs. (38th autumn meeting); *The Japan Soc. Appl. Phys.*, 258, 1977.
12. N.S. Saks, A technique for suppressing dark current generated by interface states in buried channel CCD imagers, *IEEE Electron Device Lett.*, EDL-1(7), 131–133, 1980.
13. W.F. Kosonocky and J.E. Carnes, Two-phase charge coupled devices with overlapping polysilicon and aluminum gates, *RCA Rev.*, 34, 164–202, 1973.
14. D.M. Erb, W. Kotyczka, S.C. Su, C. Wang, and G. Clough, An overlapping electrode buried channel CCD, *IEDM, Tech. Digest*, December, 24–26, 1973.
15. T. Yamada, K. Ikeda, and N. Suzuki, A line-address CCD image sensor, *ISSCC Dig. Tech. Papers*, February, 106–107, 1987.
16. M.H. White, D.R. Lampe, F.C. Blaha, and I.A. Mack, Characterization of surface channel CCD imaging arrays at low light levels, *IEEE Trans. Solid-State Circuits*, SC-9, February, 1–13, 1974.
17. T. Yamada, T. Yanai, and T. Kaneko, 2/3 inch 400,000 pixel CCD area image sensor, Toshiba Rev., No.162, 16–20, Winter 1987.
18. C.H. Séquin, F.J. Morris, T.A. Shankoff, M.F. Tompsett, and E.J. Zimany, Charge-coupled area image sensor using three levels of polysilicon, *IEEE Trans. Electron Devices*, ED-21, 712–720, 1974.
19. J. Hynecek, Virtual phase technology: a new approach to fabrication of large-area CCDs, *IEEE Trans. Electron Devices*, ED-28, 483–489, May 1981.
20. E.G. Stevens, T.-H. Lee, D.N. Nichols, C.N. Anagnostpoulos, B.C. Berkey, W.C. Chang, T.M. Kelly, R.P. Khosla, D.L. Losee, and T.J. Tredwell, A 1.4-million-element CCD image sensor, *ISSCC Dig. Tech. Papers*, 114–115, Feb. 1987.
21. G.F. Amelio, Physics and applications of charge coupled devices, *IEEE INTERCON*, New York, Digest, 6, paper 1/3, 1973.
22. K. Horii, T. Kuroda, and S. Matsumoto, A new configuration of CCD imager with a very low smear level FIT-CCD imager, *IEEE Trans. Electron Devices*, ED-31(7), 904–909, 1984.
23. A. Furukawa, Y. Matsunaga, N. Suzuki, N. Harada, Y. Endo, Y. Hayashimoto, S. Sato, Y. Egawa, and O. Yoshida, An interline transfer CCD for a single sensor 2/3 color camera, *IEDM Tech. Dig.*, December, 346–349, 1980.

24. T. Yamada, H. Goto, A. Shudo, and N. Suzuki, A 3648 element CCD linear image sensor, *IEDM Tech. Dig.*, December, 320–323, 1982.
25. H. Goto, H. Sekine, T. Yamada, and N. Suzuki, CCD linear image sensor with buried overflow drain structure, *Electron. Lett.*, 17(24), 904–905, 1981.
26. Y. Ishihara, E. Oda, H. Tanigawa, N. Teranishi, E. Takeuchi, I. Akiyama, K. Arai, M. Nishimura, and T. Kamata, Interline CCD image sensor with an anti blooming structure, *ISSCC Dig. Tech. Papers*, February, 168–169, 1982.
27. N. Teranishi, A. Kohno, Y. Ishihara, E. Oda, and K. Arai, No image lag photodiode structure in the interline CCD image sensor, *IEDM Tech. Dig.*, December, 324–327, 1982.
28. Y. Matsunaga and N. Suzuki, An interline transfer CCD imager, *ISSCC Dig. Tech. Papers*, February, 32–33, 1984.
29. J.E. Carnes and W.F. Kosonocky, Noise sources in charge-coupled devices, *RCA Rev.*, 33, 327–343, 1972.
30. M. Hamasaki, T. Suzuki, Y. Kagawa, K. Ishikawa, M. Miyata, and H. Kambe, An IT-CCD imager with electronically variable shutter speed, *ITEJ Tech. Rep.*, 12(12), 31–36, 1988.
31. T. Ishigami, A. Kobayashi, Y. Naito, A. Izumi, T. Hanagata, and K. Nakashima, A 1/2-in 380k-pixel progressive scan CCD image sensor, *ITE Tech. Rep.*, 17(16), 39–44, March, 1993.
32. S. Uya, N. Suzuki, K. Ogawa, T. Toma and T. Yamada, A 1/2 format 2.17M square pixel CCD image sensor, *ITE Tech. Rep.*, 25(28), 73–77, 2001.
33. T. Yamada, K. Ikeda, Y.G. Kim, H. Wakoh, T. Toma, T. Sakamoto, K. Ogawa, E. Okamoto, K. Masukane, K. Oda, and M. Inuiya. A progressive scan CCD image sensor for DSC applications, *IEEE J. Solid-State Circuits*, 35(12), 2044–2054, 2000.
34. H. Mutoh, 3-D optical and electrical simulation for CMOS image sensor, in Program of 2003 IEEE Workshop on CCD and AIS, Session 2, 2004.
35. H. Mutoh, Simulation for 3-D optical and electrical analysis of CCD, *IEEE Trans. Electron Devices*, 44(10), 1604–1610, October, 1997.
36. Y.D. Hagiwara, M. Abe, and C. Okada, A 380H × 488V CCD imager with narrow channel transfer gates, Proc. 10th Conf. Solid State Devices, Tokyo, 1978; JJAP, 18, Suppl. 18-1, 335–340, 1978.
37. A. Watanabe, T. Mori, S. Nagata, and K. Hiwatashi, Spatial sine-wave responses of the human visual system, *Vision Res.* 8, 1245–1263, 1968.
38. M. Tamaru, M. Inuiya, T. Misawa, and T. Yamada, Development of new structure CCD for digital still camera, *ITE Tech. Rep.*, 23(75), 31–36, 1999.
39. T. Misawa, N. Kubo, M. Inuiya, K. Ikeda, K. Fujisawa, and T. Yamada, Development of the 3,300,000 pixels CCD image sensor with 30 fps VGA movie readout function, *ITE Tech. Rep.*, 26(26), 65–70, 2002.
40. K. Oda, H. Kobayasi, K. Takemura, Y. Takeuchi, and T. Yamada, The development of wide dynamic range image sensor, *ITE Tech. Rep.*, 27(25), 17–20, 2003.

5 CMOS Image Sensors

Isao Takayanagi

CONTENTS

5.1 Introduction to CMOS Image Sensors 144
    5.1.1 Concepts of CMOS Image Sensors 144
    5.1.2 Basic Architecture 147
        5.1.2.1 Pixel and Pixel Array 147
        5.1.2.2 X–Y Pixel Addressing 148
        5.1.2.3 Fixed Pattern Noise Suppression 148
        5.1.2.4 Output Stage 149
        5.1.2.5 Other Peripherals 149
    5.1.3 Pixel-Addressing and Signal-Processing Architectures 150
        5.1.3.1 Pixel Serial Readout Architecture 150
        5.1.3.2 Column Parallel Readout Architecture 150
        5.1.3.3 Pixel Parallel Readout Architecture 151
    5.1.4 Rolling Shutter and Global Shutter 151
    5.1.5 Power Consumption 151
5.2 CMOS Active Pixel Technology 153
    5.2.1 PN Photodiode Pixels 153
        5.2.1.1 PN Photodiode Structure 154
        5.2.1.2 Photodiode Reset Noise (kTC noise) 154
        5.2.1.3 Hard Reset and Soft Reset 155
        5.2.1.4 Reset Noise Suppression in PN Photodiode Pixels 155
    5.2.2 Pinned Photodiode Pixels 156
        5.2.2.1 Low-Voltage Charge Transfer 158
        5.2.2.2 Boosting 158
        5.2.2.3 Simplification of Pixel Configuration 158
        5.2.2.4 P- and N-Type Substrates 159
    5.2.3 Other Pixel Structures for Large-Format, High-Resolution
        CMOS Image Sensors 160
        5.2.3.1 LBCAST 161
        5.2.3.2 Vertical Integration Photodiode Pixel 161
5.3 Signal Processing and Noise Behavior 161
    5.3.1 Pixel Signal Readout and FPN Suppression Circuit 162
        5.3.1.1 Column Correlated Double Sampling (CDS)
                  Scheme 162
        5.3.1.2 Differential Delta Sampling (DDS) 163

                5.3.2   Analog Front End..164
                        5.3.2.1   Serial PGA and ADC 165
                        5.3.2.2   Column Parallel PGA and ADC 165
                5.3.3   Noise in CMOS Image Sensors..165
                        5.3.3.1   Pixel-to-Pixel Random Noise 166
                        5.3.3.2   Row-Wise and Column-Wise Noise....................................167
                        5.3.3.3   Shading 167
                        5.3.3.4   Smear and Blooming 168
5.4   CMOS Image Sensors for DSC Applications 168
        5.4.1   On-Chip Integration vs. Classes of DSC Products 168
                        5.4.1.1   Toy Cameras 168
                        5.4.1.2   Middle-Class Compact DSCs..169
                        5.4.1.3   Digital SLR Cameras...169
        5.4.2   Operation Sequence for Capturing Still Images 169
                        5.4.2.1   Synchronization of Rolling Reset and Mechanical
                                  Shutter 170
                        5.4.2.2   Synchronization of Global Reset and Mechanical
                                  Shutter 170
                        5.4.2.3   Electronic Global Shutter 171
        5.4.3   Movie and AE/AF Modes 172
                        5.4.3.1   Subresolution Readout...173
                        5.4.3.2   Pixel Binning 173
                        5.4.3.3   Mode Change and Dead Frames 174
5.5   Future Prospects of CMOS Image Sensors for DSC Applications 175
References..176

This chapter discusses the motivation for developing complementary metal-oxide semiconductor (CMOS) image sensor technology and its digital still camera (DSC) applications. Basic structures, operational methods, and features of CMOS image sensors are introduced in Section 5.1 and CMOS pixel technologies are explained in Section 5.2. CMOS architectures and design techniques, along with some interesting topics related to performance limitations, are described in Section 5.3; Section 5.4 explains the requirements for DSC applications. Section 5.5 examines future prospects of CMOS image sensors.

5.1   INTRODUCTION TO CMOS IMAGE SENSORS

5.1.1   CONCEPTS OF CMOS IMAGE SENSORS

CMOS is a mainstream technology used for digital, analog, and mixed-signal applications. Its rapid growth has been driven by some huge markets, including CPUs, solid-state memories, ASIC, general-purpose logic integrated circuits, and now image sensors. The performance of CMOS analog circuit technology has also advanced to the point that most discrete analog components, except for very special, application-specific components, can be produced using recent CMOS technology. In chip circuits, a metal-oxide semiconductor field effect transistor (MOSFET) can be used for analog switches that have very high off-state impedance; this can result in excellent sample-and-hold and switched capacitor circuits. As will be shown in later sections, the sample-and-hold and switched capacitor circuits are very important building blocks in CMOS image sensors.

Another common benefit of CMOS technology is its low power consumption. The power supply voltage, V_{DD}, of CMOS devices has been reduced, with the scaling of device feature size contributing to lower power requirements. The high off-state impedance of the MOSFET can also minimize the current consumption. Based on the features just described, building image sensors using CMOS technology holds two major benefits: power reduction and on-chip functionality, which in turn lead to low-power, user-friendly, intelligent image sensors, such as highly integrated camera-on-a-chip devices.

Around 1990, two pioneering concepts in the migration to the system-on-a-chip (SOC) CMOS image sensors were proposed and developed: the camera-on-a-chip[1] and on-chip image processing.[2] In 1993, potential advantages of CMOS image sensor technology in comparison to CCD image sensors became a vigorous topic of discussion.[3,4]

Optical performance of early-stage CMOS image sensors did not compare favorably with that of CCD image sensors due to CMOS's high dark current in the photodiode and high readout noise. Therefore, the initial target applications for CMOS image sensors were limited almost entirely to low-cost cameras or special camera systems, such as those for scientific applications and industrial high-speed machine vision applications. In the late 1990s, however, some significant CCD technologies, including the charge transfer gate, pinned photodiode, and microlens, were shifted to and optimized for CMOS image sensors. Accordingly, CMOS image sensor performance improved significantly and has reached a level acceptable for DSC applications.

CMOS image sensors employ an X–Y address scheme for pixel scanning and active-type pixels. The X–Y address scanning architecture was investigated during the 1960s prior to the invention of the CCD image sensor.[5,6] This architecture was transferred to MOS-type image sensors[7,8] and CPD image sensors,[9] which were in production in the 1980s for use in video cameras. However, as CCD technology evolved, it replaced most of the MOS-type image sensors by 1990, due mainly to its larger pixel read noise.

The active pixel concept in which a photogenerated charge is amplified in a pixel and the amplified signal is read out has its roots in the phototransistor array image sensor.[5,10] One advantage of the active pixel is its suppression of noise generated and/or injected in the signal readout path, as shown in Figure 5.1(a). In contrast to active pixels, pixels without signal amplification are now referred to as passive pixels, a configuration of which is shown in Figure 5.1(b). (The 1980s era MOS-type image sensor mentioned here was a passive pixel device that transferred charge to a large readout bus connected to the pixel through an X–Y analog multiplexer constructed from N-MOS pass transistors.) As the size of the pixel array increases, the large parasitic capacitance seen on the pixel output bus degrades output signal. Also, noise injected directly onto the readout bus contributes to noise in the output signal.

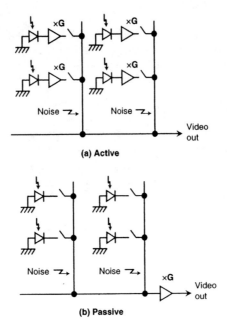

FIGURE 5.1 Active pixel and passive pixel.

CCD image sensors are categorized as passive-pixel image sensors because the photogenerated charge is transferred without amplification from a pixel to an output amplifier. As a result, CCD image quality is strongly affected by noise injection during charge transfer, which manifests as smear artifacts (see Section 4.2.4.2 in Chapter 4). In contrast, smear is rarely observed in CMOS image sensors, whose active pixel configuration, composed of a photodiode and MOS transistors, was proposed in 1968[11] and is still the foundation of today's CMOS image sensor pixels.

Differences between CMOS image sensors and CCD image sensors are:

- CMOS-based fabrication technology. The base processes and facility infrastructure for CMOS technology already exist, while the CCD technology is highly specialized.
- Readout flexibility. CMOS technology's X–Y readout scheme provides the flexibility to choose among several readout modes, such as widowing and skipping. In general, the X–Y readout has an advantage in power consumption for high-speed readout because only selected pixels are activated.
- Low power supply voltage and low power consumption. CMOS circuits can operate with lower power supply voltages than CCDs. In addition, the X–Y address scheme and on-chip functionality help reduce system power. This feature is especially important for high-speed readout.
- On chip functionality. Integrating chip control circuits, as CMOS technology does, can simplify camera electronics. Analog-to-digital conver-

CMOS Image Sensors

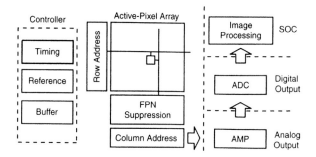

FIGURE 5.2 Typical configuration of CMOS image sensors.

sion and image processing can also be readily implemented, resulting in an image sensor with a full digital interface.

5.1.2 BASIC ARCHITECTURE

The basic configuration of a CMOS image sensor is shown in Figure 5.2. Its architecture can be categorized into three types, defined in terms of interface specifications:

- Analog output image sensors
- Digital output image sensors
- SOC type image sensors

SOC image sensors contain an imaging part, an analog front end, and a digital back end. Control functions, such as the timing generator, clock driver, and reference voltage generators, are also commonly integrated.

5.1.2.1 Pixel and Pixel Array

A CMOS image sensor pixel can be treated as a circuit consisting of a photodiode, a photodiode reset switch, signal amplification, and output circuits. A pixel with signal amplification capability is called an active pixel. Figure 5.3(a) shows a simplified architecture with a typical three-transistor active pixel and peripheral circuits. A photodiode reset transistor, M_{RS}, and a select transistor, M_{SEL}, are connected to row bus lines, and the pixel output is connected to a column signal line. The row address circuit outputs row control signals to a row to be selected. When a row-select pulse is applied at the gate of M_{SEL}, the M_{RD} and a bias current load form a source follower circuit. The photodiode voltage, V_{PIX}, is buffered by the source follower, and the buffered output voltage at a V_{PIXOUT} node is sampled on a sample-and-hold capacitor C_{SH}. The column-addressing circuit scans the sampled signals during the horizontal scanning period.

The source follower is a voltage buffer. It has a current amplification capability but has no voltage gain. However, in terms of signal charge, the photogenerated charge is calculated by $A_V \cdot (C_{SH}/C_{PIX})$, where A_V and C_{PIX} are the voltage gain of the

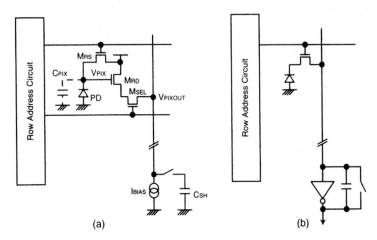

FIGURE 5.3 Pixel and peripheral circuits: (a) three-transistor active pixel with source follower readout; (b) passive pixel with amplifier readout.

source follower (<1) and the capacitance of a storage node inside a pixel, respectively. If $C_{PIX} = 5$ fF, $C_{SH} = 1$ pF, and $A_V = 0.8$, the charge gain becomes 160.

The passive pixel can also be used in CMOS image sensors,[1] the basic configuration of which is shown in Figure 5.3(b). The passive pixel simply consists of a photodiode and a select switch. The photogenerated charge is directly read out from the pixel and amplified by a charge detection amplifier located outside the pixel array. Although the passive-pixel structure makes reducing pixel size simpler and easier, the photogenerated charge is susceptible to noise injection between the pixel output and the charge detection amplifier. Because this drawback becomes significant with a large pixel array, the passive pixel is not preferable for DSC applications.

5.1.2.2 X–Y Pixel Addressing

A video signal is obtained by raster-scanning the pixel array with row (vertical) and column (horizontal) scanners. Usually, the row scanner supplies a row-select pulse and a reset pulse to pixels on the selected row once in a frame time, and the column scanner scans columns during every row period. Two common scanners used in CMOS image sensors are the shift register and the decoder. Advantages of the shift register are its simple configuration, low flip noise generation, and, in some upgraded versions, flexible readout.[12] On the other hand, a decoder allows greater scanning flexibility than the shift register and can be equipped with a window-of-interest readout and/or a skip readout.

5.1.2.3 Fixed Pattern Noise Suppression

CCDs transfer photogenerated charge to a charge-detection amplifier located at the end of the CCD register so that all signals are read out through the same amplifier. Therefore, the amplifier offset is inherently constant. On the other hand, each pixel amplifier in a CMOS image sensor has an offset variation, which results in the generation of fixed pattern noise (FPN). Threshold voltage variation of M_{RD} is the

CMOS Image Sensors

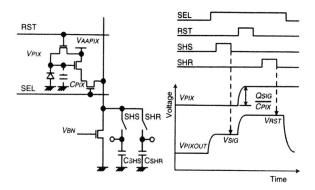

FIGURE 5.4 Principle of fixed pattern noise suppression in a three-transistor active pixel.

largest source of noise and usually falls in the range of several tens of millivolts. Therefore, a noise suppression circuit must be implemented in CMOS image sensors to suppress the offset variations.

The principle of noise suppression is shown in Figure 5.4. First, after the integration time, a pixel output, V_{SIG}, that contains the photoinduced signal and the amplifier offset is read out and stored in a memory. After the photodiode is reset, the pixel output, V_{RST}, that contains only the amplifier offset is read out again and stored in a different memory. By subtracting one output from the other, the amplifier offset can be cancelled. It should be noted that offsets caused by dark current variations cannot be suppressed. Details of the noise suppression circuits are described in Section 5.3.

5.1.2.4 Output Stage

As described at the beginning of this section, the three types of signal output are analog, digital, and SOC-type outputs. Some variations exist in the output schemes of analog output image sensors, such as a single-ended output scheme (with the on-chip FPN suppression) and a differential dual-output scheme (one corresponds to V_{SIG} and the other for V_{RST}). The single-ended output allows the use of conventional analog front-end (AFE) integrated circuits (ICs) developed for CCD image sensors.

Digital output by an on-chip analog-to-digital converter (ADC) has become the most popular scheme for consumer applications because it can make a camera-head design extremely simple by eliminating the analog interface from the image sensor and the AFE circuit (Actually, the AFE is no longer needed). In an SOC-type image sensor, a digital signal processor is implemented. Basic image processing, such as exposure control, gain control, white balance, and color interpolation, is implemented in the on-chip processor.

5.1.2.5 Other Peripherals

In most CMOS image sensors, reference voltage/current generators, drivers, and the timing controller are integrated on chip. The on-chip timing generator and clock buffers reduce external control signals, and the on-chip reference generators allow the image sensor to operate on a single power supply voltage. These on-chip integrations reduce interface power consumption and eliminate the possible noise mixture that could be introduced through chip-to-chip interface.

5.1.3 Pixel-Addressing and Signal-Processing Architectures

Several architectures for pixel addressing and signal processing, such as FPN suppression and analog-to-digital conversion, are available in CMOS image sensors. They can be classified by how many pixels are simultaneously read out and processed, as shown in Figure 5.5. As described in Section 3.1.3 in Chapter 3, the integration timing between pixels or rows differs in the X–Y address scheme, unless a global shutter (see Section 5.4.2.3) is introduced. The integration timing for each pixel depends on the processor architecture shown in Figure 5.5.

5.1.3.1 Pixel Serial Readout Architecture

In a pixel serial-processing architecture, shown in Figure 5.5(a), row- and column-select pulses choose one pixel at a time, then read out and process them sequentially. This scheme achieves complete X–Y addressing. The integration time is shifted pixel by pixel. The transversal signal line (TSL) image sensor[13] and noise/signal sequential readout image sensor[14] use the pixel serial readout.

5.1.3.2 Column Parallel Readout Architecture

The column parallel readout architecture, shown in Figure 5.5(b), is very popular and is used in most CMOS image sensors. Pixels in a row are read out simultaneously and processed in parallel. The processed signals are stored in a line memory, then read out sequentially. The integration time is shifted row by row. A pixel needs only a row select pulse in this architecture, which reduces the number of bus lines carrying control pulses to a pixel. Therefore, a relatively simple pixel configuration that is suitable for pixel size reduction can be realized. Because one row time can be assigned for a row processing, column parallel processors operate at relatively low frequencies. The lower frequency operation allows the use of low-power analog

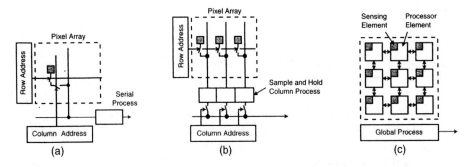

FIGURE 5.5 Pixel readout and processing schemes: (a) pixel serial readout and processing; (b) column parallel readout and processing; (c) pixel parallel readout and processing.

CMOS Image Sensors

circuits, which is especially effective in high-resolution, high-pixel readout operations, such as those required in DSC applications.

5.1.3.3 Pixel Parallel Readout Architecture

The pixel parallel, or frame simultaneous readout, is used for special applications, such as very high-speed image processing.[15,16] A processor element (PE) is located in each pixel and performs image processing in parallel, as shown in Figure 5.5(c). Compressed signal or only a signal of interest is output through the global processor. Compared to other architectures, the pixel parallel readout is especially advantageous for the high-speed operation required for machine vision applications because of its parallel processing and possible relaxation of the bottleneck in pixel readout rate. The disadvantage is that the pixel configuration becomes much more complex, resulting in a large pixel size and low fill factor and making it unsuitable for DSC applications in the foreseeable future.

5.1.4 ROLLING SHUTTER AND GLOBAL SHUTTER

To control the exposure time, CMOS image sensors require an additional reset scan in which the shutter pulses scan the pixel array prior to a readout scanning,[17] as shown in Figure 5.6. The interval between the shutter pulse and the readout pulse determines the exposure time. This process is similar to the operation of a mechanical rolling shutter (known as a focal-plane shutter). Therefore, this shutter operation is called an electronic rolling shutter, in contrast to the global shutter of CCD image sensors.

Video applications reproduce consecutive images on a display at a video rate, and the difference between the rolling shutter and global shutter is hardly visible. Prior to the advent of imaging semiconductors, imaging tubes operated in the rolling shutter mode. However, when capturing still pictures, the shifted exposure times distort the image of moving objects, as shown in Figure 5.7. Ways of addressing the distortion problem include using a mechanical shutter, increasing the frame rate, and implementing an on-chip frame memory (see Section 5.4.2.3). Shutter control for still imaging will be discussed in Section 5.4.

5.1.5 POWER CONSUMPTION

Because several circuit blocks, such as a chip controller, an FPN suppression circuit, a gain stage, and an ADC, are integrated into a CMOS image sensor chip, it is fair to take the corresponding independent parts in a CCD camera system into account when comparing power consumption of the two devices. Given this comparison, a camera head with a CMOS image sensor consumes one half to one tenth of the power of a CCD-based camera head. Along with the inherent low-power needs of the CMOS device, the following factors contribute to its low power consumption:

- Lower power supply. CMOS image sensors operate on a power supply voltage of 2.5 to 3.3 V, considerably lower than the CCD operation voltages, typically −8, 3, and 15 V. A CCD image sensor's output buffer also requires a high supply voltage (see Section 4.2.5 in Chapter 4).

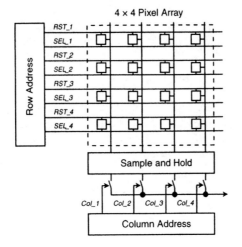

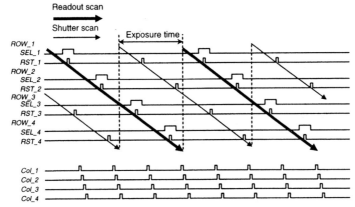

FIGURE 5.6 Pixel scanning with rolling shutter.

- X–Y address readout. Because of a CMOS sensor's X–Y address scanning scheme, only selected pixels consume power at any given time. Conversely, to transfer charge on a CCD sensor, all its registers should be driven in parallel continuously. Therefore, power consumption of an X–Y address image sensor is much lower than that of a CCD image sensor.
- Single power supply and on-chip integration. A CMOS image sensor can operate with a single power supply. In addition, integration of a timing controller and reference generators reduces the number of external devices needed at a chip-to-chip interface and, consequently, the power they demand.
- On-chip signal processing. Recently, digital output CMOS image sensors have become popular in consumer applications. Performing analog gain and analog-to-digital conversion in an image sensor can eliminate an external AFE and high-speed analog I/Os, which consume more power than the digital interface.

CMOS Image Sensors

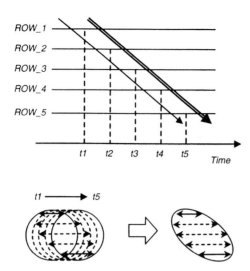

FIGURE 5.7 Shape distortion of moving objects in rolling shutter.

In CMOS image sensors, in principle, an increase in chip size does not affect power consumption because of their X–Y address schemes. On the other hand, in CCD image sensors, the larger area of CCD registers directly increases the load capacitance of clock drivers. In addition, the larger chip size requires higher charge transfer efficiency and makes it difficult to reduce drive pulse height. Therefore, the difference in power consumption between CMOS image sensors and CCD image sensors grows as chip size and frame rate increase.

5.2 CMOS ACTIVE PIXEL TECHNOLOGY

A CMOS image sensor's active pixel consists of a photodiode and a readout circuit; thus, many variations of pixel configurations have been proposed and demonstrated. However, the pixel configurations for DSC applications should be as simple as possible to achieve excellent fill-factor of a photodiode and a high-resolution pixel array. In this section, two major photodiode structures will be explained with associated readout schemes.

5.2.1 PN Photodiode Pixels

PN photodiode pixels were popular in early generations of CMOS image sensors. They can be incorporated into a CMOS process/design technology with relatively simple modifications, which can be represented in a circuit schematic, thus enabling image sensor design within a general-purpose IC design environment. Therefore, the PN photodiode pixel is still a cost-effective solution for low-cost image sensors and/or a small amount of custom image sensors.

Although the noise performance is commonly worse than the pinned photodiode pixels described in the next section, PN photodiode pixels possess the noticeable advantage of offering a larger full-well capacity. Full-well capacity can be increased

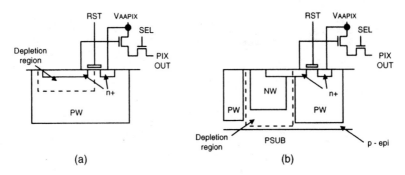

FIGURE 5.8 PN photodiode structure: (a) n⁺/PW photodiode structure; (b) NW/PSUB photodiode structure.

through photodiode capacitance in the PN photodiode pixels, but that of a pinned photodiode pixel is limited by the pinned potential of a photodiode.

5.2.1.1 PN Photodiode Structure

An n⁺/p-well (PW) or an n-well (NW)/p-type substrate (PSUB) junction is commonly used for photodiode formation, as shown in Figure 5.8. In an n⁺/PW photodiode, a high-concentration shallow n⁺ region is formed in a PW region, and photoconversion is performed at the depletion region of the junction. Although the photodiode is very simple, the increased dopant concentration of the PW region in recent highly integrated CMOS technologies reduces the thickness of the depletion layer and affects sensitivity.[18] In terms of realistic sensitivity, n⁺/PW photodiode is suitable for above 0.5 ~ 0.8 μm rule CMOS technology.

For processes below 0.35 ~ 0.5 μm, an NW/PSUB junction is used for a photodiode. The surface NW region is formed in a low-concentration epitaxial layer, and peripherals of the photodiode are isolated by PW regions. Because the dopant concentration of the epitaxial layer is very low, the depletion layer reaches the P-type substrate edge. A larger photoconversion volume can thus be obtained even with a highly integrated CMOS process.

Fundamental problems with the PN photodiode are dark current due to surface generation and thermal noise associated with the photodiode reset. To reduce the surface generation, a buried photodiode or pinned diode structure has been introduced[19] to the NW/PSUB photodiode. However, the source–bulk junction of the reset transistor still causes thermal leakage.

5.2.1.2 Photodiode Reset Noise (kTC noise)

As described in Section 3.3.3.4 in Chapter 3, the photodiode reset circuit forms an RC low-pass filter circuit and the reset noise, or kTC noise, appears to be associated with the reset operation. The reset noise is given by

$$v_n = \sqrt{\frac{kT}{C_{PD}}} \qquad (5.1)$$

CMOS Image Sensors

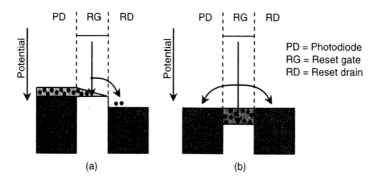

FIGURE 5.9 Potential diagram of reset operation: (a) soft reset; (b) hard reset.

In the FPN suppression operation described in Section 5.1.2.3, a pixel output signal that contains the photoinduced signal and offset of the pixel source follower is read out first, followed by the offset signal readout after the pixel reset. These signals are then subtracted to obtain the photoinduced signal only. However, the reset noise components in the two signals are different because the reset for the first signal is different from the one for the second signal and, of course, they are not correlated at all. Therefore, the reset noise in the resulting signal becomes $\sqrt{2}$ times the value obtained from Equation 5.1.

5.2.1.3 Hard Reset and Soft Reset

Actual reset noise depends on the operation modes of the reset transistor. When the reset transistor operates in its saturation region, electrons stored on a photodiode are drained to the reset drain during the reset period. On the other hand, when it operates in its linear region, electrons stored on a photodiode can move back and forth between the photodiode and the reset drain. Potential diagrams of these two reset operations are illustrated in Figure 5.9. The photodiode reset in the saturation mode of the reset transistor is called "soft reset," and that in the linear mode is called "hard reset," the level of which is given by Equation 5.1. The reset noise in the soft reset mode is reduced to approximately $1/\sqrt{2}$, which is associated with the current rectification effect during the reset. Detailed analysis of reset noise has been reported.[20,21]

A drawback of the soft reset is that it introduces image lag caused by the dependence of the photodiode voltage after the reset on the exposure, which means it does not have time to settle sufficiently to a certain fixed level within the limited reset period. Injecting a bias charge before the soft reset operation, called "flushed reset," can eliminate the image lag.[22]

Reset noise significantly affects the temporal noise floor. For example, with a photodiode capacitance of 5 fF and a photodiode voltage swing of 1 V, the kTC noise charges in the soft reset and full-well capacity are 28 e$^-$ and 32 ke$^-$, respectively, meaning the reset noise limits the maximum dynamic range to 61dB.

5.2.1.4 Reset Noise Suppression in PN Photodiode Pixels

Reset noise is a significant problem in PN photodiode pixels. In fact, a noise floor of 28 e$^-$ in the above example is relatively high compared to those of CCD image sensors or CMOS image sensors with a pinned photodiode pixel, which is around 10 e⁻. Several schemes have been proposed to suppress reset noise.

5.2.1.4.1 Reset Noise Reduction Using Nondestructive Readout

In the case of the three-transistor pixel shown in Figure 5.4, pixel signals can be read out nondestructively multiple times if the row select pulses are applied several times without supplying reset pulses. Using this feature, reset noise can be suppressed through reset frame subtraction.[23] Just after the reset operation, pixel outputs are read out nondestructively and saved in an external offset frame memory, at which point exposure starts. After the exposure time, pixel signals are read out again, and those pixel offsets saved in the frame memory are subtracted from the second signals. Because the offset and the reset noise of each pixel is contained in the offset frame, reset noise can also be cancelled. This scheme has been utilized for scientific applications with noise reductions up to one third reported.[24]

5.2.1.4.2 Active Feedback for Reset Noise Correction

The reset noise appears as a voltage fluctuation at the photodiode. Several methods have been proposed to suppress this voltage fluctuation using a circuit design approach. A common principle of the reset noise suppression circuits is to form a negative feedback onto the photodiode voltage. The pixel readout circuit operates in an inverter amplifier mode in which the feedback mechanism forces the photodiode node to settle at a fixed reset level during the reset period.[25,26] A column-based feedback circuitry was also proposed in which a negative feedback loop from the column signal line to the reset gate inside a pixel is formed during the reset period. A reduction in reset noise of almost an order of magnitude was reported using this technique.[27]

These methods are intriguing in terms of improving pixel performance through circuit design. Although DSC image sensors utilizing active feedback pixels are not yet reported, the active readout methods have potential for use in future image sensors, e.g., stacked photodiode image sensors[28] and global shutter pixels. In stacked photodiode pixels, a metal contact is commonly needed between the photodiode and detection node. Therefore, the complete charge transfer is barely achieved, which in turn generates reset noise. On the other hand, a global shutter pixel needs a memory capacitor in the pixel (see a memory-in-pixel configuration in Section 5.4.2.3), and resetting the memory capacitor should generate reset noise, as seen with the PN photodiode pixel.

5.2.2 Pinned Photodiode Pixels

Two significant challenges exist in conventional PN photodiode pixels: dark current and reset noise reduction. To address these problems, the pinned photodiode structure[29] was introduced and has become popular in recent CMOS image sensors. The basic pinned photodiode pixel configuration is shown in Figure 5.10(a). The pixel consists of a pinned photodiode and four transistors that include a transfer gate (M_{TX}), a reset transistor (M_{RS}), an amplifier transistor (M_{RD}), and a select transistor (M_{SEL}). Thus, the pixel structure is often called a four-transistor pixel in contrast to the conventional three-transistor pixel.

CMOS Image Sensors

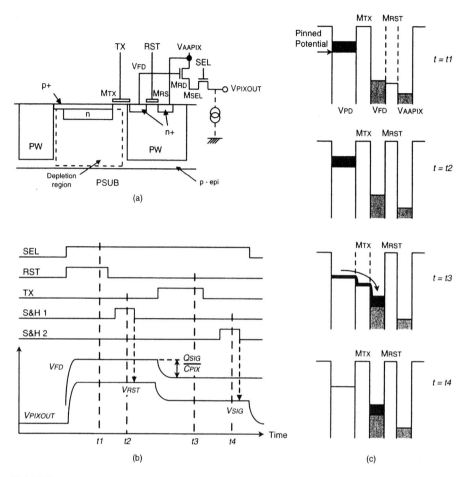

FIGURE 5.10 Pinned photodiode pixel (four-transistor pixel): (a) pixel structure; (b) timing diagram; (c) potential diagram.

The operation sequence and corresponding potential diagrams are illustrated in Figure 5.10(b) and (c), respectively. At the beginning of the pixel readout, the floating diffusion (FD) node is reset by the reset transistor, and then the pixel output (V_{PIXOUT}) is read out once as a reset signal (V_{RST}). V_{RST} includes the pixel offset and reset noise at the FD node. Next, the transfer gate, M_{TX}, turns ON so that the accumulated charge in the pinned photodiode is transferred completely to the FD node. Because of the complete charge transfer, no reset noise is generated during the transfer operation. The transferred charge drops the potential at the FD node, and V_{PIXOUT} decreases. After the transfer operation, V_{PIXOUT} is read out again as V_{SIG}. By subtracting V_{RST} from V_{SIG}, the pixel FPN and the reset noise can be removed, which results in a reduced readout noise of less than 10 e-.[30] Another important benefit of the pinned photodiode is low dark current. Because the surface of the pinned photodiode is shielded by a p+ layer, dark current can be reduced as low as 50 pA/cm^2 at 60°C.[31]

Next, a few important issues and topics regarding the pinned photodiode pixel will be introduced.

5.2.2.1 Low-Voltage Charge Transfer

Although the pinned photodiode and transfer gate structure has been used in CCD image sensors, the power supply voltage in CMOS image sensors is much lower than the CCD transfer pulse height. Incomplete charge transfer from the photodiode to the FD node causes excess noise, image lag, and a nonlinear response. Therefore, the pinned photodiode and the transfer gate in CMOS image sensors should be optimized for low-voltage operation.

The reset level of the FD node is around 2 to 2.5 V when using a power supply voltage of 2.5 to 3.3 V. Taking the FD voltage drop at the charge transfer into account, the acceptable pinned potential of the pinned photodiode is then 1.0 to 1.5 V. Therefore, very accurate dopant control is required in fabrication. High transfer efficiency with a low-voltage transfer pulse requires optimizing the structure of the region between the photodiode edge and the transfer gate[32] as well as optimizing photodiode depth. In terms of pixel layout, forming an L-shaped transfer gate also improves transfer efficiency due to a three-dimensional potential effect,[33] whose high-speed charge transfer is as fast as 2 ns with a 3.3-V power supply.

5.2.2.2 Boosting

A higher voltage setting for the initial FD voltage helps improve efficiency of the charge transfer from PD to FD and the dynamic range of the pixel output swing. Increasing the pulse height of the RST pulse through on-chip power boosting is one suitable solution.[34]

However, in the case of the N-type substrate (see Section 5.2.2.4), positive power boosting is difficult to realize because of the common power supply voltage connecting through the N-type substrate. To achieve higher FD initial voltage, in-pixel boosting,[35] the pixel configuration shown in Figure 5.11, has been introduced. The formation of M_{RD} and M_{SEL} is reversed, in contrast to the conventional three-transistor configuration. M_{RS} turns ON while M_{SEL} is kept OFF, then M_{SEL} turns ON. When M_{SEL} is shorted, the source node voltage of M_{RD} increases. This voltage increase is fed to V_{FD} through the gate capacitance of M_{RD} because the inversion layer is formed beneath the gate and is coupled with the source node voltage, V_{PIXOUT}. In this way, the boosted V_{FD} is attained, which contributes to the generation of a fringing field at the transfer gate, improving transfer efficiency and the dynamic range of the pixel source follower.

5.2.2.3 Simplification of Pixel Configuration

Reducing the number of transistors in a pixel is effective for pixel size reduction. In the pinned photodiode pixel, a pinned photodiode is isolated from the readout transistors by a transfer gate. This feature can be used to reduce the number of transistors per pixel so that one readout circuit can be shared by multiple photodiodes, as shown in Figure 5.12, which reduces the size of the pixel. This shared pixel configuration has several variations. The FD is shared by four consecutive pixels in the row,[36] two horizontal pixels,[35] 2 × 2 pixels.[37] The shared-pixel archi-

CMOS Image Sensors

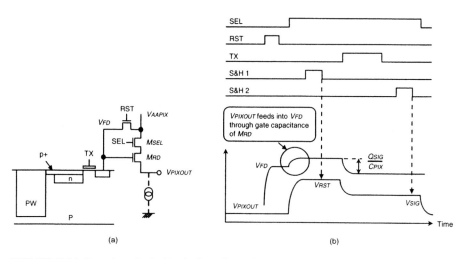

FIGURE 5.11 Boosting pixel: (a) pixel configuration; (b) timing diagram.

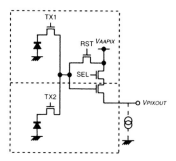

FIGURE 5.12 Shared pixel configuration (one readout circuit shared by two pixels).

tecture is suitable for creating a small pixel, but care needs to be taken in its design to ensure uniform characteristics among the pixels that share one readout circuit.

To reduce the number of transistors further, the select transistor, M_{SEL}, is eliminated using a pixel reset voltage control, as shown in Figure 5.13. After the pixel readout, pixel reset voltage, V_{RST}, is lowered and sampled on the FD node. Because the FD voltage maintains at its low voltage, the pixel remains inactive until the next readout operation. A pixel select transistor can thus be removed. Sharing the V_{RST} bus with V_{AAPIX} bus[35] or V_{PIXOUT} bus[36] realizes further pixel simplification.

5.2.2.4 P- and N-Type Substrates

In CMOS image sensors, n-type and p-type substrates, shown in Figure 5.14, can be used because the vertical shutter (see Section 4.3.3 in Chapter 4) is not employed in the pixel. However, for two reasons, the p-type substrate is more commonly used in general-purpose CMOS processes. From a circuit design point of view, the p-type substrate has an advantage because of the large amount of CMOS intellectual property (IP) that has been developed for it. Also, using a p-type substrate requires only minor modifications from the conventional CMOS process. On the other hand,

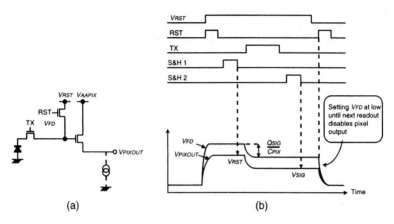

FIGURE 5.13 Pulsed V_{RST} pixel: (a) pixel configuration; (b) timing diagram.

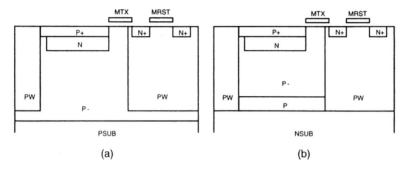

FIGURE 5.14 Pinned photodiode in (a) P-type and (b) N-type substrates.

carriers generated in a p-type substrate and/or p-well region can diffuse into neighboring pixels, which results in pixel-to-pixel crosstalk. Also, diffusion dark current from the substrate may exist. Therefore, process improvements are necessary to suppress carrier diffusion from the p-type substrate.

Better pixel-to-pixel isolation can be expected with an N-type substrate because photogenerated carriers in the deep substrate and/or P-well isolation regions are mostly drained through the N-type substrate. To avoid reach-through between the N-type substrate and N photodiode, a deep PW blocking region is introduced below the photodiode.[35] The depth and profile of the deep PW block region determine the photoconversion layer thickness.

5.2.3 Other Pixel Structures for Large-Format, High-Resolution CMOS Image Sensors

The PN photodiode pixel and the pinned photodiode pixel are very common pixel technologies for CMOS image sensors; however, two others are used in image sensors for digital single lens reflex (DSLR) cameras at the time of this writing. One is an active pixel with a junction field effect transistor (JFET) amplifier transistor; the other is a vertically integrated photodiode pixel.

CMOS Image Sensors

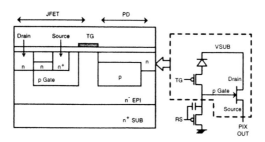

FIGURE 5.15 LBCAST.

5.2.3.1 LBCAST

A lateral buried charge accumulator and sensing transistor (LBCAST) array consists of a pinned photodiode; a junction field effect transistor (JFET); a charge transfer transistor; and a reset transistor, as shown in Figure 5.15.[38] During the accumulation period, RS is kept at low and the P gate of the n-channel JFET is in its OFF state. During pixel readout, RS is turned to high and the voltage of the P gate increases through a coupling capacitor between the P gate and RS, which activates the JFET. PIXOUT is connected to a constant current load so that the P gate of the JFET acts as a source follower input transistor. Then, the accumulated signal charge on the photodiode is transferred to the P gate by turning TG ON. A photo signal can be obtained from the voltage change at the PIXOUT node using a column CDS circuit (not shown). In addition to the low dark current and the complete charge transfer due to the pinned photodiode, its low noise nature is enhanced by the JFET, which exhibits lower temporal noise characteristics than MOSFETs.

5.2.3.2 Vertical Integration Photodiode Pixel

Absorption length in a semiconductor has the wavelength dependence of incident light. As illustrated in Figure 3.2 in Chapter 3, shorter wavelength photons such as violet light are absorbed near the photodiode surface and longer wavelengths such as red light can reach deeper regions. This feature can be used for color separation.[39–41]

As shown in Figure 5.16, stacking three photodiodes and detecting a photogenerated charge for individual photodiodes provides red/green/blue (RGB) color information at the same location.[42] In contrast to image sensors with on-chip color filters, the vertical integration pixel provides a larger collection area for generated carriers and color interpolation-free image sensors. Although the fundamental problems of a PN photodiode pixel — namely, reset noise and dark current — are present, the vertical integration of photodiodes is a fairly interesting approach.

5.3 SIGNAL PROCESSING AND NOISE BEHAVIOR

In this section, two typical noise suppression schemes and signal processor architectures will be introduced, followed by descriptions of noise generation mechanisms in CMOS image sensors. The following explanation will be based on the pinned

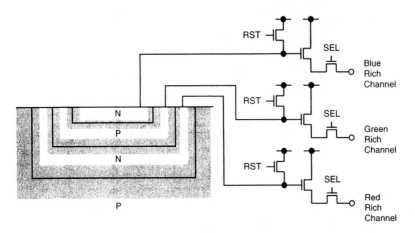

FIGURE 5.16 Vertical integration photodiode pixel.

photodiode pixel most commonly incorporating the four-transistor pixel technology. Although several types of pixels are used in image sensors built for DSCs, as shown in the Section 5.2, the principles discussed are common to other CMOS image sensors, as well.

5.3.1 Pixel Signal Readout and FPN Suppression Circuit

The photogenerated charge is amplified in the pixel and transferred to the noise suppression circuit, which makes a peculiar but very important contribution to the image quality of CMOS image sensors. In recent CMOS pixels, pixel temporal noise has already been reduced to less than 500 µV. Therefore, the noise suppression circuit is requested to suppress pixel offset variations less than the temporal noise, namely, 100 to 200 μV_{rms}. As the temporal noise is reduced by pixel/signal chain improvements, residual FPN, which is the output of the FPN suppression circuit, must be much lower than the temporal noise.

Voltage-domain source follower readout is the most common readout scheme. As described in Section 5.1, the FPN caused by variations of the MOS threshold voltage should be suppressed. Except for the two representative schemes illustrated next, some unique configurations have been reported and realized.[14,46]

5.3.1.1 Column Correlated Double Sampling (CDS) Scheme

The combination of a pixel source follower and a column CDS circuit,[43] shown in Figure 5.17, is the most basic configuration. A bias current source is connected at the pixel output node to compose a source follower circuit that generates buffered output following the voltage at the FD.

When FD is initialized, the sample and hold (S/H) capacitor is clamped at V_{CLP}. After turning off the clamp switch, the charge transfer gate TX is opened and the signal charge is transferred from the photodiode to the FD, dropping the V_{FD} due to the charge injection. The S/H capacitor then samples the voltage drop, and the pixel's

CMOS Image Sensors

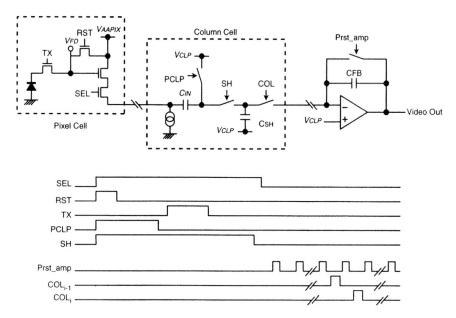

FIGURE 5.17 Correlated double sampling (CDS) for noise suppression.

DC offset is suppressed by the CDS operation. The memorized signal voltage on the S/H capacitor C_{SH} can be expressed as

$$V_{SH} = V_{CLP} - A_V \cdot \frac{Q_{PH}}{C_{FD}} \cdot \frac{C_{IN}}{C_{IN} + C_{SH}} \tag{5.2}$$

where A_V, Q_{PH}, C_{IN}, and C_{FD} are the source follower gain, signal charge, input capacitance of the CDS circuit, and capacitance at the FD, respectively. Following the column CDS, signals stored in the column S/H capacitors are scanned.

Although the circuit is relatively simple, its performance is affected by a voltage gain loss of $C_{IN}/(C_{IN} + C_{SH})$, threshold voltage variations of the clamp switches, and less common mode noise rejection. Using a switched capacitor (SC) amplifier and multiple SC stages can address these problems.[44]

5.3.1.2 Differential Delta Sampling (DDS)

The differential delta sampling (DDS) noise suppression circuit was introduced to eliminate the characteristic mismatch between two signal paths in a differential FPN suppression circuit.[45] Figure 5.18 shows an example of an FPN suppression circuit using DDS. S/H capacitors, C_{SHS} and C_{SHR}, sample V_{SIG} and V_{RST}, respectively. The output of each PMOS source follower buffer is connected to a differential input of a differential amplifier through an offset capacitor.

After the two signals are sampled and held on the S/H capacitors, horizontal scanning starts. When the ith column is selected by a COL_i pulse from the column

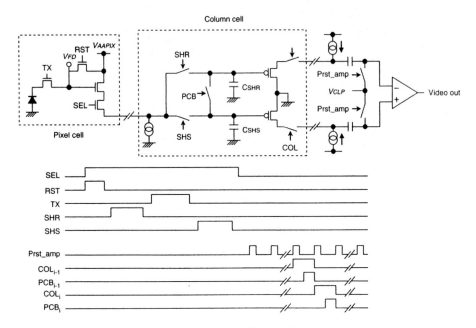

FIGURE 5.18 Differential delta sampling (DDS) for noise suppression.

address circuit, two parallel PMOS source follower circuits are activated and generate differential outputs that correspond to V_{SIG} and V_{RST}. Inputs of the differential amplifier are initialized at V_{CLP}, canceling the offset voltages of the PMOS source followers. A crowbar operation initiated by turning PCB pulse high then follows. During the first half of the column signal readout, the differential amplifier outputs an offset of its own. When the two S/H capacitors are shorted by the crowbar switch, the outputs change, depending on voltage differences between V_{SIG} and V_{RST}.

The differential signal occurring after the crowbar operation can be expressed as

$$\{(\alpha_1 + \alpha_2)/2\} \cdot (V_{RST} - V_{SIG})$$

where α_1 and α_2 are the gains of each respective channel. Because the two gain factors are averaged into one, the gain mismatch can be suppressed along with the offset mismatch between channels.

5.3.2 ANALOG FRONT END

The analog processing in CMOS image sensors is divided into column parallel and serial segments. Usually, the noise suppression circuit is implemented in column parallel. An analog programmable gain stage and an ADC stage follow the noise suppression circuit. Gain settings of the programmable gain amplifier (PGA) determine the camera's ISO speed and are controlled so that the input range of the ADC is effectively utilized.

CMOS Image Sensors

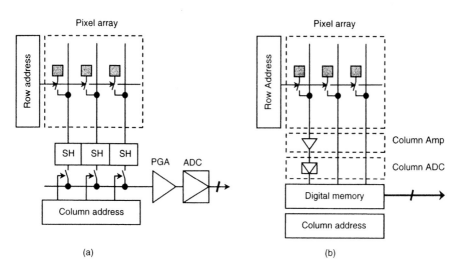

FIGURE 5.19 (a) Serial and (b) column parallel processors.

5.3.2.1 Serial PGA and ADC

Configuration of the programmable gain amplifier (PGA) and ADC, shown in Figure 5.19(a), is quite similar to the AFE device for CCD signal processing. The noise-suppressed signals of one row are stored at once in the S/H memory array, and then the memorized analog signals are scanned and transferred to PGA and ADC in serial. The combination of a multiple-stage, programmable-gain, video-bandwidth amplifier and a pipeline ADC is the most popular configuration for serial processing.

5.3.2.2 Column Parallel PGA and ADC

When a gain amplifier is implemented in the column parallel part, the amplifier block is called a "line-amplifier" or "column amplifiers." Similarly, when multiple ADCs are implemented in the column, the ADC structure is called "column ADCs." In the column ADC architecture shown in Figure 5.19(b), noise suppression, gain, and ADC are performed in parallel, the resulting digital data is stored in a line buffer memory, and the column decoder scans the line buffer memory.

The required sampling rate in the column parallel ADC is defined as an inverse of one row cycle and commonly is as slow as 10 to 100 kHz. Therefore, slow but very low-power ADC schemes, such as ramp ADC[47–49] or successive approximation ADC,[50] are employed. Compared to the serial approach, the column parallel ADC approach has the advantage of high-speed operation because of its highly integrated parallel processing.[51,52]

5.3.3 NOISE IN CMOS IMAGE SENSORS

In addition to pixel-related noise, implemented circuits also generate noise in the output image. In order to estimate noise source and improve performance, it is important to understand the relationship between noise sources and observed noise in output images. In still imaging, temporal noise and FPN affect image quality equally because the temporal noise is latched in the still image.

5.3.3.1 Pixel-to-Pixel Random Noise

When noise in an image is completely random from pixel to pixel, the noise is caused by individual pixel noise and/or pure temporal noise in the signal path. Pixel random noise can be categorized into pixel FPN and pixel temporal noise. The pixel FPN is caused by variations in dark current, photodiode sensitivity, the storage capacitor, and source follower characteristics (offset as well as gain). On the other hand, pixel temporal noise is affected by photodiode reset noise, photon shot noise, dark current shot noise, and pixel source-follower circuit noise. In the case of pinned photodiode pixels, reset noise (see Section 5.2.1.2) can be neglected by the CDS operation.

The shot noise can be expressed by:

$$v_{n\_shot} = \frac{q}{C_{FD}} \sqrt{N_{sig} + N_{dark}} \quad (5.3)$$

where N_{sig} and N_{dark} are the number of photogenerated electrons and thermally generated leakage electrons, respectively.

Assuming an ideal bias current source and sufficient settling time, the noise power and noise bandwidth of the pixel follower circuit are expressed as $\gamma \cdot 4kT/g_m$ and $(1/4)g_m/C_{LV}$, respectively, where g_m and C_{LV} denote transconductance of M_{RD} and load capacitance of the source follower output, respectively. Gamma is the MOS channel noise factor and is estimated to be ~2/3 for long-channel transistors. C_{LV} includes sample-and-hold capacitance and a total parasitic capacitance of the pixel signal bus line. Supposing noise suppression by double sampling, thermal noise will be increased by approximately $\sqrt{2}$ times. Accordingly, the thermal noise component of the pixel source-follower after noise suppression can be expressed by

$$v_{n\_therm} \sim \sqrt{\frac{4}{3}kT/C_{LV}} \quad (5.4)$$

In typical CMOS image sensors, C_{LV} has a value of several picofarads. When C_{LV} is 5 pF, Equation 5.4 estimates the thermal noise component at the FD node to be 33 μV_{rms}.

Flicker noise power spectral density can be approximated by $S_n(f) = K_f'/f$, where K_f' is a coefficient that depends on device parameters, such as transistor size, surface trap state density, etc. Because flicker noise has time-domain correlation, transfer function of the CDS operation is used to estimate the contribution from the M_{RD} flicker noise. The CDS transfer function can be expressed by Equation 5.5, assuming each sampling operation is expressed by the δ-function;

$$H(j2\pi f) = 1 - e^{-j2\pi f \Delta T} \quad (5.5)$$

CMOS Image Sensors

where ΔT is the interval between two samples of CDS.

The resulting output referred flicker noise component is estimated by

$$v_{n\_flicker}^2 = \int_0^\infty |H_{CDS}(j2\pi f)|^2 \cdot |H_{SF}(j2\pi f)|^2 \cdot \frac{K_f{'}}{f} \cdot df \tag{5.6}$$

where $H_{SF}(j2\pi f)$ is the transfer function of the source follower. Assuming $H_{SF}(j2\pi f)$ is represented by a single-pole, low-pass filter characteristic with the low-frequency voltage gain A_V and the cut-off frequency of f_c, Equation 5.6 can be rewritten as

$$v_{n\_flicker}^2 = 2K_f{'} \cdot \int_0^\infty \frac{A_V^2}{1+(f^2/f_c^2)} \cdot \frac{1-\cos(2\pi f \Delta T)}{f} \cdot df \tag{5.7}$$

Therefore, the flicker noise coefficient, K_f', and the interval between two samples, ΔT, should be examined carefully when designing the CDS circuit.

5.3.3.2 Row-Wise and Column-Wise Noise

Because stripe-wise noise is four to five times more visible than pixel random noise, reducing stripe noise is one of the most important issues in CMOS image sensor development. Performance variation in the column parallel segment, such as the noise suppression circuit, often causes vertical stripe noise. With column parallel AGC and ADC architectures, gain and residual offset variations also cause vertical stripe noise. Cyclic noise injection at a particular timing of the horizontal (column) scanning period also causes vertical stripe noise. Major causes of such kinds of noise injection are flip noise from the binary counter, flip-noise in the column decoder, and charge injection of the column select switch. Most of these are affected by column operation frequency.

On the other hand, common noise injection during pixel readout or column parallel operation causes a horizontal stripe noise because all signals in a row are affected by the noise at a same time. For example, a fluctuating power supply voltage for a pixel's signal sample-and-hold circuit in the noise suppression block generates a horizontal stripe noise.

5.3.3.3 Shading

Two factors cause shading in CMOS image sensors. The first is optical nonuniformity; the second is electrical in nature and is best described as follows. The principle of optical shading in CMOS sensors is similar to that in CCDs. How light angle dependence of sensitivity affects the optical shading is influenced by the lens F-number. Optical shading can be addressed through microlens optimization, pixel layout optimization, and thinning of the material between the silicon and the microlens. Electrical shading is due to voltage drop, pulse delay, and the distributed parasitic effect inside the array block. Dual-row driver architecture is often employed to supply pixel drive pulses from both sides of the array, which can minimize the pulse delay. To reduce the voltage drop of power supplies, tree structure layout of the bias line[53] or implementing local bias generators for every column[54] are reported.

5.3.3.4 Smear and Blooming

Active-pixel image sensors feature very high smear protection. Usually, the smear in CMOS image sensors is much lower than the acceptable limit. Antiblooming performance of CMOS image sensors depends on the substrate type and schemes of excess charge overflow in a pixel.

The dark noise floor has been addressed through a buried pinned photodiode. The basic performance of a CMOS image sensor's pixel is very close to that of a CCD. Improvements in the optical performance, pixel size, and noise reduction in CMOS image sensors are also in progress. However, their intermetal layer is thicker than that of CCDs, affecting their optical crosstalk and degrading resolution and angle tolerance in small pixels. Decreasing the interlayer thickness is a major focus of present CMOS image sensor design and will improve transmittance and optimize the structure of the microlens. One method being explored is designing image sensors with fewer metal layers than conventional CMOS chips.[34–37]

5.4 CMOS IMAGE SENSORS FOR DSC APPLICATIONS

5.4.1 ON-CHIP INTEGRATION VS. CLASSES OF DSC PRODUCTS

DSC products can be grouped into several classifications as described in Section 1.3 in Chapter 1. In this chapter, the focus is on DSC cameras, including toy cameras, midclass compact cameras, and digital SLR cameras. An examination is made of three types of each, their requirements, and how much on-chip integration is appropriate in a CMOS image sensor for each DSC category. Appropriate on-chip integration levels for each class of DSC are suggested in Figure 5.20.

5.4.1.1 Toy Cameras

Toy cameras and low-end DSCs offer simple features, small sizes, and low prices of approximately $100 U.S. Low-end DSCs are similar to the cameras used in mobile phones, in which low cost is the highest priority. Therefore, a fixed focal length lens is usually used, and a mechanical shutter and an LCD display seldom are. On-chip

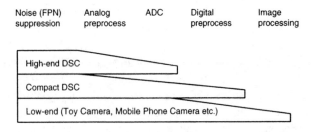

FIGURE 5.20 On-chip function for signal processing and DSC applications.

integration in image sensors is very effective in reducing the number of electronic parts, which lowers system cost and can reduce the camera's size. In fact, SOC type CMOS image sensors are becoming common in mobile phone applications and are expected to appear in future low-end DSCs.

5.4.1.2 Middle-Class Compact DSCs

Because this class of cameras demands relatively high-performance image processing, specially developed image processors are commonly used. This suggests that the required on-chip functionality of CMOS image sensors — including ADC and early-stage preprocessing such as noise suppression — comes basically from the sensor's analog front end. Because view finding, auto exposure (AE), and auto white balance (AWB) controls as well as auto focus (AF) adjustments are performed using the image sensor output, a video rate readout mode with subresolution is necessary.

At the time of this writing, CCD image sensors with pixel sizes of 2.2 to 3.5 μm dominate this class of DSC products. However, as explained earlier, CMOS image sensor performance is approaching that of CCDs, even with its small pixel sizes, which are decreasing rapidly to levels below 3 μm thanks to the shared-type pixels discussed in Section 5.2.2. Once CMOS and CCD sensors are relatively equal, the advantages of CMOS image sensors, such as low power consumption even at high frame rates, can be fully utilized to give DSCs functions that they do not have now.

5.4.1.3 Digital SLR Cameras

Achieving the highest possible image quality is everything for DSLR cameras. Therefore, the emphasis in image sensor design, whether CMOS or CCD, should be improving optical performance. At the same time, because separate AE/AWB/AF sensors and an optical viewfinder are implemented in DSLRs, subresolution video readout may not be a necessary sensor function. However, the high-speed readout capability of CMOS image sensors is considered very important for consecutive multiframe image capture, and the advantages of CMOS image sensors described in Section 5.1.1 can be fully utilized in applications that require a large-format image sensor with a high-frame-rate capability.[55,56] As described in Section 5.1.5, the difference in power consumption between CMOS and CCD image sensors grows as chip size and frame rate increase, so power consumption is much lower in CMOS image sensors. On-chip, column-parallel analog-to-digital conversion offers potentially higher speed readout and lower power consumption when high resolution and high frame rates are required.[55]

5.4.2 OPERATION SEQUENCE FOR CAPTURING STILL IMAGES

In this section, operation sequences are explained, with emphasis on operational timing between a CMOS image sensor and a mechanical shutter. As described in Figure 5.6, a row-by-row reset/readout operation is usually employed in CMOS image sensors; thus, the pixel-integration period shifts by a row period between

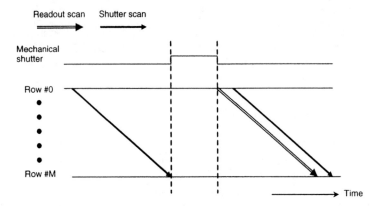

FIGURE 5.21 Mechanical shutter timing with rolling reset.

every row. This is a constraint of mechanical shutter operation and strobe light timing.

5.4.2.1 Synchronization of Rolling Reset and Mechanical Shutter

When the image sensor operates in a rolling reset mode, opening a mechanical shutter is only valid during the vertical blanking period. Mechanical shutter timing and the internal rolling shutter operation are shown in Figure 5.21. First, the reset scan should be completed while the mechanical shutter is closed. Pixels on each row start integration after the reset. Then, the mechanical shutter opens for a pre-determined period of time, which corresponds to the light exposure time. Pixel readouts start after the mechanical shutter is closed. If the mechanical shutter is open, overlapping the reset or readout scanning durations, a different exposure time for a part of the rows results.

This scheme has the advantage of avoiding the offset shading caused by dark current because the integration time is identical for all rows. However, it is not without its drawbacks: namely, long shutter-lag due to the pre-reset scanning and difficulty in using a normally open type of mechanical shutter.

5.4.2.2 Synchronization of Global Reset and Mechanical Shutter

In order to use a normally open type of mechanical shutter, simultaneous reset for all pixels — called global reset — or a fast reset scanning is necessary. Figure 5.22 shows the shutter sequence with internal reset/readout timing. When all the pixels are reset simultaneously or a reset scan is completed within a negligibly short duration compared to the exposure time, exposure begins. After a mechanical shutter is closed, readout starts. As shown in the figure, this scheme is similar to the sequence for interlaced scan CCD image sensors. A normally open shutter can be used with the focal plane AE/AF followed by still imaging. A possible drawback to this scheme

CMOS Image Sensors

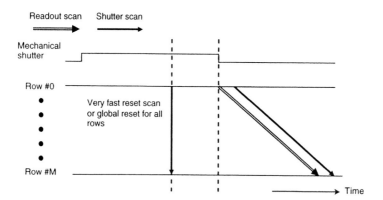

FIGURE 5.22 Mechanical shutter timing with global reset.

is that the dark current integration time is different for each row and that difference could generate shading.

5.4.2.3 Electronic Global Shutter

In order to make the integration period identical for all pixels, an electronic global shutter is necessary as is seen in interline CCD image sensors. This scheme has not yet been achieved in CMOS image sensors for DSC applications, but it is greatly desired because, without the electronic global shutter, a mechanical shutter must be used. A mechanical shutter determines the intervals between image capturing periods in a consecutive imaging mode but is not as fast as the full electronic exposure control provided by an electronic global shutter.

The speed at which consecutive images are captured should improve up to a frame rate that an image sensor can achieve without the use of a mechanical shutter. (Note that, as with exposure control, the readout speed of a CMOS image sensor can be faster than that of a CCD image sensor.) Meanwhile, the rolling reset/readout can capture consecutive images without a mechanical shutter, but it is not preferred in DSC applications because of the distortion it creates in still images, as illustrated in Figure 5.7.

Implementing the global shutter on CMOS image sensors has two approaches. One is a memory-in-pixel approach[57]; the other is a frame-memory approach.[58] Although these have not yet been adopted in DSC applications, they are of future interest.

The memory-in-pixel scheme has been used in CMOS image sensors for high-frame-rate applications.[51] In this scheme, photogenerated charge accumulation and readout operations are performed individually. As shown in Figure 5.23, a PG pulse initiates the start of the exposure; then, TG is turned on during the vertical blanking, which is followed by the readout scanning. This operation is very similar to the interline-transfer CCD (IT CDD) operation. The pulse PG corresponds to the electronic shutter in CCD image sensors. Photo leakage from a bright light at the pixel memory during the storage time, t_s, affects the stored signal, as is the case in CCD image sensors. In contrast to the vertical smear of CCD image sensors, in CMOS

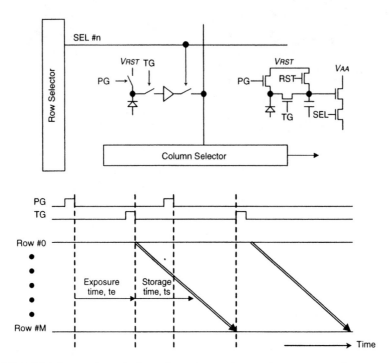

FIGURE 5.23 Memory-in-a-pixel configuration and operation sequence for consecutive imaging.

image sensors, it causes a lag-wise artifact because the photo leakage affects only the pixels exposed by the bright light. For DSC applications, further technology improvements are required to reduce the photo leakage, thermal leakage at the memory node, and readout noise reduction, especially kTC noise suppression.

The frame memory approach, shown in Figure 5.24, is similar to the concept of frame transfer CCD (FT CCD). The pixel signal is read out as quickly as possible and transferred into a frame memory row by row; then, the memorized signals are scanned for outputting. The frame memory should be completely covered by a metal light shield to avoid photo leakage at the memory cell. This photo-leakage-free configuration is a big advantage compared to the memory-in-pixel approach. Issues that must be addressed when using the frame-memory scheme are increased chip costs due to the additional on-chip memory; the difference in exposure times between the rows (which would be significant in a large pixel array); and how to achieve a low-noise signal readout from the pixel array.

5.4.3 Movie and AE/AF Modes

In recent DSCs, movie mode is becoming a very popular function. This requires the fast frame readout to perform AE/AF adjustments using the video output. In this section, movie imaging and subresolution operation will be described with some features relating to X–Y addressed readout image sensors.

CMOS Image Sensors

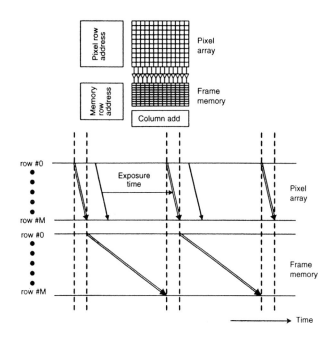

FIGURE 5.24 Frame memory CMOS imager configuration and operation sequence for consecutive imaging.

5.4.3.1 Subresolution Readout

Subresolution, the readout that enables faster frame rates for view finding, AE/AF operation, and movie imaging, are very common features in recent DSC image sensors. The two schemes for subresolution readout are pixel skipping and signal binning. Pixel-skip readout is easily achieved in CMOS image sensors because of the scanning flexibility of the X–Y address readout. In Figure 5.25, two contrasting examples are shown for a three-to-one skip in color image sensors. In example (a), the distance between pixels is kept constant; a set of four color pixels is treated as a pseudopixel in the case of (b). The constant pitch readout provides better spatial resolution but enhances color aliasing (also called a "color moiré" artifact). On the other hand, the pseudopixel pitch readout minimizes color aliasing but provides less spatial resolution. Therefore, choosing a skipping manner depends on the purpose of the subresolution readout and the interpolation algorithm in the subsequent back-end process.

5.4.3.2 Pixel Binning

Like reduced fill factor, skipping readout generates aliasing artifacts (see Section 3.5.2 in Chapter 3). Aliasing can be reduced by adding pixel signals in the subresolution unit,[59] as shown in Figure 5.26. This method is called pixel binning, which, in the case of CCDs, is performed by charge mixing in CCD registers, as described in Chapter 4, Section 4.3.5. In CMOS image sensors, pixel binning can be performed

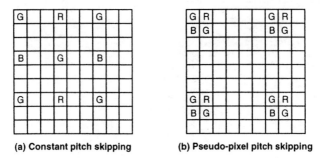

FIGURE 5.25 Color pixel skipping: (a) constant pitch skipping; (b) pseudopixel pitch skipping.

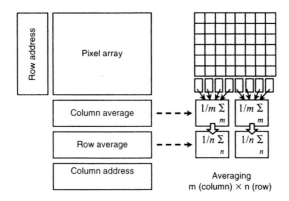

FIGURE 5.26 Pixel binning.

by circuit operation except for special pixels (such as shared architecture pixels introduced in Section 5.3).

In the CCD charge domain binning, the signal charge increases N times, where N is the number of binned pixels. Sensitivity becomes N times higher and saturation exposure decreases to $1/N$ because the charge handling capacity of the CCD register is limited, meaning that the charge binning has no effect on the dynamic range. In CMOS image sensors, a variety of binning schemes is possible[58,59] due to the technology's ability to implement on-chip signal processing circuits, which may achieve higher performance than the charge domain binning seen in CCD imagesensors.

5.4.3.3 Mode Change and Dead Frames

Changing scanning modes and/or exposure times is often required in a consecutive imaging AE/AF or movie mode. Changing modes in consecutive imaging is somewhat complex using the rolling reset scheme. As an example, Figure 5.27 shows an internal operation timing where the exposure time is lengthened from t_{int1} to t_{int2}, forcing a dead frame, in which exposure control is not completed, to appear. Similar situations occur when a window size and/or a skipping step is changed. Users of

CMOS Image Sensors

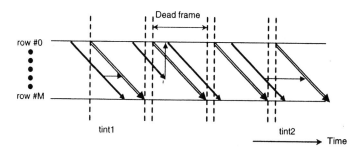

FIGURE 5.27 Generation of a dead frame at operation mode change.

CMOS image sensors should be aware of the possibility that a dead frame can be associated with mode changes, although its appearance depends on how the array control logic operates in each CMOS image sensor product.

5.5 FUTURE PROSPECTS OF CMOS IMAGE SENSORS FOR DSC APPLICATIONS

As the movie mode becomes a more accepted function in a DSC, increasing its memory requirements and improving the real-time processing performance, the differences between DSC and digital video (DV) technologies will be reduced or eliminated. The operational advantages of CMOS image sensors, including lower power consumption and higher performance in fast-operation mode, will make DSCs more attractive. In addition, a CMOS image sensor's readout flexibility enables the development of multiple-mode cameras, in which DSC and DV features are combined.

High-resolution movie CMOS image sensors have also been developed.[56,60] An 8-Mpixel, digital-output, ultrahigh-definition television (UDTV) image sensor[56] that can operate at 60 frames per second (fps) progressive readout consumes only 760 mW, including internal and output buffering power components, and enables a camera head with approximately one-tenth the power consumption of a CCD-based camera head. Although this is an extreme case, the potential CMOS image sensors have for high-speed operation will have an impact on future DSC designs.

Moreover, movie imaging in DSC applications is not limited to a legacy video format. Many interesting options can also be considered, such as high-frame-rate imaging. For example, a 5000-fps VGA CMOS image sensor has been demonstrated.[51] Its power consumption is only 500 mW, making it a realistic device for DSCs.

Finally, as evidenced by their many recent advancements, CMOS image sensors have great potential, and many of their future innovations will be unique to the technology. Three-dimensional imaging and high-dynamic-range imaging, for example, may become very familiar technologies to DSCs in the near future.

REFERENCES

1. D. Renshaw et al., ASIC vision, *Proc. IEEE CICC* (Custom Integrated Circuit Conf.), 7.3.1–7.3.4, 1990.
2. K. Chen et al., PASIC: a processor–A/D converter–sensor integrated circuit, *Proc. IEEE ISCAS* (Int. Symp. Circuits Syst.), 3, 1990.
3. E.R. Fossum, Active pixel sensors: are CCDs dinosaurs?, *Proc. SPIE, Charge-Coupled Devices Solid State Optical Sensors III*, 1900, 2–14, 1993.
4. E.R. Fossum, CMOS image sensors: electronic camera-on-a-chip, *IEEE Trans. Electron Devices*, 44(10), 1689–1698, 1997.
5. M.A. Schuster and G. Strull, A monolithic mosaic of photon sensors for solid-state imaging applications, *IEEE Trans. Electron Devices*, ED-13(12), 907–912, 1966.
6. P.K. Weimer et al., A self-scanned solid-state image sensor, *Proc. IEEE*, 55(9), 1591–1602, 1967.
7. M. Aoki et al., MOS color imaging device, *ISSCC Dig. Tech. Papers*, 26–27, February, 1980.
8. M. Aoki et al., 2/3-inch format MOS single-chip color imager, *IEEE Trans. Electron Devices*, ED-29(4), 745–750, 1982.
9. S. Terakawa et al., A CPD image sensor with buried-channel priming couplers, *IEEE Trans. Electron Devices*, ED-32(8), 1490–1494, 1985.
10. R.H. Dyck and G.P. Weckler, Integrated arrays of silicon photodetectors for image sensing, *IEEE Trans. Electron Devices*, ED-15(4), 196–201, 1968.
11. P.J.W. Noble, Self-scanned silicon image detector arrays, *IEEE Trans. Electron Devices*, ED-15(4), 202–209, 1968.
12. T. Nomoto et al., A 4M-pixel CMD image sensor with block and skip access capability, *IEEE Trans. Electron Devices*, 44(10), 1738–1746, 1997.
13. M. Noda et al., A solid state color video camera with a horizontal readout MOS imager, *IEEE Trans. Consumer Elecron.*, CE-32(3), 329–336, 1986.
14. K. Yonemoto and H. Sumi, A CMOS image sensor with a simple fixed-pattern-noise-reduction technology and a hole accumulation diode, *IEEE Trans. Electron Devices*, 35(12), 2038–2043, 2000.
15. K. Aizawa et al., Computational image sensor for on sensor compression, *IEEE Trans. Electron Devices*, 44(10), 1724–1730, 1997.
16. M. Ishikawa and T. Komuro, Digital vision chips and high-speed vision system, *Dig. Tech. Papers of 2001 Symp. VLSI Circuits*, 1–4, 2001.
17. T. Kinugasa et al., An electronic variable-shutter system in video camera use, *IEEE Trans. Consumer Electron.*, CE-33, 249–258, August 1987.
18. H.S. Wong, Technology and device scaling considerations for CMOS imagers, *IEEE Trans. Electron Devices*, 43(12), 2131–2142, 1996.
19. H.-C. Chien et al., Active pixel image sensor scale down in 0.18 μm CMOS technology, *IEDM Tech. Dig.*, 813–816, December, 2002.
20. B. Pain et al., Analysis and enhancement of low-light-level performance of photodiode-type CMOS active pixel imagers operated with sub-threshold reset, *IEEE Workshop CCDs Adv. Image Sensors*, R13, 140–143, 1999.
21. H. Tian, B. Fowler, and A.E. Gamal, Analysis of temporal noise in CMOS photodiode active pixel sensor, *IEEE J. Solid-State Circuits*, 36(1), 92–101, 2001.
22. B. Pain et al., An enhanced-performance CMOS imager with a flushed-reset photodiode pixel, *IEEE Trans. Electron Devices*, 50(1), 48–56, 2003.
23. J.E.D. Hurwitz et al., An 800k-pixel color CMOS sensor for consumer still cameras, *Proc. SPIE*, 3019, 115–124, 1997.

24. I. Takayanagi et al., Dark current reduction in stacked-type CMOS-APS for charged particle imaging, *IEEE Trans. Electron Devices*, 50(1), 70–76, 2003.
25. B. Fowler et al., Low noise readout using active reset for CMOS APS, *Proc. SPIE*, 3965, 126–135, 2000.
26. I. Takayanagi et al., A four-transistor capacitive feedback reset active pixel and its reset noise reduction capability, *IEEE Workshop CCDs Adv. Image Sensors*, 118–121, 2001.
27. B. Pain et al., Reset noise suppression in two-dimensional CMOS photodiode pixels through column-based feedback-reset, *IEDM Tech. Dig.*, 809–812, December, 2002.
28. T. Watabe et al., New signal readout method for ultrahigh-sensitivity CMOS image sensor, *IEEE Trans. Electron Devices*, 50(1), 63–69, 2003.
29. R.M. Guidash et al., A 0.6 µm CMOS pinned photodiode color imager technology, *IEDM Tech. Dig.*, 927–929, 1997.
30. A. Krymski, N. Khaliullin, and H. Rhodes, A 2e⁻ noise 1.3 Megapixel CMOS sensor, *IEEE Workshop CCDs Adv. Image Sensors*, 2003.
31. S. Inoue et al., A 3.25M-pixel APS-C size CMOS image sensor, *IEEE Workshop CCDs Adv. Image Sensors*, 16–19, 2001.
32. I. Inoue et al., Low-leakage-current and low-operating-voltage buried photodiode for a CMOS imager, *IEEE Trans. Electron Devices*, 50(1), 43–47, 2003.
33. K. Yonemoto and H. Sumi, A numerical analysis of a CMOS image sensor with a simple fixed-pattern-noise-reduction technology, *IEEE Trans. Electron Devices*, 49(5), 746–753, 2002.
34. H. Rhodes et al., CMOS imager technology shrinks and image performance, *Proc. IEEE Workshop Microelectron. Electron Devices*, 7–18, April 2004.
35. K. Mabuchi et al., CMOS image sensor using a floating diffusion driving buried photodiode, *ISSCC Dig. Tech. Papers*, 112–113, February 2004.
36. H. Takahashi et al., A 3.9 µm pixel pitch VGA format 10b digital image sensor with 1.5-transistor/pixel, *ISSCC Dig. Tech. Papers*, 108–109, February 2004.
37. M. Mori et al., A ¼ in 2M pixel CMOS image sensor with 1.75-transistor/pixel, *ISSCC Dig. Tech. Papers*, 110–111, February 2004.
38. T. Isogai et al., 4.1-Mpixel JFET imaging sensor LBCAST, *Proc. SPIE*, 5301, 258–262, 2004.
39. S. Mohajerzadeh, A. Nathan, and C.R. Selvakumar, Numerical simulation of a p–n–p–n color sensor for simultaneous color detection, *Sensors Actuators A: Phys.*, 44, 119–124, 1994.
40. M.B. Choikha, G.N. Lu, M. Sedjil, and G. Sou, A CMOS linear array of BDJ color detectors, *Proc. SPIE*, 3410, 46–53, 1998.
41. K.M. Findlater et al., A CMOS image sensor with a double-junction active pixel, *IEEE Trans. Electron Devices*, 50(1), 32–42, 2003.
42. A. Rush and P. Hubel, X3 sensor characteristics, *J. Soc. Photogr. Sci. Technol. Jpn.*, 66(1), 57–60, 2003.
43. J. Hynecek, BCMD – an improved photosite structure for high density image sensors, *IEEE Trans. Electron Devices*, 38(5), 1011–1020, 1991.
44. N. Kawai and S. Kawahito, Noise analysis of high-gain, low-noise column readout circuits for CMOS image sensors, *IEEE Trans. Electron Devices*, 51(2), 185–194, 2004.
45. R.H. Nixon et al., 256 × 256 CMOS active pixel sensor camera-on-a-chip, *IEEE J. Solid-State Circuits*, 31(12), 2046–2050, 1996.
46. Y. Matsunaga and Y. Endo, Noise cancel circuit for CMOS image sensor, *ITE Tech. Rep.*, 22(3), 7–11, 1998 (in Japanese).

47. W. Yang et al., An integrated 800 × 600 CMOS imaging system, *ISSCC Dig. Tech. Papers*, 304–305, February, 1999.
48. U. Ramacher et al., Single-chip video camera with multiple integrated functions, *ISSCC Dig. Tech. Papers*, 306–307, February 1999.
49. T. Sugiki et al., A 60mW 10b CMOS image sensor with column-to-column FPN reduction, *ISSCC Dig. Tech. Papers*, 108–109, February, 2000.
50. Z. Zhou, B. Pain, and E.R. Fossum, CMOS active pixel sensor with on-chip successive approximation analog-to-digital converter, *IEEE Trans. Electron Devices*, 44(10), 1759–1763, 1997.
51. A.I. Krymski and N. Tu, A 9-V/lux-s 5000-frames/s 512 × 512 CMOS sensor, *IEEE Trans. Electron Devices*, 50(1), 136–143, 2003.
52. R. Johansson et al., A multiresolution 100 GOPS 4 Gpixels/s programmable CMOS image sensor for machine vision, *IEEE Workshop CCDs Adv. Image Sensors*, 2003.
53. B. Pain et al., CMOS digital imager design from a system-on-a-chip perspective, *Proc. 16th Int. Conf. VLSI Design*, VLSI '03, 395–400, 2003.
54. S. Iversen et al., An 8.3-megapixel, 10-bit, 60 fps CMOS APS, *IEEE Workshop CCDs Adv. Image Sensors*, 2003.
55. A.I. Krymski et al., A high-speed, 240-frames/s, 4.1-Mpixel CMOS sensor, *IEEE Trans. Electron Devices*, 50(1), 130–135, 2003.
56. I. Takayanagi et al., A 1-1/4 inch 8.3M pixel digital output CMOS APS for UDTV application, *ISSCC Dig. Tech. Papers*, 216–217, 2003.
57. M. Wäny and G.P. Israel, CMOS image sensor with NMOS-only global shutter and enhanced responsivity,*IEEE Trans. Electron Devices*, 50(1), 57–62, 2003.
58. Z. Zhou, B. Pain, and E.R. Fossum, Frame-transfer CMOS active pixel sensor with pixel binning, *IEEE Trans. Electron Devices*, 44(10), 1764–1768, 1997.
59. R. Panicacci et al., Programmable multiresolution CMOS active pixel sensor, *Proc. SPIE*, 2654, 72–79, 1996.
60. R.M. Iodice et al., Broadcast quality 3840 × 2160 color imager operating at 30 frames/s, *Proc. SPIE*, 5017, 1–9, 2003.

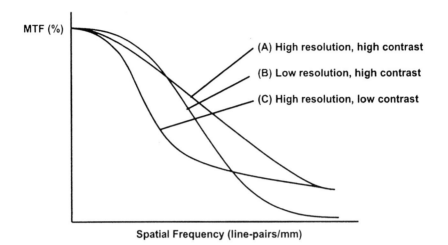

COLOR FIGURE 2.6 Three patterns of MTF spatial frequency characteristics.

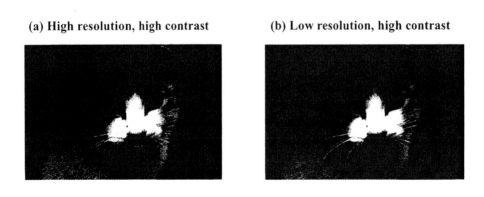

COLOR FIGURE 2.7 Images for three patterns of MTF spatial frequency characteristics: high resolution, high contrast; low resolution, high contrast; high resolution, low contrast.

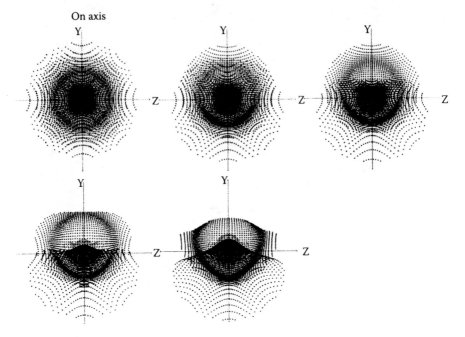

COLOR FIGURE 2.8 Example of a color spot diagram.

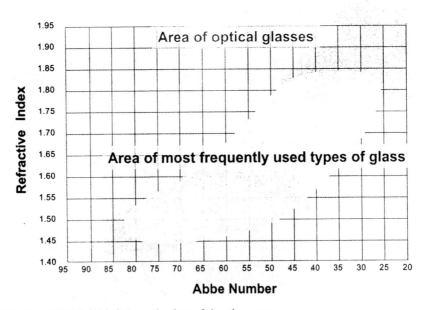

COLOR FIGURE 2.14 Schematic view of the glass map.

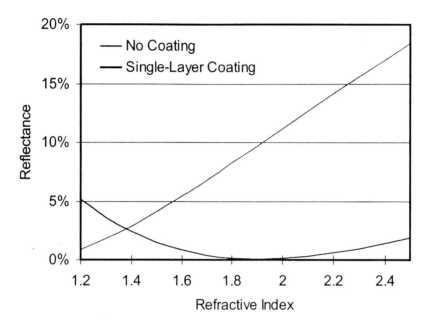

COLOR FIGURE 2.16 Correlation between the refractive index of glass and its reflectance.

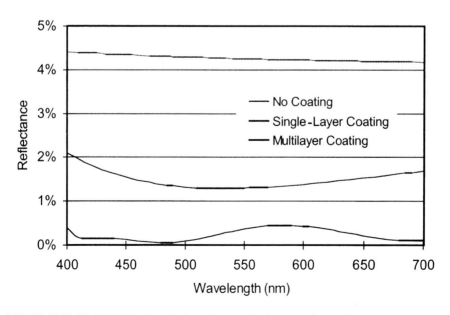

COLOR FIGURE 2.17 Wavelength dependency of reflectance for coatings.

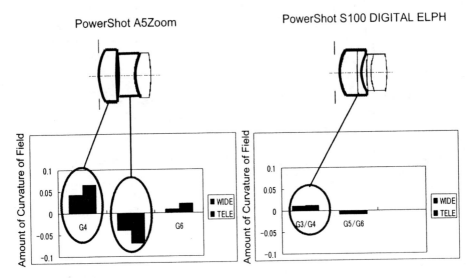

COLOR FIGURE 2.19 Comparison of decenter sensitivity.

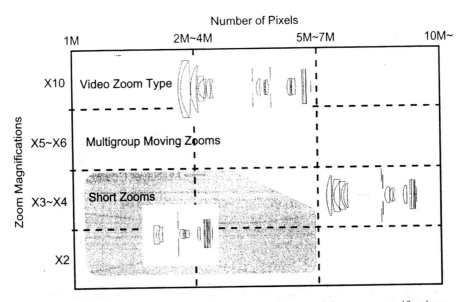

COLOR FIGURE 2.20 The correlation between zoom types and the zoom magnifications and their supported pixel counts.

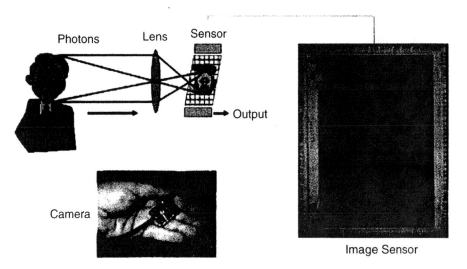

COLOR FIGURE 3.1 Image sensor.

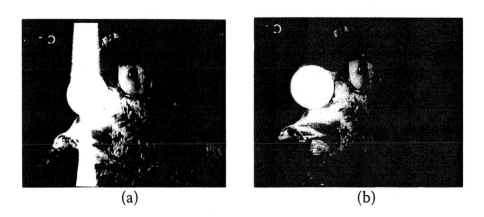

COLOR FIGURE 4.24 A blooming image and an improved image using VOD: (a) a spurious image caused by the blooming; (b) right image brought by VOFD.

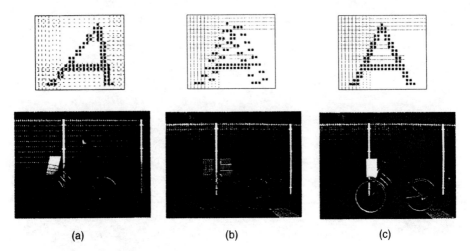

COLOR FIGURE 4.35 Shot images: (a) a shot image taken by raster (or line address) scan CMOS imager; (b) a shot image taken by interlace scan CCD imager; (c) a shot image taken by progressive scan CCD imager.

COLOR FIGURE 4.39 A picture shot with 1/10,000 sec electronic shutter.

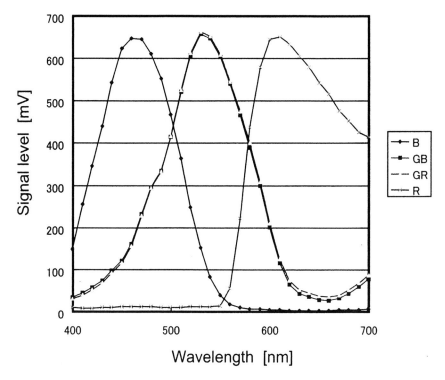

COLOR FIGURE 6.5 Example of spectral response.

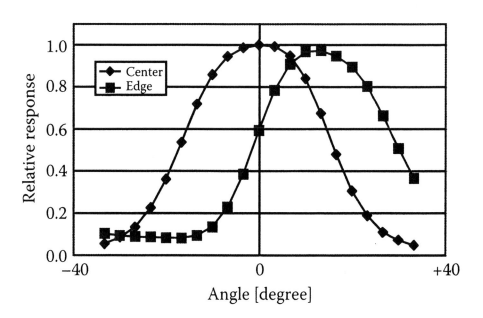

COLOR FIGURE 6.8 Example of angular response data.

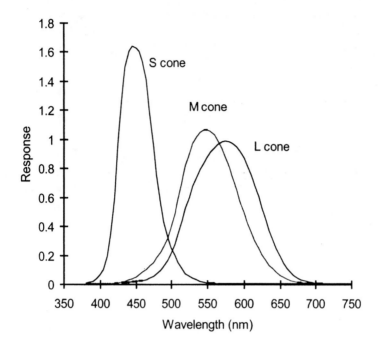

COLOR FIGURE 7.1 Human cone sensitivity (estimated).

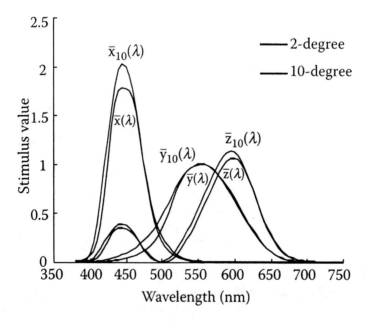

COLOR FIGURE 7.2 CIE color-matching functions.

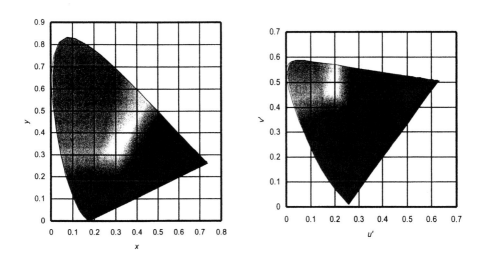

COLOR FIGURE 7.3 Diagram of xy and $u'v'$ chromaticity: CIE 1931 chromaticity diagram (left) and CIE 1976 UCS diagram (right).

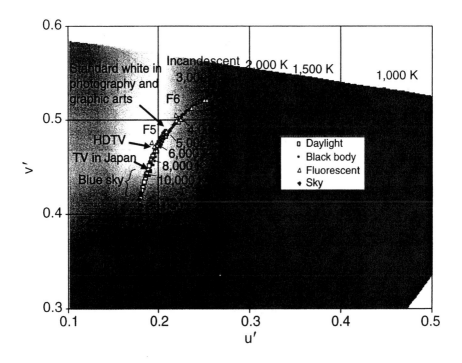

COLOR FIGURE 7.4 Light sources on the $u'v'$ chromaticity diagram.

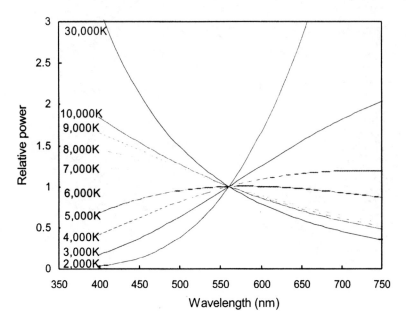

COLOR FIGURE 7.5 Spectral distribution of black body.

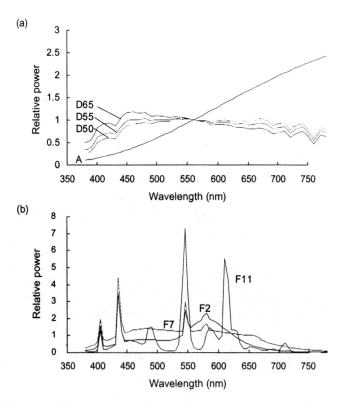

COLOR FIGURE 7.6 Spectral distribution of CIE standard illuminants (a) and fluorescent lamps (b).

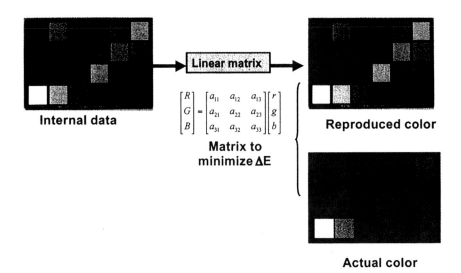

COLOR FIGURE 7.7 Characterization by linear matrix.

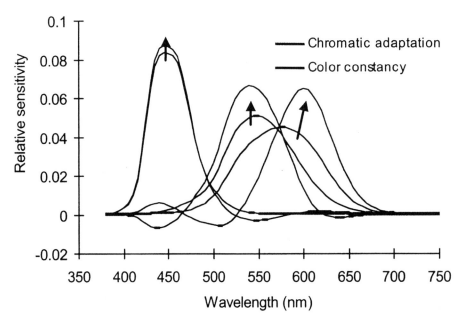

COLOR FIGURE 7.8 Equivalent color sensitivity in white balancing optimized for chromatic adaptation and color constancy.

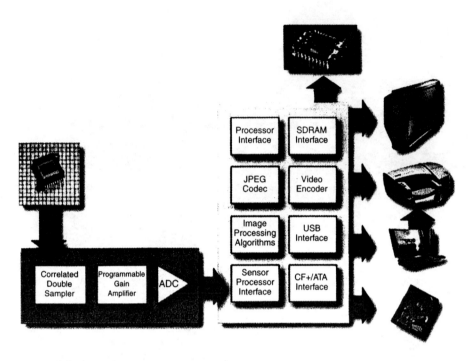

COLOR FIGURE 9.1 Camera block diagram.

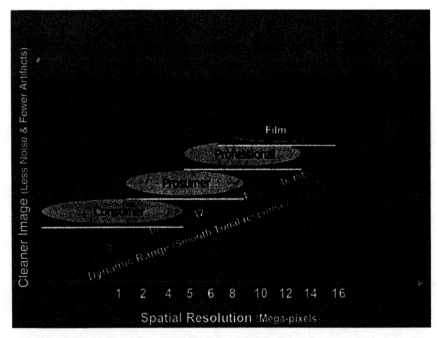

COLOR FIGURE 9.2 Conceptual mapping of imaging products.

COLOR FIGURE 9.10 Reproduced image comparison. The upper image is reproduced by the NDX-1260 device and the lower image is obtained by other AFE.

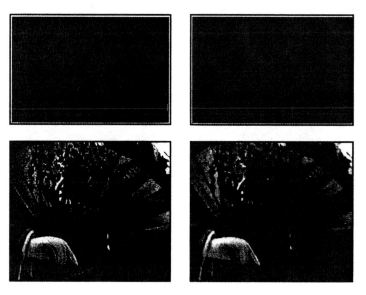

COLOR FIGURE 10.3 Difference of frequency response.

COLOR FIGURE 10.4 Normal (left) and noisy (right) images.

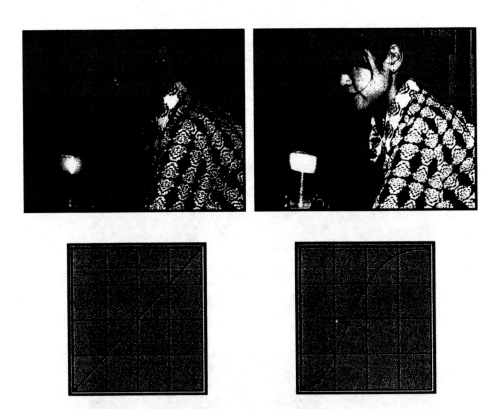

COLOR FIGURE 10.5 Original (left) and tone-enhanced (right) images.

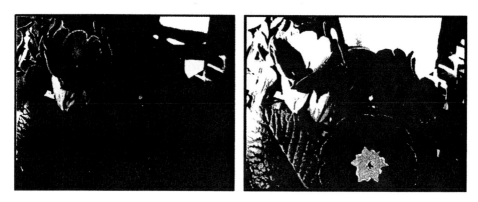

COLOR FIGURE 10.7 Images with enough (left) and insufficient (right) D-range.

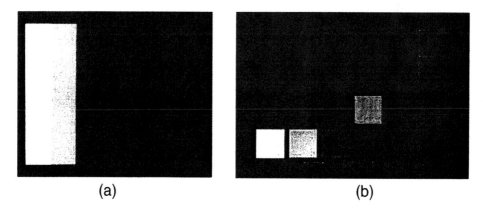

COLOR FIGURE 10.8 Test chart for measuring color reproduction: (a) color bar chart; (b) Macbeth's color chart.

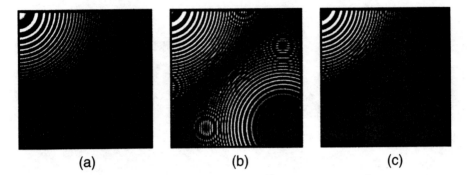

COLOR FIGURE 10.9 CZP pattern and examples of generated color moiré: (a) CZP pattern; (b) taken without OLPF; (c) taken with OLPF.

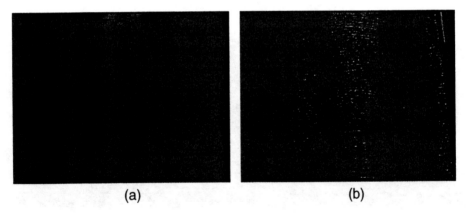

COLOR FIGURE 10.10 An example of color moiré in an actual picture: (a) moiré is not observed (out of focus); (b) color moiré generated (in focus).

 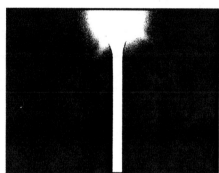

COLOR FIGURE 10.11 Examples of smear.

COLOR FIGURE 10.12 Example pictures: (a) with white clipping; (b) wide D-range; (c) with toneless black.

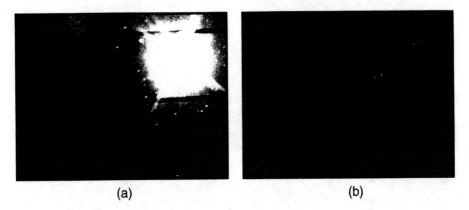

COLOR FIGURE 10.14 Ghost images with (a) a light source in the subject area and (b) a light source out of the subject area.

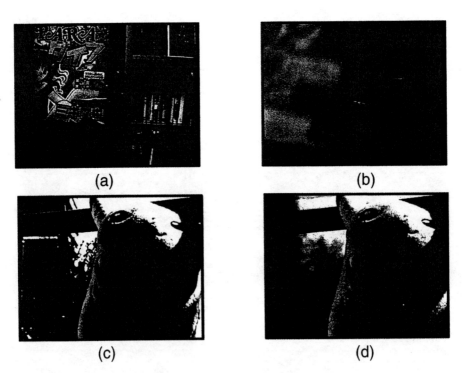

COLOR FIGURE 10.15 Changes of depth of field: (a) wide angle (short focal length); (b) narrow angle (long focal length); (c) narrow aperture; (d) open aperture.

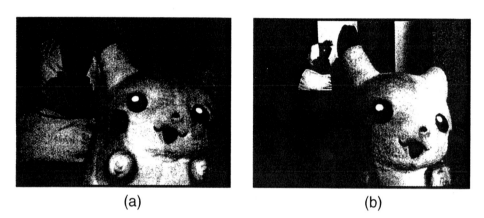

COLOR FIGURE 10.16 Difference of perspective taken with (a) a narrow-angle lens and (b) a wide-angle lens.

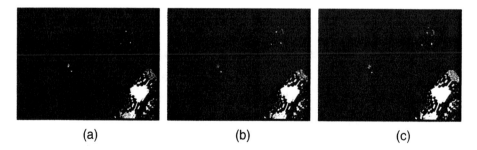

COLOR FIGURE 10.17 Difference of quantization bit number: (a) 8-b/color (full color); (b) 4-b/color; (c) 3-b/color.

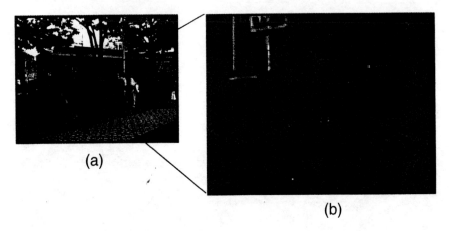

COLOR FIGURE 10.18 An example of compression noise (block distortion): (a) highly compressed picture; (b) expansion of a part.

COLOR FIGURE 11.1 An illustration of "being there."

6 Evaluation of Image Sensors

Toyokazu Mizoguchi

CONTENTS

6.1 What Is Evaluation of Image Sensors? 180
    6.1.1 Purposes of Evaluation 180
    6.1.2 Evaluation Parameters for Image Quality and an Image Sensor 180
    6.1.3 Environment for Image Sensor Evaluation 181
6.2 Evaluation Environment 181
    6.2.1 Evaluation of an Image Sensor and Evaluation Environment 181
    6.2.2 Basic Configuration of Image Sensor Evaluation Environment 181
        6.2.2.1 Light Source (Light Box, Pulsed LED Light Source, High-Irradiance Light Source) 182
        6.2.2.2 Imaging Lens 183
        6.2.2.3 Optical Filters (Color Conversion Filters, Band-Pass Filters, IR-Cut Filters) 183
        6.2.2.4 Jig for Alignment of Image Sensor's Position (x, y, z, Azimuth, Rotation) 183
        6.2.2.5 Temperature Control 183
        6.2.2.6 Evaluation Board 184
        6.2.2.7 Image Acquisition Board and PC 184
        6.2.2.8 Evaluation/Analysis Software 185
    6.2.3 Preparation 185
        6.2.3.1 Measurement of Light Intensity 185
        6.2.3.2 Adjustment of Optical Axis 186
6.3 Evaluation Methods 186
    6.3.1 Photoconversion Characteristics 186
        6.3.1.1 Sensitivity 189
        6.3.1.2 Linearity 189
        6.3.1.3 Saturation Characteristic 189
    6.3.2 Spectral Response 190
    6.3.3 Angular Response 193
    6.3.4 Dark Characteristics 194
        6.3.4.1 Average Dark Current 194
        6.3.4.2 Temporal Noise at Dark 195
        6.3.4.3 FPN at Dark (DSNU) 197

6.3.5	Illuminated Characteristics 198
	6.3.5.1	Temporal Noise under Illumination 198
	6.3.5.2	Conversion Factor 198
	6.3.5.3	FPN under Illumination (PRNU) 198
6.3.6	Smear Characteristics 199
6.3.7	Resolution Characteristics 200
6.3.8	Image Lag Characteristics 201
6.3.9	Defects 201
6.3.10	Reproduced Images of Natural Scenes 202

6.1 WHAT IS EVALUATION OF IMAGE SENSORS?

6.1.1 Purposes of Evaluation

Evaluation of image sensors, also referred to as characterization, is the process of determining if a particular sensor's performance parameters meet the final product's specifications. This evaluation is necessary for:

- *Development of a product that uses an image sensor.* Evaluation parameters and judgment criteria are determined by the specifications of the final product where the image sensor is to be used. Therefore, the judgment criteria change on a product-to-product basis. For example, evaluation parameters and results should be different for digital still cameras and video cameras. Even for DSC applications, although the evaluation parameters are almost identical, it is likely that evaluation results are different for single lens reflex (SLR) cameras, compact type cameras, and toy cameras. However, the primary specifications of a final product are determined by the evaluation results of the image sensor.
- *Development of an image sensor.* The purpose of evaluation of the image sensor is to improve and/or optimize the design performance by feeding the results back to the design and fabrication process engineers. Thus, it is important to know the relationship between the evaluation results for the application and the image sensor design. For example, if an evaluation result shows that the sensitivity is not high enough to meet the specification for a particular product, further investigation of each component that determines the sensitivity, such as performance of the microlens and/or readout electronics, is required.

In this chapter, the evaluation of image sensors for DSC product applications will be described.

6.1.2 Evaluation Parameters for Image Quality and an Image Sensor

Evaluation parameters for an image sensor are closely linked to the evaluation parameters for image quality of a picture taken by a DSC (discussed in Chapter 10).

TABLE 6.1
Relationship between Evaluation Items for an Image Sensor and for Image Quality

Evaluation parameters for image sensor	Evaluation parameters for image quality
Resolution	Sharpness
Spectral response	Moiré, false color
Characteristics under illumination	Color reproduction
Defects	Noise
Characteristics at dark	Shading
Angular response	Tone curve
Photo-conversion characteristics	Dynamic range
Smear	Other artifacts
Image lag	

The relationship between the evaluation parameters for an image sensor and for image quality is shown in Table 6.1.

6.1.3 ENVIRONMENT FOR IMAGE SENSOR EVALUATION

The environment for image sensor evaluation consists of evaluation methodology (software) that specifies how each item is measured or quantified and associated equipment (hardware). It is impossible to obtain correct and reproducible results if any part of the experimental setup is not clearly defined for each test.

6.2 EVALUATION ENVIRONMENT

6.2.1 EVALUATION OF AN IMAGE SENSOR AND EVALUATION ENVIRONMENT

Image sensor performance is obtained by measuring an input–output relationship under a certain operating condition. Input factors and environmental factors that affect the image sensor performance for DSC applications are: (1) input component: light (wavelength, intensity, incident angle, polarization) and its two-dimensional distribution over time; electric inputs (drive pulses, bias voltages, etc.); and (2) environmental factor: temperature. The evaluation environment provides a method that allows us to control these factors and to obtain numerical results with necessary accuracy.

It is less meaningful to compare evaluation results when they are obtained under different evaluation environments because the image sensor performance is highly sensitive to the factors mentioned previously. Factors that affect image sensor performance are shown in Figure 6.1.

6.2.2 BASIC CONFIGURATION OF IMAGE SENSOR EVALUATION ENVIRONMENT

Figure 6.2 shows a basic configuration of an image sensor evaluation environment. Each component is described in the following subsections.

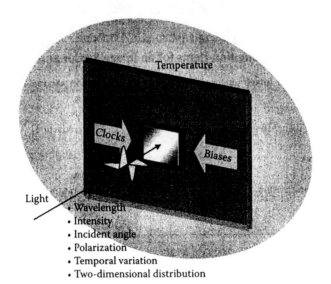

FIGURE 6.1 Factors that affect image sensor performance.

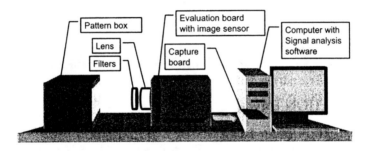

FIGURE 6.2 A basic configuration of the image sensor evaluation environment.

6.2.2.1 Light Source (Light Box, Pulsed LED Light Source, High-Irradiance Light Source)

Most of the evaluation parameters, such as photon-conversion characteristics, are measured with a light box. The color temperature of the light emitted from a light box is preferably 3200K, with which many image sensor manufacturers specify performance of their image sensor products. In cases when other color temperatures are needed, a color conversion filter is inserted in the light source or in an optical system in the evaluation equipment.

For evaluation of image lag, a light source that is capable of turning on and off with short rise time/fall time (a few microseconds), such as an LED, is used; it is synchronized with driving pulses to the image sensor. In addition to these light sources, a high-irradiant light source should be used for evaluation of a reproduced image of the sun, which is one of most severe scenes for image sensors to capture. It is very important to measure the irradiance, color temperature, spectral irradiance, and uniformity of the light source prior to a measurement, without relying on the

Evaluation of Image Sensors 183

specification sheet of the light source. This is important to assess the results based on the actual measured performance of the light source.

6.2.2.2 Imaging Lens

The image circle of a lens to be used should be larger than the optical format of an image sensor under test. If the position of the exit pupil for the image sensor under test is known, a suitable lens should be selected. It is also important to obtain data on spectral transmittance and degradation in light intensity near the perimeter of the image area at the minimum F-number of the lens.

Another important lens parameter is resolution. An imaging lens must have adequate horizontal and vertical resolution compared to the image sensor pixel resolution to avoid a blurred image. For example, an image sensor with 3-μm pixel pitch requires a much higher resolution lens than an image sensor with 7-μm pixel pitch.

Aperture mechanisms of lenses for SLR cameras and TV cameras are not suitable for precise evaluation of image sensors because they have errors between their nominal F-value and an actual value. Thus, reproducibility of the F-number setting is not guaranteed.

6.2.2.3 Optical Filters (Color Conversion Filters, Band-Pass Filters, IR-Cut Filters)

Spectral transmittance of every filter should be measured before performing evaluation of image sensors. A simple way to measure the spectral transmittance of the filters is to compare two data sets measured with a spectral radiometer. One data set corresponds to the radiation spectrum of the light source and the other is that of the light source with a filter placed in front. It is preferable that the center wavelengths of the transmission of a band-pass filter match that of an on-chip color filter of the image sensor.

6.2.2.4 Jig for Alignment of Image Sensor's Position (x, y, z, Azimuth, Rotation)

A mechanical apparatus for arranging the image sensor's position with high accuracy and high reproducibility is needed to measure characteristics of an image sensor that depend on incident light ray angle, such as angular dependence and shading of luminous/chroma signals associated with the angular response. Because the position of an image sensor chip assembled in a package has some small error, it is preferable to have an additional position arrangement mechanism when considering a need to change samples under test.

6.2.2.5 Temperature Control

The most important environmental factor for image sensors for DSC application is chip temperature. In order to design appropriate signal-processing algorithms, it is extremely important to obtain temperature dependence of defect pixels that have extraordinarily large dark current, as well as average dark current. Determining the temperature dependence of the dark current and defect pixels also allows designers to reduce dark current in the image sensor development.

In addition, temperature affects noise performance, full-well capacity, and charge transfer efficiency in CCD image sensors. Therefore, it is important to prepare equipment capable of controlling temperature in illuminated conditions as well as in dark condition. Because monitoring the temperature of an image sensor chip is very difficult, the relationship in temperature between the measurement point and the image sensor chip should be obtained. When an image sensor chip under test is large, special care should be taken for its temperature distribution across the die and/or package.

6.2.2.6 Evaluation Board

To perform evaluation of image sensors, it is necessary to acquire the raw data from each pixel with necessary resolution. For an image sensor provided by an image sensor manufacturer, an evaluation board made by the manufacturer, which is supposed to offer the best sensor performance, is used. For an image sensor that is being developed and for which a special evaluation board is not available, a board must be developed to digitize the image sensor's output signal without deteriorating its performance. Points to note are:

- Noise injection to an image sensor output should be avoided (board noise << image sensor noise).
- All the pixel data, including dummy pixels and optical black (OB) pixels, should be obtained.
- No signal processing, such as OB clamp, edge enhancement, etc. should be applied.
- A signal interface that is required by a data acquisition board, such as LVDS, etc., should be implemented.
- A sufficiently wide frequency bandwidth and low noise floor in the analog front-end circuitry should be guaranteed.
- A sufficiently high linearity in analog circuitry (including an analog-to-digital converter, ADC) should be guaranteed.
- When settings of an image sensor's operation modes are needed, an associated interface should be implemented.

6.2.2.7 Image Acquisition Board and PC

An image acquisition board and a PC are needed to process and analyze digital data that come from the evaluation board mentioned earlier. An image acquisition board available in the market can be used. Important points to note when selecting an acquisition board include:

- It should be capable of handling required image size, data rate, and ADC resolution.

- It should be capable of acquiring still images synchronized with other hardware, such as a mechanical shutter.
- It should be capable of acquiring consecutive frame capture.
- Multichannel input ports should be equipped, when necessary.
- Stable data acquisition should be guaranteed.

6.2.2.8 Evaluation/Analysis Software

The following sections describe examples of procedures to obtain particular evaluation results using raw data from an image sensor. To derive the evaluation results, image-processing software is used. Functions required for this type of software include:

- A region of pixel data to be processed can be selected.
- A skipped image with a designated unit (e.g., 4 × 4 pixels) can be generated.
- An average and a standard deviation of pixel data can be obtained.
- Addition, subtraction, multiplication, and division can be performed for image data.
- Addition and subtraction between a set of image data can be performed.
- Filter operations, such as smoothing or median filtering, can be performed.

6.2.3 PREPARATION

6.2.3.1 Measurement of Light Intensity

Exposure, the input factor for an image sensor, is a value of illuminance at the image sensor plane (face-plate illumination) multiplied by the integration time. However, a direct measurement of the face-plate illuminance requires complicated procedures and is usually difficult. Therefore, in most cases, the face-plate illuminance is obtained using the following equation with measuring the object illuminance, E_p:

$$E_p = \frac{E_0 R T_L}{4F^2(m+1)^2} \quad [\text{lux}] \quad (6.1)$$

where E_0, R, T_L, F, and m denote the object illuminance; reflectance of the object; transmittance of an imaging lens; F-number of an imaging lens; and the magnification factor, respectively.

For most cases in the evaluation described in this chapter, the object is a light box. In this case, the object illuminance is calculated using the following equation, with an assumption that the light-emitting plane of the light box is considered as a perfect diffusive plate:

$$E = \pi B \quad (6.2)$$

where B is the irradiance of the light-emitting plane of the light box in [nit].

As a common note for each evaluation described later, it is important to use the light source in its stable performance condition. That is, stability from turn-on and dependence on power supply voltage should be measured and the light source should be used under such control.

6.2.3.2 Adjustment of Optical Axis

The position of an image sensor under test is adjusted using a laser beam in such a way that a laser beam from an optical system in which the image sensor is installed is reflected back to the emitting point of the laser.

6.3 EVALUATION METHODS

6.3.1 Photoconversion Characteristics

Photoconversion characteristics or photon/electron transfer characteristics are very important, together with noise characteristics mentioned later to determine ISO speed of a DSC. Here, methods to obtain the photoconversion characteristics and how to derive associated parameters are described.

Measurement method. The output signal of an image sensor is measured under a constant optical condition using an electronic shutter function of the image sensor. A graph showing a relationship between the face-plate exposure and image sensor output should be created.

Procedures.

1. Images of the light box are acquired using several integration time settings. For each integration time, an image at dark condition is captured in order to eliminate dark current components.
2. A dark image is subtracted from an illuminated image (the integration time is the same for two images). Then, an average from a center portion with more than 10×10 pixels is calculated. For an image sensor with on-chip color filter array, averages for each color plane are obtained.
3. The face-plate exposure is obtained from the irradiance of the light box and the lens setting using Equation 6.1 and Equation 6.2. A graph showing a relationship between the face-plate exposure and image sensor output is plotted.

Evaluation condition. An example of the evaluation conditions is shown in the corresponding column of Table 6.2. It should be modified based on the operating conditions of a DSC in which the image sensor is to be equipped.

Note that, as described later, image sensor performance depends upon incident light ray angle. In general, the dependence becomes larger in an image sensor with a smaller pixel size and the image sensor output from the periphery of its imaging array differs from a level corresponding to the face-plate illuminance expressed by Equation 6.1. Therefore, F-number of an imaging lens should be noted as an important optical parameter.

TABLE 6.2
Evaluation Conditions (I)

			Photoconversion characteristics	Temporal noise under illumination	FPN under illumination	Smear	Resolution	Image lag
Equipment to use	Light source			Light box	Standard condition	Cold spot	Light box	LED
	Lens		Lens for SLR cameras (with/without a lens cap)					
	Filter			IR cut filter		IR cut filter ND filters	IR cut filter R/G/B filters	IR cut filter
	Other		—	—	—	V/10 mask	—	—
Evaluation condition	Optical condition	Brightness of the light source			Standard condition			Adjust
		Lens F-number			Open			
		Filter	If necessary, an IR filter is inserted			An IR filter and ND filters are inserted	An IR filter and R/G/B filters are inserted appropriately	An IR filter is inserted
	Temperature			25°C/60°C			25°C	25°C/60°C
	Integration time		Variable (0 to 3× saturation equivalent)	Variable (a linear region of the conversion characteristics should be used)	Variable	1/1000 sec	Variable for adjusting exposure	Appropriately set

Evaluation of Image Sensors

TABLE 6.2
Evaluation Conditions (I) (continued)

		Photoconversion characteristics	Temporal noise under illumination	FPN under illumination	Smear	Resolution	Image lag
	Drive condition			Standard drive condition			
Data to acquire	Region of pixels	More than 10 × 10 pixels at the center of the imaging array	All effective pixels in the imaging array	All effective pixels in the imaging array	Optical black pixels on the same rows and columns as those of V/10 spot	All effective pixels in the imaging array	
	Data processing	Averaging of a set of image data (dark image data subtracted)	Standard deviation of image data obtained by subtracting two images under illumination	Appropriate filtering	Averaging of a set of image data (dark image data subtracted)	Averaging of sequential frames	Averaging over the area of an image of the LED light source

Evaluation of Image Sensors

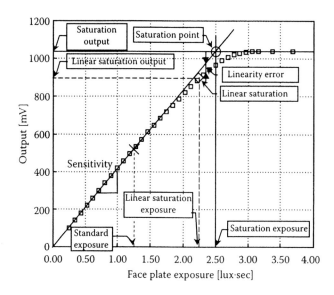

FIGURE 6.3 Photoconversion characteristic showing the linear saturation point.

6.3.1.1 Sensitivity

The slope of a straight line that is determined by an output level for zero exposure and an output level for a standard exposure* gives the sensitivity in [output unit/lux-sec].

6.3.1.2 Linearity

A straight line that is determined by an output level for zero exposure and an output level for a standard exposure is regarded as an ideal characteristic. Linearity error is defined as a deviation of a measurement characteristic from the ideal characteristic.

6.3.1.3 Saturation Characteristic

A few measurement points that are saturated at a constant value determine the "saturation output level" of an image sensor. The exposure value at the intersection of the saturation level and the ideal characteristic mentioned in Section 6.3.1.2 gives the "saturation exposure." The saturation output and saturation exposure are not meaningful for DSCs. What is meaningful is the maximum exposure that can be used for tone representation of an image. Here, we define it as a "linear saturation exposure." Also, we define the maximum output level that corresponds to the linear saturation exposure as "linear saturation output." These definitions are seen in Figure 6.3 and Figure 6.4.

The saturation exposure does not have a significant meaning for DSCs as described previously, but it is used as a reference point for high-light level characteristics of an image sensor, such as smear characteristics for CCD sensors. A

---

* Differs by specifications of a DSC.

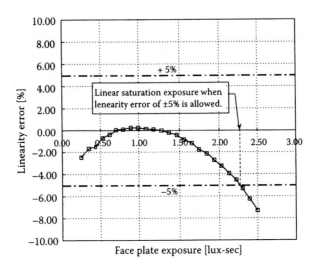

FIGURE 6.4 Definition of the linear saturation exposure.

difference between the linear saturation exposure and the saturation exposure is treated as a margin in high-light level performance.

Assessment of photoconversion characteristics. Measurement results of sensitivity, saturation characteristics, and linearity are judged based on the requirement of ISO speed for a DSC product.

6.3.2 SPECTRAL RESPONSE

Spectral response data are very important in identifying an IR-cut filter, color reproduction algorithm, and other parameters to be used in a DSC. Spectral response represents an image sensor output per incident light energy per wavelength in the range of spectrum where a DSC system operates. When spectral response is measured, the following should be taken into account:

- Use of a linear region of image sensor output levels
- Positional dependence of output light energy from a monochromator and the output angle of light rays: the output light beam from a monochromator has a certain cross-section but the whole cross-section area is not necessarily uniform. Also, the output angle of the beam should be small because an image sensor response has an angular dependence, as discussed in Section 6.3.3.
- Polarization: some image sensors respond differently depending on the polarization of light. The polarization dependence should be measured with a polarization filter.

Measurement method. There are two ways to measure spectral response using a monochromator. Image sensor outputs are measured while the monochromator is controlled so that (1) the light energy per unit wavelength is maintained at a constant value; or (2) the image sensor outputs are maintained at a constant value. With the

Evaluation of Image Sensors

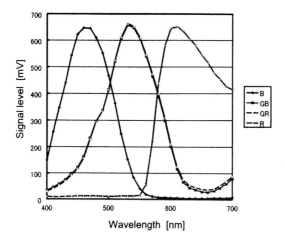

FIGURE 6.5 (See **Color Figure 6.5** following page 178.) Example of spectral response.

first method, an evaluation board for an image sensor can be used and the spectral response is directly obtained from the measurement results. However, its measurement accuracy depends on the accuracy of light energy control and signal-to-noise ratio (S/N) of the image sensor output. (When the sensor output is small, the measurement accuracy degrades.) With the second method, a feedback mean is required from the sensor output level to the light energy control. However, measurement accuracy does not depend on S/N and/or nonlinearity of the image sensor outputs.

Procedures.

1. An image sensor is irradiated by monochrome light from a monochromator and a set of image data is acquired. At each monochrome light level, an image at dark should be captured, which should be subtracted to avoid the dark current contribution.
2. A subtraction is performed between an image of the monochrome light and dark images. An average is obtained from 10 × 10 pixels in a center portion of the imaging array. For image sensors with on-chip color filter array, responses from each color plane are obtained.
3. The relationship between wavelength and the ratio of the image sensor output to the light energy is plotted. An example is shown in Figure 6.5.

Evaluation condition. Evaluation conditions are summarized in the corresponding column in Table 6.3.

Assessment of spectral response. Assessment of spectral response cannot be done alone; it should be done by taking color signal-processing algorithms and a way of setting parameters in the color signal-processing algorithms into account. For example, if the blue spectral response is lower than green and red responses for pixels using a Bayer color filter array, the increase in blue gain for white balance will introduce unacceptable noise levels.

TABLE 6.3
Evaluation Conditions (II)

			Spectral response	Angular response	Dark characteristics	Dark temporal noise
Equipment to use	Light source		Spectrometer (monochromater)	R, G, B LEDs	—	—
	Lens		—		Lens for SLR cameras (with a lens cap)	
	Filter		ND filters		—	
	Other		—			
Evaluation condition	Optical condition	Wavelength	380 ~ 780 nm	LED	—	
		Step	10 nm			
		Filter	If necessary, ND filters are inserted		—	
	Temperature		25°C		25°C/60°C	
	Integration time		Constant (a linear region of the conversion characteristics should be used)		Variable (0 ~ several tens of seconds)	Set depending on the purpose
	Drive condition				Standard drive condition	
Data to acquire	Region of pixels		More than 10 × 10 pixels at the center of the imaging array		All of effective pixels in the imaging array	
	Data processing		Averaging of a set of image data (dark image data subtracted)		Averaging	Standard deviation of a image data obtained by subtracting two dark images

Quantum efficiency. Quantum efficiency (QE) is obtained by converting the vertical axis of the spectral response into the level of signal charge in electrons divided by the number of incident photons.

Evaluation of Image Sensors

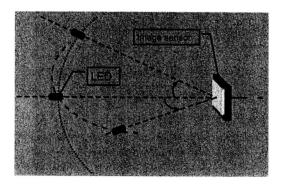

FIGURE 6.6 A method to control the incident light angle (1).

6.3.3 Angular Response

Image sensor performance parameters that depend upon the angle of incident light include the following:

- Light collection characteristic of the microlens and the shape of photodiode aperture
- Structure of the reflective interface material and dielectric layers, and their reflectance/refraction indices
- Incident light angle on the Si detector

Regarding first factor, the detector size shrinks with the scaling of pixel sizes but it is not easy to reduce the height of the structure above the detector because it is determined by electrical considerations, process geometry, and process layer thicknesses. Therefore, the angular response degrades with the pixel size scaling. For the second factor, light passing through a color filter experiences multiple reflections and passes light to neighboring pixels, which may cause color mixture. The third factor is a cause of color mixture where angled light directly absorbs near the charge transfer channel of the neighboring pixel.

Evaluation method. Two methods can be used to control the incident light angle illuminating an image sensor: (1) an image sensor is fixed and the location of a point light source located far from the image sensor is changed, as shown in Figure 6.6; or (2) a parallel light beam is used and the angle between the beam and an image sensor surface is changed, as shown in Figure 6.7. It is relatively easy to perform measurements using method 1 by using an LED. However, drawbacks include limited freedom in setting various wavelengths and difficulty in controlling the angle accurately. With method 2, precise data can be acquired using a collimator. However, special equipment that controls the angle needs to be prepared. The angular response tends to be significant along the diagonal of an imaging array where the incident angle is largest. Therefore, the evaluation should be done for each color plane up to the maximum angle of the imaging lens used in the camera system.

Evaluation condition. Evaluation conditions are summarized in the corresponding column in Table 6.3.

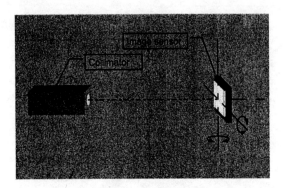

FIGURE 6.7 A method to control the incident light angle (2).

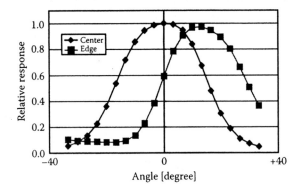

FIGURE 6.8 (See Color Figure 6.8 following page 178.) Example of angular response data.

Assessment of angular response. Assessment should be done by considering the characteristics of the imaging lens used and allowable levels of luminance shading and chroma shading. Also, F-number dependence of the angular response should be examined. Figure 6.8 and Figure 6.9 show an example of angular response data and a reproduced image, respectively.

6.3.4 Dark Characteristics

Characteristics of an image sensor with no light exposure are referred to as dark characteristics. A reproduced image of the dark must provide a uniform black image that should be a reference of a gray scale representation in principle. However, the outputs from pixels fluctuate temporally and spatially due to several factors. In this section, methods to evaluate the dark characteristics are presented.

6.3.4.1 Average Dark Current

Thermally generated charge, in addition to photogenerated charge, is accumulated over time at the charge storage region of a pixel and the charge transfer region (in CCD image sensors). This thermally generated current is called "dark current." In the evaluation of dark current, temperature should be controlled very accurately. The

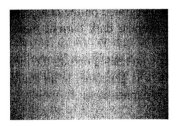

FIGURE 6.9 Example of a reproduced image showing the angular response.

evaluation is most often performed at a die temperature of 60°C, which considers the worst case of operating conditions of DSCs, in which the highest ambient temperature and the heat generation by the image sensor are taken into account.

Measurement method. Image sensor output levels are measured with integration time using an electronic shutter as a parameter, while the die temperature is kept constant. Then, a slope of the output change with the integration time is obtained.

Procedures.

1. Completely block all light from the optical system using a lens cap. Also, it is best to ensure that no stray light enters the system by enclosing the system using optical shielding cloth and/or reducing ambient room lighting.
2. Acquire a set of image data, each with a different integration time.
3. Obtain an average from the selected pixel window for each image obtained at each integration time and plot the relationship between the integration time and the output.
4. Obtain the rate of output change in [output unit/s] from the plot. An average dark rate in [e⁻/s] can be obtained using a conversion factor (described in Section 6.3.5.2).

Evaluation condition. An example of the evaluation conditions for the dark current measurements is shown in the corresponding column in Table 6.3. The conditions should be identified specifically for operating conditions of a DSC product.

6.3.4.2 Temporal Noise at Dark

Components in the output signal that fluctuate over time are called random noise or temporal noise. Temporal noise in number of electrons at dark n_d includes the dark current shot noise, n_{sd}, reset noise, n_{rs}, and read noise, n_{rd}, and is given by

$$n_d^2 = n_{sd}^2 + n_{rs}^2 + n_{rd}^2 \tag{6.3}$$

The dark current shot noise is given by

$$n_{sd} = \sqrt{N_d} \tag{6.4}$$

where N_d is the dark charge integrated on the charge storage region of a pixel. N_d can be obtained from the average dark current measurement mentioned in the previous section.

In case of image sensors with analog output, the measured noise should contain a noise component, n_{sys}, which is generated in an evaluation board. Therefore, the measured dark noise, n_{meas}, is expressed as

$$n_{meas}^2 = n_d^2 + n_{sys}^2 \qquad (6.5)$$

Measurement method. Main noise sources in image sensors are thermal noise and shot noise. These noise components have flat frequency distribution and, thus, are white noise. Because the probability distribution of white noise follows the Gaussian distribution, the root-mean squared (rms) random noise can be estimated by the standard deviation, σ, of samples taken in the time domain. Noise in each output signal from a pixel does not have any spatial and temporal correlations; image data captured at a certain time instances can be used to obtain temporal noise.

However, each output signal from a pixel does have an offset component that does not change in time, which results in FPN. Therefore, this offset component should be removed before calculating temporal noise. The simplest way is to perform subtraction between two frames of image data, which eliminates the FPN component. Note that the temporal noise component becomes $\sqrt{2}$ times larger because the two frames of image data have no correlation. Again, as noted before, the system noise component should be eliminated. It can be measured from image data obtained by feeding a zero signal with same output impedance as that of an image sensor under test to the input of the evaluation board.

Evaluation condition. Evaluation conditions for temporal noise at dark are summarized in the corresponding column in Table 6.3.

Assessment of temporal noise evaluation. Temporal noise at dark affects image quality when operating in low-light conditions. Therefore, temporal noise should be evaluated as a factor that determines S/N when a camera ISO speed is set at high values.

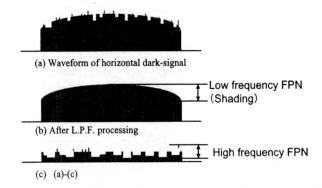

FIGURE 6.10 (a) Dark FPN; (b) a procedure to extract a shading component; (c) a high-frequency component.

6.3.4.3 FPN at Dark (DSNU)

Nonuniformity in image data that do not change over time is called fixed pattern noise (FPN). FPN and nonuniformity at dark are called "dark FPN" and "dark signal nonuniformity" (DSNU), respectively. The dark FPN consists of two components: a spatially, slowly varying component (shading) and a high-frequency component, as shown in Figure 6.10(a). Origins of FPN vary by device structures.

FPN should be evaluated in its magnitude and pattern. Unusual patterns may be caused by a device structure and/or back-end image-processing algorithms. Thus, evaluation should be done on a case-by-case basis for such unusual FPN patterns. Here, we focus on stripe-like FPN because it is usually caused by device structures and is very noticeable compared to pixel noise sources.

Measurement method. In order to obtain FPN, temporal noise components in the image data should be eliminated by averaging several frames. If we take an average from N images frames, S/N is improved by a factor of $\sqrt{N}$. The number of frames used for the FPN measurements to meet a determined accuracy depends on the magnitude of the random noise with respect to the FPN level:

- For a low-frequency component (shading) — after applying a median filtering, smoothing filtering, or both, which can suppress the high-frequency component, a peak-to-peak (p–p) value of the image data is used as a measure of shading, as shown in Figure 6.10(b).
- For a high-frequency component — the image data obtained using the preceding procedure is subtracted from the N-time averaged image data, as shown in Figure 6.10(c). A histogram of the resulting data and/or a p–p value is used as a measure of the high-frequency component. Note that a use of standard deviation is less meaningful in case the distribution is not Gaussian.
- For stripe FPN — obtain average values along a row or column direction with consecutive pixels. Differences in magnitude of the average values are considered as stripe FPNs. Pixel outputs that are significantly different from a normal output level should be removed by a median filter because they tend to affect the magnitude of the stripe FPN.

Evaluation condition. The same condition as that for temporal noise at dark should be used.

Assessment of dark FPN evaluation. Together with the temporal noise at dark, the dark FPN affects S/N of a reproduced image at high ISO speed settings. The pattern of FPN, as well as its magnitude, is important because, in general, stripe FPN in columns or rows is visible in an image if the p–p difference is approximately the same as the rms value of the pixel noise (FPN + temporal). Visual assessment under low-light conditions (high ISO) should be performed.

6.3.5 ILLUMINATED CHARACTERISTICS

Image sensor characteristics under uniform illumination are referred to as "illuminated characteristics" here.

6.3.5.1 Temporal Noise under Illumination

Temporal noise under illumination n_{illum} is given by

$$n_{illum}^2 = n_d^2 + n_s^2 \qquad (6.6)$$

where n_d and n_s are the random noise at dark and the photon shot noise, respectively.

6.3.5.2 Conversion Factor

The amount of charge collected in a pixel is converted to an output in an image sensor with a conversion factor, ζ, in [output unit/e⁻]. The output of an image sensor may be a voltage, current/charge, or digits. Also, the relationship between the integrated charge and the output varies depending on circuit configurations and parameters. Therefore, when comparing different image sensor devices, the comparison should be done by the input referred numbers that are measured in e⁻ (electrons) at the charge detection node inside a pixel.

Measurement method. From Equation 3.51 and Equation 3.52 in Chapter 3, we have obtained the relationship of Equation 3.53. The conversion factor, ζ, can be obtained from the equivalent formula as follows:

$$n_s'^2 = \zeta \cdot S_s' \qquad (6.7)$$

where n_s' and S_s' are the photon shot noise component in the output and the output of the image sensor in [output unit], respectively. The measurement procedures are the same as those for random noise at dark, except that the image sensor is uniformly illuminated. The photon shot noise component is obtained through Equation 6.6.

Evaluation condition. Evaluation conditions for temporal noise under illumination are summarized in the corresponding column in Table 6.2.

Assessment of conversion factor estimation. The conversion factor is not a parameter for evaluation. However, a true comparison of each evaluation parameter with input referred values is only possible by knowing the conversion factor.

6.3.5.3 FPN under Illumination (PRNU)

Ideally, the image sensor output from each color channel is uniform when an image sensor is illuminated by uniform white light. However, this is not usually the case for several reasons, such as variations of pixel performance and angular response of a microlens. Also, nonuniformity in saturation exposure and saturation output are seen due to variation of pixel structure caused by fabrication process variation.

Evaluation of Image Sensors

Measurement method. Acquire image data of a light box without a test chart. Variation in spatial irradiance of the light box and transmittance of the imaging lens should be corrected to avoid error in pixel measurements. The high- and low-frequency components are evaluated using the same manner as that for dark FPN. Create images from each color channel (R, G_R, G_B, B). The low-frequency component of a G channel is treated as luminous shading. Color shading is evaluated by a low-frequency component of images of (B – G) and (R – G), or B/G and R/G. Because the luminous and color (chroma) shading depend on the color signal-processing algorithms, they should preferably be evaluated with the same algorithms. Also, shading is usually caused by the angular response, so F-number dependence needs to be measured and examined. Variation of the saturation output is evaluated as a *p–p* value under two to three times higher illumination than the nominal saturation exposure.

Evaluation condition. Evaluation conditions for FPN under illumination are summarized in the corresponding column in Table 6.2.

Assessment of FPN under illumination. The high-frequency component of FPN under illumination should be examined from a color S/N point of view. Also, the luminous/color shadings should be judged by taking into account the performance of the optical system used. Variation of saturation output should be evaluated, considering the fact that the minimum value determines the dynamic range of a DSC, because the saturation output level correlates with the linear saturation level.

6.3.6 Smear Characteristics

Although smear and blooming can be distinguished by their generation mechanisms, we will refer to smear as any artifacts in an image of very bright objects.

Measurement method. Take an image of a bright light with its size V/10, where V denotes the height of effective imaging pixels at the center of the imaging array, and measure outputs from optical black pixels on the same rows and columns, as shown in Figure 6.11. The light exposure is usually set for a high-speed electronic shuttering condition at approximately 1000 times the saturation exposure of the image sensor under test. For example, this value should be set as follows: first, the light intensity is set so that an image sensor outputs its saturation level while it operates in the electronic shutter mode at 1/1000 sec and receives the light through an ND (neutral density) filter with 1/1000 attenuation. Then, the ND filter is removed to produce 1000 times saturation exposure.

Smear suppression ratio is obtained using the following formula:

$$Smear\ suppression\ ratio = 20 \cdot \log \frac{average\ output\ of\ OB\ pixels}{saturation\ output \times 1000}\ [dB] \quad (6.8)$$

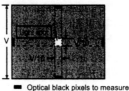

■ Optical black pixels to measure

FIGURE 6.11 Measurement of smear.

Evaluation condition. Evaluation conditions for smear are summarized in the corresponding column in Table 6.2.

Assessment of smear evaluation. For DSC applications, a smear suppression ratio of −100 to120 dB is a critical value for acceptance. Also, the pattern associated with the smear artifacts is an important factor. Even if the smear suppression ratio is acceptable, generation of a visually unnatural pattern is not acceptable.

6.3.7 Resolution Characteristics

Camera system resolution characteristics are determined by image-processing algorithms and imaging lens characteristics in addition to the resolution characteristics of the image sensor. Thus, measurement of the amplitude response, AR, of the image sensor is needed rather than only obtaining the limiting resolution of the system. Responses near the limiting resolution (i.e., the Nyquist frequency) and the associated aliasing are determined by the shape of a photodiode. Another factor that affects the AR is pixel-to-pixel cross-talk that has become more significant as a result of the pixel size scaling. This is caused by the pixel structure and occurs optically and electrically. Here, a simple method to obtain the response is introduced.

Measurement method. Obtain a monochrome sensor without an on-chip color filter array. A CZP (circular zone plate) chart is set on the light box. Obtain AR along the horizontal/vertical directions and any other directions of interest from an envelope of the image data taken. Repeat measurements with and without R/G/B filters. Figure 6.12 shows an example of AR.

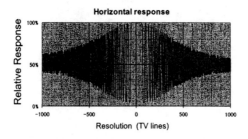

FIGURE 6.12 Example of amplitude response (AR).

In case the currently available charts are not good enough for recent high-resolution image sensors to evaluate the resolution characteristics, the image height on a focal plane should be adjusted so that the corresponding spatial frequency can be obtained. Some points to be noted are:

- Nonuniformity of brightness of the light box should be obtained without inserting a test chart.
- MTF of the imaging lens should be known. The lens MTF data are used for correction of resolution data.
- An image sensor under test should operate in its linear region of the photoconversion characteristic.
- The in-phase and out-of-phase responses should be measured near the limiting resolution.

Evaluation conditions. These are summarized for resolution characteristics in the corresponding column in Table 6.2.

Assessment of evaluation of resolution. Usually, the measurement results are used to determine parameters of an optical low-pass filter and signal processing of a back-end processor in a DSC. The resolution should be evaluated by the total resolution of the DSC, as given by Equation 3.58 in Chapter 3. Therefore, the measurement results provide data for optimizing parameters for signal processing in a DSC. If measured data with R/G/B filters are different, it is possible that crosstalk between pixels exists.

6.3.8 Image Lag Characteristics

Image lag degrades image quality of pictures taken in consecutive shooting modes and movie modes of a DSC. Lag occurs when some residual charge remains in a pixel after the signal is transferred. This charge remains in the pixel during the next consecutive exposed frame and may corrupt the image — especially images of bright objects in low-light scenes.

Measurement method. An LED is turned on and off to illuminate an image sensor under test at a time before the signal readout that is synchronized with the image sensor operation, as shown in Figure 6.13. The image lag is evaluated using the following equation:

$$Lag = \frac{S2}{S1} \times 100 \ [\%] \qquad (6.9)$$

where *S1* and *S2* are the lag component in the first frame and that of the second frame after the LED is on, respectively, as shown in Figure 6.13. Evaluation conditions for image lag are summarized in the corresponding column in Table 6.2.

Assessment of evaluation of image lag. Image lag is not acceptable in an image sensor if it is visually detectable for all operating modes. The allowable level is determined by the highest ISO speed of the intended DSC product.

6.3.9 Defects

A "defect" pixel is (1) a pixel that does not respond to incident light (a white/ black spot defect, etc.); or (2) a pixel that responds to incident light but with significantly different characteristics from those of normal pixels. For the latter type of pixel,

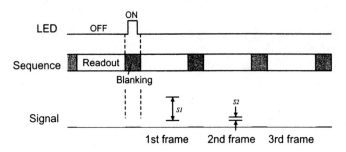

FIGURE 6.13 Measurement of image lag.

evaluation conditions and an allowable level vary, depending on specifications of a DSC. Also, patterns of cluster positions that are to be accepted or rejected are determined by information of positions of defects that include both types of defect pixels and defect correction algorithms in digital signal processing.

Measurement method. Images that are generated to evaluate the high-frequency components of FPN (see Section 6.3.4.3 and Section 6.3.5.3) may be used. By applying a particular threshold level to the image data, defect pixels are detected.

Evaluation conditions. The conditions are basically identical for measurements of DSNU and PRNU. Note that the dark current, which is a major contribution to defects, is highly sensitive to increase of temperature. Dark current is doubled at 5 to 10°C in the die temperature range from 25 to 60°C. Therefore, defects and the associated threshold levels are usually defined for a specific operating temperature.

6.3.10 REPRODUCED IMAGES OF NATURAL SCENES

Images captured of very bright light sources, such as the Sun, which has the highest irradiance in nature, can cause unpredictable artifacts. These artifacts are not necessarily from a direct incidence of strong light. Common artifacts include flare-like artifacts that appear at the periphery of an image of an object having strong irradiance and a phenomenon in which a high-light spot turns to a black spot in some CMOS image sensors. These artifacts may be seen when performing smear measurements (described in Section 6.3.6).

Evaluation method. To confirm if these artifacts occur, scenes summarized in Table 6.4 are created. Pictures are taken and examined in all the operation modes

TABLE 6.4
Scenes to Confirm if Artifacts Caused by Bright Light Sources Occur

Scene	Shooting condition
Scenes in which the sun is behind a building (position of the sun: at the top, most right, and most left in an image)	F-number of a lens and the electronic shutter speed of an image sensor variable
Scenes in which the sunlight is reflected by a side mirror, windshield, and hood of a car (the reflection point is at the bottom of an image)	

Evaluation of Image Sensors

used in a DSC. For example, if a movie function and electronic view finder (EVF) function are to be implemented in a DSC in addition to still picture mode with a mechanical shutter, movie pictures should be taken at the required frame rate.

Assessment of reproduced images of natural scenes. It is difficult to quantify the evaluation criteria to the reproduced images of natural scenes. A qualitative approach must be used to evaluate reproduced images of natural scenes. In general, images must appear natural and not contain visually unappealing artifacts that cause the viewer to question the appearance objects in the scene.

7 Color Theory and Its Application to Digital Still Cameras

Po-Chieh Hung

CONTENTS

7.1 Color Theory 206
    7.1.1 The Human Visual System 206
    7.1.2 Color-Matching Functions and Tristimulus Values 206
    7.1.3 Chromaticity and Uniform Color Space..207
    7.1.4 Color Difference...210
    7.1.5 Light Source and Color Temperature 211
7.2 Camera Spectral Sensitivity..212
7.3 Characterization of a Camera 214
7.4 White Balance 215
    7.4.1 White Point 215
    7.4.2 Color Conversion 216
        7.4.2.1 Chromatic Adaptation 216
        7.4.2.2 Color Constancy 217
7.5 Conversion for Display (Color Management) 218
    7.5.1 Colorimetric Definition 218
    7.5.2 Image State...219
    7.5.3 Profile Approach 220
7.6 Summary 220
References..221

Color is one of the most important aspects to the design of digital still cameras because the first impression of image quality is mostly affected by the quality of color. In conventional photography, sensitometry, which is based on dye amount, is often used. Unlike conventional photography, which uses only silver halide film and silver halide paper, digital photography systems output to many other types of media such as CRT (cathode ray tube), LCD (liquid crystal display), and inkjet printing. This means that we must consider a basic principle applicable to any media: colorimetry. In this chapter, the basic knowledge of color used for the design of a DSC is described.

7.1 COLOR THEORY

7.1.1 THE HUMAN VISUAL SYSTEM

The human eye has three types of light receptors: L (long), M (middle), and S (short) cones. The cones have the spectral sensitivities illustrated in Figure 7.1. The reason why human eyes sense color is that these three sensitivities are different from each other. Humans recognize color after processing the light information from the cones through nerves to the brain. The stimuli from the cones are converted into luminance, red–green, and yellow–blue signals. It should be noted these three signals are imitated by the YCbCr signals used in image compression such as JPEG. These signals are transmitted to the brain for recognition as another set of color attributes, such as lightness, chroma, and hue.

In addition to the cones, an eye has another type of receptor, which is called rod. Because it is deactivated in a bright environment, it does not contribute to the recognition of color. It is only active in a dark environment — approximately from sunset to sunrise, during which the three cones are deactivated. A single receptor does not give spectral information, so color is not sensed in such a dark environment.

7.1.2 COLOR-MATCHING FUNCTIONS AND TRISTIMULUS VALUES[1]

In industrial use, it is desired that color information be quantified and exchangeable. CIE (International Commission on Illumination) specifies the methodology, which is called colorimetry. It essentially traces the process in the human visual system with some modifications, approximations, and simplification for easy use in industry.

Instead of the cone sensitivities, a set of equivalent sensitivity curves is defined as a linear transformation of the set of cone sensitivities. The sensitivity curves, depicted in Figure 7.2, are called the CIE color-matching functions. (Note that the set of cone sensitivities mentioned previously is historically estimated from the color-

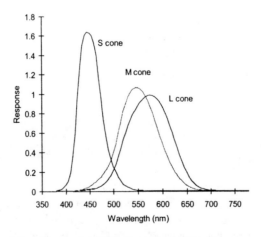

FIGURE 7.1 (See Color Figure 7.1 following page 178.) Human cone sensitivity (estimated).

Color Theory and Its Application to Digital Still Cameras

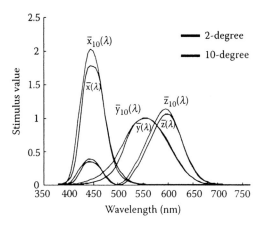

FIGURE 7.2 (See Color Figure 7.2 following page 178.) CIE color-matching functions.

matching functions.) These functions are noted as $\bar{x}(\lambda)$, $\bar{y}(\lambda)$, $\bar{z}(\lambda)$, where $\bar{y}(\lambda)$ is adjusted to represent luminance to fulfill industrial demands.

Two sets of color matching functions are defined by CIE. The first one, determined in 1931, was obtained for a view angle of less than 2°. In 1964, the other set was obtained for a view angle of between 4° and 10° and is denoted by $\bar{x}_{10}(\lambda)$, $\bar{y}_{10}(\lambda)$, and $\bar{z}_{10}(\lambda)$. Although some small errors and individual differences have been reported, these are still believed to be a typical representation of the sensitivity of normal human vision.

When an arbitrary object with spectral reflectance, $R(\lambda)$, is illuminated by a light source with spectral radiance, $L(\lambda)$, tristimulus values, which are also a linear transformation of the cone sensitivities, can be obtained as follows:

$$X = \int L(\lambda)R(\lambda)\bar{x}(\lambda)d\lambda$$

$$Y = \int L(\lambda)R(\lambda)\bar{y}(\lambda)d\lambda \qquad (7.1)$$

$$Z = \int L(\lambda)R(\lambda)\bar{z}(\lambda)d\lambda$$

This is called the CIE 1931 (1964) standard colorimetric system. Tristimulus values calculated by Equation 7.1 form the basis of quantification of color. Here, the Y value provides luminance and the X and Z values physically have no meaning.

7.1.3 CHROMATICITY AND UNIFORM COLOR SPACE

Tristimulus values certainly quantify a color. However, it is hard for us to imagine the color from these values. Two forms of representing color have become prevalent:

chromaticity coordinates and uniform color spaces. The former is a simplified way of ignoring the luminance axis and may be easily understood by a two-dimensional representation. The latter attempts to imitate human color recognition in three dimensions. Unfortunately, there is no unique way to represent color due to trade-offs between the desire for simple equations and the complexity within the actual human visual system. As a result, every representation has some deficiencies, such as geometrical homogeneity. The ideal geometrical homogeneity is that the geometrical distance of two colors in a color space is equal to the apparent color difference all over the place.

The two following representations of chromaticity are the most widely used. The simplest one, xy chromaticities, is calculated from CIE coordinates by the following equations:

$$x = \frac{X}{X+Y+Z}, y = \frac{Y}{X+Y+Z} \qquad (7.2)$$

Although this representation tends to give a large geometric distance for a unit color difference in the green region compared with unit differences in other colors, it is still widely used due to its simplicity. To provide a better representation in terms of geometric uniformity, the following $u'v'$ chromaticity is also widely accepted. This is also used for the calculation of the CIE 1976 $L^*u^*v^*$ uniform color space described later.

$$u' = \frac{4X}{X+15Y+3Z}, v' = \frac{9Y}{X+15Y+3Z} \qquad (7.3)$$

These two chromaticity diagrams are illustrated in Figure 7.3. The horseshoe arch is the locus of monochromatic lights and represents the range of color that eyes can recognize.

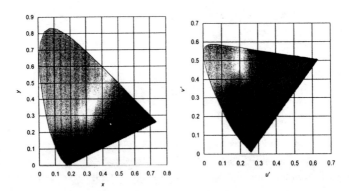

FIGURE 7.3 (See Color Figure 7.3 following page 178.) Diagrams of xy and $u'v'$ chromaticity: CIE 1931 chromaticity diagram (left) and CIE 1976 UCS diagram (right).

Color Theory and Its Application to Digital Still Cameras

For uniform color spaces, the CIE 1976 $L*a*b*$ color space and the CIE 1976 $L*u*v*$ color space recommended by CIE in 1976 are widely used. These are designed to simulate color recognition. Because they use orthogonal coordinates, additional equations are necessary in order to convert them into the cylindrical coordinates of lightness, chroma, and hue. These are used to represent a color analogous to the color recognition.

These equations require the tristimulus values of the white point for normalization. Usually, a virtual perfect reflecting diffuser is used for the measurement of a white point, but obviously the choice of a white point is not easy in an actual scene. The uniform color space is developed for object colors, not for self-emitting light sources. Therefore, the choice of a white point may be somewhat arbitrary. In practice, the estimation of a perfect reflecting diffuser or the brightest white in the scene is often used. The best practice may be to provide the definition of the white point along with $L*a*b*$ or $L*u*v*$ values. It should be noted that the color spaces were developed for a specific observing condition — a color patch with a neutral gray background ($L* = 50$). If the background does not match the condition, the uniformity of the color spaces is not guaranteed.

The equations to compute these values are as follows:
$L*$ (common for $L*a*b*$ and $L*u*v*$):

$$L* = \begin{cases} 116\left(\dfrac{Y}{Y_0}\right)^{\frac{1}{3}} - 16 & \dfrac{Y}{Y_0} > 0.008856 \\ 903.29\left(\dfrac{Y}{Y_0}\right) & \dfrac{Y}{Y_0} \leq 0.008856 \end{cases} \quad (7.4)$$

$a*b*$:

$$X_n = \begin{cases} \left(\dfrac{X}{X_0}\right)^{\frac{1}{3}} & \dfrac{X}{X_0} > 0.008856 \\ 7.787\left(\dfrac{X}{X_0}\right) + \dfrac{16}{116} & \dfrac{X}{X_0} \leq 0.008856 \end{cases}$$

$$Y_n = \begin{cases} \left(\dfrac{Y}{Y_0}\right)^{\frac{1}{3}} & \dfrac{Y}{Y_0} > 0.008856 \\ 7.787\left(\dfrac{Y}{Y_0}\right) + \dfrac{16}{116} & \dfrac{Y}{Y_0} \leq 0.008856 \end{cases}$$

$$Z_n = \begin{cases} \left(\dfrac{Z}{Z_0}\right)^{\frac{1}{3}} & \dfrac{Z}{Z_0} > 0.008856 \\ 7.787\left(\dfrac{Z}{Z_0}\right) + \dfrac{16}{116} & \dfrac{Z}{Z_0} \leq 0.008856 \end{cases}$$

$$a^* = 500(X_n - Y_n)$$
$$b^* = 200(Y_n - Z_n)$$
(7.5)

u*v*:

$$u^* = 13L^*(u' - u'_0)\,,\ v^* = 13L^*(v' - v'_0) \tag{7.6}$$

C*(chroma), h (hue angle):

$$C_{ab}^* = (a^{*2} + b^{*2})^{\frac{1}{2}},\ h_{ab} = \dfrac{180°}{\pi}\tan^{-1}\left(\dfrac{b^*}{a^*}\right) \tag{7.7}$$

where X_0, Y_0, and Z_0 denote the tristimulus values of white point, and u'_0 and v'_0 are its $u'v'$ chromaticity.

These were intended to use simple equations and the normalization by white point does not well model the visual adaptation; therefore, color appearance models such as CIECAM02[2] have been developed. Color appearance models take into account the adaptation of the visual system into their equations in addition to uniform color space. The model is designed to predict lightness, chroma, hue, etc. in arbitrary observing conditions.

7.1.4 COLOR DIFFERENCE

The geometrical difference of two colors in a uniform color space should be proportional to the apparent, or perceived, color difference. The color difference is denoted by ΔE^* (delta E). In the L*a*b* color space, ΔE is written as ΔE^*_{ab} and is computed as follows (For ΔE^*_{uv}, a^* and b^* are substituted by u^* and v^*):

$$\Delta E^*_{ab} = \left[(L^*_1 - L^*_2)^2 + (a^*_1 - a^*_2)^2 + (b^*_1 - b^*_2)^2\right]^{\frac{1}{2}} \tag{7.8}$$

The color difference is a typical measure to evaluate color quality against a target color. Although the CIE 1976 L*a*b* color space and the CIE 1976 L*u*v* color space are geometrically different, it is thought that a color difference value of two to three in these color spaces may be a target color difference when the original and the reproduced pictures are observed side by side. Because ΔE^*_{ab} and ΔE^*_{uv} are calculated by simple formulae, the color differences are not homogenous. For more accuracy, ΔE^*_{94}, ΔE^*_{2000}, and other formulae are recommended.[3,4]

7.1.5 LIGHT SOURCE AND COLOR TEMPERATURE

Chromaticity of light sources observed in life can be approximately plotted along the black body locus as shown in Figure 7.4. However, it is necessary to pay attention to the fact that the actual spectral distributions of light sources with the same chromaticity are not similar. The spectral distributions of black body sources, several natural light sources, and artificial lamps are shown in Figure 7.5 and Figure 7.6.

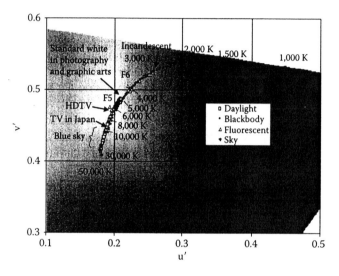

FIGURE 7.4 (See Color Figure 7.4 following page 178.) Light sources on the $u'v'$ chromaticity diagram.

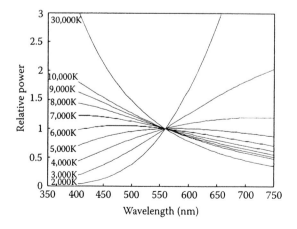

FIGURE 7.5 (See Color Figure 7.5 following page 178.) Spectral distribution of black body.

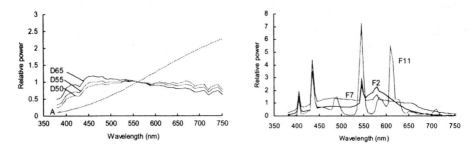

FIGURE 7.6 (See Color Figure 7.6 following page 178.) Spectral distribution of CIE standard illuminants (left) and fluorescent lamps (right).

Human beings have been evolving under natural light sources, which are close to the black body locus in chromaticities and spectral distributions; thus, the human visual system gives natural color recognition under these light sources. On the other hand, artificial light sources such as fluorescent lamps do not have similar characteristics to black body light sources as shown in Figure 7.6. Nevertheless, the colors of objects under such artificial light sources would be recognized as natural because most objects do not have steep spectral changes.[5] As tristimulus values are calculated with different types of such light sources having the same chromaticity, there would be no significant differences.

When a color is carefully observed, however, color changes are sometimes noticed. For example, under a low-cost fluorescent lamp, skin color might appear dark and yellowish. This is caused by the spectral difference between artificial and natural light sources, and this deficiency can be evaluated by the CIE 13.3 color rendering index.[6] Table 7.1 depicts the indices of typical light sources. *Ra* indicates an index for averaged color, and *Rn* indicates an index for specific color. Some types of fluorescent lamps with a low *Ra* index (average color rendering index) tend to cause the previously mentioned phenomenon. For instance, illuminant F2 produces a low index value for skin color at R15.[7] The result is that object colors cannot be observed satisfactorily under such a light source, and neither can a digital still camera, in general.

Correlated color temperature is an index to represent the chromaticity of light sources based on the fact that the chromaticities of most light sources lie along the black body locus. A residual small difference from the black body locus is approximated to be the nearest point in terms of color difference. Therefore, if two different light sources have the identical correlated temperature, the color of an object under each light source will not be identical in terms of spectral distribution and chromaticity. Because the small difference of chromaticity can be compensated by visual adaptation and may be neglected, it is suggested that the color rendering index be noted along with color temperature to avoid such misunderstanding.

7.2 CAMERA SPECTRAL SENSITIVITY

The simplest target color of a camera system is the scene. (Targeting color will be discussed later.) In order to realize this, the camera sensitivity curves must be a linear transformation of the color-matching functions; otherwise, two objects

TABLE 7.1
Color Rendering Index for Some Light Sources

Sample light source			F2	F7	F11	A	D65	D50	D55
Chromaticity		x	0.3721	0.3129	0.3805	0.4476	0.3127	0.3457	0.3324
		y	0.3751	0.3292	0.3769	0.4074	0.3290	0.3585	0.3474
Reference light source (P: black body; D: daylight)			P	D	P	P	D	D	D
Correlated color temperature			4200	6500	4000	2856	6500	5000	5500
Average color rendering index	Ra		64	90	83	100	100	100	100
Special color rendering index	R1	7.5 R 6/4	56	89	98	100	100	100	100
	R2	5 Y 6/4	77	92	93	100	100	100	100
	R3	5 GY 6/8	90	91	50	100	100	100	100
	R4	2.5 G 6/6	57	91	88	100	100	100	100
	R5	10 BG 6/4	59	90	87	100	100	100	100
	R6	5 PB 6/8	67	89	77	100	100	100	100
	R7	2.5 P 6/8	74	93	89	100	100	100	100
	R8	10 P 6/8	33	87	79	100	100	100	100
	R9	4.5 R 4/13	−84	61	25	100	100	100	100
	R10	5 Y 8/10	45	78	47	100	100	100	100
	R11	4.5 G 5/8	46	89	72	100	100	100	100
	R12	3 PB 3/11	54	87	53	100	100	100	100
	R13	5 YR 8/4	60	90	97	100	100	100	100
	R14	5 GY 4/4	94	94	67	100	100	100	100
	R15	1 YR 6/4	47	88	96	100	100	100	100

observed identically by human eyes will result in different color signal outputs from the same camera system. This phenomenon is called sensitivity (or observer) metamerism. Unless the characteristics of the objects are known in priority, it is impossible to exactly estimate the tristimulus values of the original scene in such a case. This criterion of camera sensitivity is called the Luther condition.[8] In practice, however, it is difficult to adjust a set of spectral sensitivity curves to conform to the Luther condition due to the practical production of filters, sensors, and optical lenses.

Real reflective objects have limited characteristics; the spectral reflectance of an object in general does not change steeply with respect to wavelength. This characteristic allows cameras to estimate tristimulus values for the object even though they do not satisfy the Luther condition. Theoretically speaking, as long as the spectral reflections of an object are always composed of three principal components, a three-channel camera having any three types of sensitivity curves can exactly estimate the tristimulus values of the object. (Images observed on TV displays have such characteristics.) For the evaluation of sensitivity metamerism, the DSC/SMI (digital still camera/sensitivity metamerism index) was proposed in ISO/CD 17321-1. Because it takes into account the spectral reflectance of normal objects, the index correlates well with subjective tests.[9] Unfortunately, at the time of this writing, the document is not publicly available. It is expected to be published in 2006.

7.3 CHARACTERIZATION OF A CAMERA

A typical methodology to characterize a camera with a linear matrix colorimetrically is to use test patches whose spectral responses are similar to those of real objects. Suppose that the target tristimulus values for color targets are given by:

$$T = \begin{bmatrix} X_1 & \cdots & X_i & \cdots & X_n \\ Y_1 & \cdots & Y_i & \cdots & Y_n \\ Z_1 & \cdots & Z_i & \cdots & Z_n \end{bmatrix} \quad (7.9)$$

and the estimated tristimulus values are given by:

$$\hat{T} = \begin{bmatrix} \hat{X}_1 & \cdots & \hat{X}_i & \cdots & \hat{X}_n \\ \hat{Y}_1 & \cdots & \hat{Y}_i & \cdots & \hat{Y}_n \\ \hat{Z}_1 & \cdots & \hat{Z}_i & \cdots & \hat{Z}_n \end{bmatrix} = \begin{bmatrix} a_{11} & a_{12} & a_{13} \\ a_{21} & a_{22} & a_{23} \\ a_{31} & a_{32} & a_{33} \end{bmatrix} \cdot \begin{bmatrix} r_1 & \cdots & r_i & \cdots & r_n \\ g_1 & \cdots & g_i & \cdots & g_n \\ b_1 & \cdots & b_i & \cdots & b_n \end{bmatrix} = A \cdot S$$

$$(7.10)$$

where matrix S is measurement data through the camera.

To obtain 3×3 matrix A, simple linear optimization or recursive nonlinear optimization can be applied. The simple linear solution can be calculated by:

$$A = T \cdot S^T \cdot (S \cdot S^T)^{-1} \quad (7.11)$$

However, the resultant approximation tends to yield large visual errors in dark areas due to the nature of optimization in the linear domain and to visual characteristics; cubic roots are approximately proportional to one's recognition, as seen in Equation 7.4 and Equation 7.5.

An alternate method minimizes the total visual color difference, J, using a recursive conversion technique. Delta E may be calculated in a CIE uniform color space described earlier.

$$J = \sum_{i=1}^{n} w_i \Delta E(X_i, Y_i, Z_i, \hat{X}_i, \hat{Y}_i, \hat{Z}_i) \quad (7.12)$$

where w_i is a weight coefficient for each color patch. Matrix A may be optimized mainly for important colors. It should be noted that any minimization techniques converge to the identical result as long as the Luther condition is satisfied.

In practice, preparing a test chart is most problematic. A typical misconception is to use a test chart made by a printer that uses three or four colorants. When the

Color Theory and Its Application to Digital Still Cameras

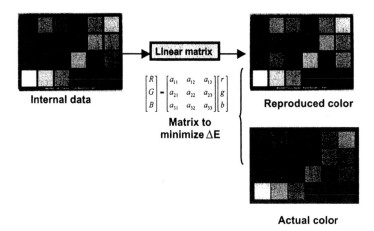

FIGURE 7.7 (See Color Figure 7.7 following page 178.) Characterization by linear matrix.

test chart is produced by such a device, the principal components of the spectral reflectances of the color patches will be limited to the characteristics of the printer, which are not the appropriate representation of a real scene. Figure 7.7 illustrates the color characterization using the Gretag Macbeth color checker, which is designed to simulate the spectral reflectances of real objects.[10]

The reflectances of most reflective objects can be accurately composed of five to ten principal components. Thus, the set of color patches should cover these characteristics but does not necessarily have many color patches. It should be noted that, because the resultant characterization is suitable for reflective objects and not for self-emitting light sources such as a neon bulb, color LEDs, and color phosphors, there is room for improvement.

7.4 WHITE BALANCE

One of the most challenging processes in a digital camera is to find an appropriate white point and to adjust color. As described in Section 7.1.5, real scenes contain many light sources. In such a situation, the human visual system adapts to the circumstances and recognizes the objects as if they were observed in typical lighting conditions, while a camera's sensor still outputs raw signals. For instance, they recognize white paper as white in the shadow of a clear sky even though its tristimulus values give a bluish color because the paper is illuminated by sky blue. It is known that the major adjustment is performed in the retina by adjusting each cone's sensitivity. This process is called chromatic adaptation.

7.4.1 WHITE POINT

A digital camera needs to know the adapted white that the human vision system determines to be achromatic on site. Primarily, three types of approaches are used to estimate the white point, or chromaticity, of a light source in a camera.

- *Average of the scene.* The first approach assumes that the average color of the entire scene is a middle gray, typically at 18% reflectance. This approach was traditionally used in film-processing labs to adjust the color balance of negative film for prints. Many movie cameras for consumers do not accumulate color over just one image, but rather over the entire sequence of images spanning a couple of minutes in the recent past. When the averaged scene is supposed to be achromatic, the resultant color should be that of the light source.
- *Brightest white.* The second approach presumes that the brightest color is white. Because the light source should be brightest in the scene by its nature, brighter objects are expected to encompass more characteristics of the light source than others. Typically, the brightest point is presumed to be the same color as the light source. However, the real scene may have self-emitting objects such as traffic signals, so the wrong light source may be estimated by mistake. To reduce the chance of this happening, only bright colors with chromaticity near the black body locus may be chosen; the other colors, even though brighter, will be eliminated from consideration.
- *Color gamut of the scene.* The last approach is to estimate the light source from the distribution of colors captured by a camera. This theory supposes that the scene has several color objects that, when illuminated by a light source, are supposed to cover all of the objects' spectral distributions statistically, thus producing the observed color gamut of the scene. That is, the light source would be estimated by comparing the correlations between the color distribution of a captured image and the entries in a color gamut database built from expected scene spectral reflectances and typical light sources.

A practical algorithm will be constructed with a mixture of these approaches along with other statistical information. Optimization should be performed for typical scenes and typical lighting conditions encountered by typical users of the camera system.

7.4.2 COLOR CONVERSION

Once the white point is found, the next step is to convert all colors into the desired ones; this includes transforming the estimated white to be achromatic. Two policies are typically used: chromatic adaptation and color constancy.

7.4.2.1 Chromatic Adaptation

Chromatic adaptation in the eye is mostly controlled by the sensitivity of the cones. Analogously, camera signals can be approximated into the tristimulus values of the cones, and the camera can control the gain of these signals. The typical calculation is performed as follows:

$$\begin{bmatrix} r' \\ g' \\ b' \end{bmatrix} = A^{-1} B^{-1} \cdot \begin{bmatrix} \dfrac{L'_W}{L_W} & 0 & 0 \\ 0 & \dfrac{M'_W}{M_W} & 0 \\ 0 & 0 & \dfrac{S'_W}{S_W} \end{bmatrix} \cdot B \cdot A \cdot \begin{bmatrix} r \\ g \\ b \end{bmatrix} \qquad (7.13)$$

where

$$\begin{bmatrix} L \\ M \\ S \end{bmatrix} = B \cdot A \cdot \begin{bmatrix} r \\ g \\ b \end{bmatrix} \qquad (7.14)$$

and r, g, and b are original camera signals. L_W, M_W, S_W, and L'_W, M'_W, L'_W denote cone tristimulus values of white points at the original scene and at the RGB color space, respectively.

This adaptation is called the von Kries model. Matrix A may be calculated by Equation 7.11. Matrix B is used to transform CIE tristimulus values into cone responses. An example of matrix B is shown as follows:[11]

$$\begin{bmatrix} L \\ M \\ S \end{bmatrix} = \begin{bmatrix} 0.4002 & 0.7076 & -0.08081 \\ -0.2263 & 1.16532 & 0.04570 \\ 0. & 0. & 0.91822 \end{bmatrix} \cdot \begin{bmatrix} X \\ Y \\ Z \end{bmatrix} \qquad (7.15)$$

7.4.2.2 Color Constancy

It is known that the color of an object under a variety of light sources is recognized as if it were under day light as long as the Ra of the light source is high,[12] although no information about the spectral distribution of the light sources and the spectral reflections of the objects is available. This is called color constancy. A simple linear approach is to use Equation 7.13 with matrix B optimized in terms of color constancy; this would estimate the corresponding color under a standard light source. It is known that the resulting equivalent sensitivity curve will be negative in places. Figure 7.8 illustrates this characteristic.

In a nonlinear approach, more than one matrix optimized for each light source may be prepared. The camera could choose one of them, depending upon the type of light source. However, this approach increases the risk that an incorrect matrix is chosen.

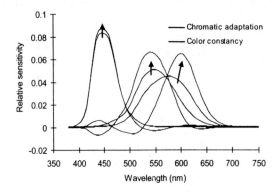

FIGURE 7.8 (See Color Figure 7.8 following page 178.) Equivalent color sensitivity in white balancing optimized for chromatic adaptation and color constancy.

7.5 CONVERSION FOR DISPLAY (COLOR MANAGEMENT)

Because the digital values of image data explicitly do not contain color definition, it is necessary that the image data be interpreted by a receiver without any ambiguity. Therefore, the relationship between color signal values (digital count) and their physical meanings (i.e., colorimetry) should be well defined. A typical approach to color management in digital cameras is to use a standard color space scheme. Two key points should be considered: colorimetric definition and its intent.

7.5.1 COLORIMETRIC DEFINITION

Image data from a digital camera may be sent to a display without any conversions. This means that the color of RGB data is defined based on the color appearance on a display. Therefore, the color space based on a specific display is used as the standard color space. The most popular standard color encoding is sRGB (standard RGB), which is defined for an average CRT with a certain observing condition. The following is the definition of sRGB for an 8-bit system.[13]

$$R'_{sRGB} = R_{8bit} \div 255$$
$$G'_{sRGB} = G_{8bit} \div 255 \qquad (7.16)$$
$$B'_{sRGB} = B_{8bit} \div 255$$

$$R'_{sRGB}, G'_{sRGB}, B'_{sRGB} \leq 0.04045$$

$$R_{sRGB} = R'_{sRGB} \div 12.92$$
$$G_{sRGB} = G'_{sRGB} \div 12.92 \qquad (7.17)$$
$$B_{sRGB} = B'_{sRGB} \div 12.92$$

Color Theory and Its Application to Digital Still Cameras

$$R'_{sRGB}, G'_{sRGB}, B'_{sRGB} \geq 0.04045$$

$$R_{sRGB} = [(R'_{sRGB} + 0.055)/1.055]^{2.4}$$
$$G_{sRGB} = [(G'_{sRGB} + 0.055)/1.055]^{2.4} \quad (7.18)$$
$$B_{sRGB} = [(B'_{sRGB} + 0.055)/1.055]^{2.4}$$

$$\begin{bmatrix} X \\ Y \\ Z \end{bmatrix} = \begin{bmatrix} 0.4124 & 0.3576 & 0.1805 \\ 0.2126 & 0.7152 & 0.0722 \\ 0.0193 & 0.1192 & 0.9505 \end{bmatrix} \begin{bmatrix} R_{sRGB} \\ G_{sRGB} \\ B_{sRGB} \end{bmatrix} \quad (7.19)$$

The white point of sRGB is the chromaticity of D65, and the primaries of sRGB are the ones of HDTV (high-definition television). Because negative values are not allowed and the maximum value is limited, the colors represented by the standard are limited to the color gamut of the virtual display.

To overcome the gamut problem, two more color spaces are often used: sYCC and Adobe RGB. (In the specification, Adobe RGB is formally defined as the "DCF optional color space.") The sYCC[14] color space is used for image compression to improve the efficiency for better compression ratio at equivalent image quality. Because the sYCC is a superset of sRGB, it can represent more chromatic colors. The Adobe RGB color space has a wider color gamut compared with sRGB by about 15% in the $u'v'$ chromaticity diagram. For further information, please refer to Exif v2.21[15,16] and DCF.[17]

7.5.2 IMAGE STATE

The important concept to clarify in digitization of images is the image state. Because sRGB defines only the observing condition including the display, there is no guideline for the kind of color image that should be encoded into sRGB color space, no matter how beautiful or dirty the colors appear. At first sight, readers may think that the tristimulus values or the white balanced tristimulus values of the real scene are the good target. However, most users desire pleasing colors rather than correctly reproduced colors.

When the color is intended to reproduce the color of the real scene, the data are called scene-referred image data. In all other cases, for example, when the color is intended to adjust for the user's memory or preferred colors, they are called output-referred image data. ISO 22028-1 defines this concept.[18] Figure 7.9 illustrates the flow of signals and their image state in a digital camera.

The process of converting scene-referred image data to output-referred image data is called color rendering. TV standards use scene-referred image data, although the term was not defined at the time that these standards were published. This kind of color image can be seen on display monitors in TV broadcasting studios. Once the image is transmitted to a home, the image is seen with some color adjustments

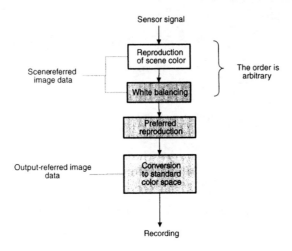

FIGURE 7.9 Conceptual flow of color image processing in a digital camera.

— typically, increasing contrast and chroma, manually by the consumer. Thus, TV is enjoyed with preferred colors.

Digital cameras using sRGB should calculate output-referred colors internally and encode them into image files. Thus, the user enjoys preferred colors on monitor displays without any further adjustment. It is obvious that the preferred colors are not unique. The desired result can be influenced by cultural background such as region, race, profession, and age. The method of color rendering is ill defined, and further study is encouraged to determine the best conversion — perhaps, for specific users.

7.5.3 PROFILE APPROACH

Another way to control color is to use a profile to establish the relationship between digital data and colorimetric values. The ICC (International Color Consortium)[19] offers the specification of a profile format. This approach is especially effective when it is used for the conversion of RAW data output by the camera because each camera has its own sensor characteristics that, in turn, determine the color characteristics as well.

7.6 SUMMARY

In this chapter, the basics of color theory and the concepts necessary for the design of digital cameras were described. Colorimetry is the key theory to quantify color. In order to reproduce colors as our eyes see them, the Luther condition must be considered. For the exact reproduction of a scene under a standard light source, there is a systematic way to characterize a camera. For nonstandard light sources, a white point must be found and color should be converted accordingly. Many camera users wish to see preferred colors rather than true colors; in order to generate preferred colors, we still need empirical adjustments and studies of users' preferences. Finally, for the transmission of color image data captured by a digital camera, standard color spaces like sRGB are typically used.

In conventional cameras, these kinds of processing were done by silver halide film and the photofinishing process. A digital camera must handle all of the processing inside. The knowledge of color described in this chapter is essential to improving the image quality of digital cameras.

REFERENCES

1. Publication CIE 15.2-1986, *Colorimetry*, 2nd ed., 1986.
2. Publication CIE 159: 2004 *A Colour Appearance Model for Colour Management Systems*: CIECAM02, 2004.
3. CIE 116-1995, Industrial colour-difference evaluation, 1995.
4. M.R. Luo, G. Cui, and B. Rigg, The development of the CIE 2000 colour difference formula, *Color Res. Appl.*, 25, 282–290, 2002.
5. J. Tajima, H. Haneishi, N. Ojima, and M. Tsukada, Representative data selection for standard object colour spectra database (SOCS), *IS&T/SID 11th Color Imaging Conf.*, 155–160, 2002.
6. CIE 13.3-1995, Method of measuring and specifying colour rendering properties of light sources, 1995.
7. JIS Z 8726: 1990, Method of specifying colour rendering properties of light sources, 1990.
8. R. Luther, Aus dem Gebiet der Farbreizmetrik, *Zeitschrift fur Technische Physik*, 12, 540–558, 1927.
9. P.-C. Hung, Sensitivity metamerism index for digital still camera, *Color Sci. Imaging Technol., Proc. SPIE*, 4922, 1–14, 2002.
10. C.S. McCamy, H. Marcus, and J.G. Davidson, A color-rendition chart, *J. Appl. Photogr. Eng.*, 2, 3, 95–99, 1976.
11. Y. Nayatani, K. Hashimoto, K. Takahama, and H. Sobagaki, A nonlinear color-appearance model using Estevez–Hunt–Pointer primaries, *Color Res. Appl.*, 12, 5, 231–242, 1987.
12. P.-C. Hung, Camera sensitivity evaluation and primary optimization considering color constancy, *IS&T/SID 11th Color Imaging Conf.*, 127–132, 2002.
13. IEC 61966-2-1 Multimedia systems and equipment – Colour measurement and management – Part 2.1: Colour management – Default RGB colour space – SRGB, 1999.
14. IEC 61966-2-1 Amendment 1, 2003.
15. JEITA CP-3451, Exchangeable image file format for digital still cameras: Exif Version 2.2, 2002.
16. JEITA CP-3451-1 Exchangeable image file format for digital still cameras: Exif Version 2.21 (Amendment Ver 2.2), 2003.
17. JEITA CP-3461, Design rule for camera file system: DCF Version 2.0, 2003.
18. ISO 22028-1: 2004, Photography and graphic technology – extended colour encodings for digital image storage, manipulation and interchange – part 1.
19. http://www.color.org, specification ICC. 1:2003-09, file format for color profiles (version 4.1.0).

8 Image-Processing Algorithms

Kazuhiro Sato

CONTENTS

8.1 Basic Image-Processing Algorithms 224
    8.1.1 Noise Reduction 225
        8.1.1.1 Offset Noise 225
        8.1.1.2 Pattern Noise 226
        8.1.1.3 Aliasing Noise 226
    8.1.2 Color Interpolation 226
        8.1.2.1 Rectangular Grid Sampling 228
        8.1.2.2 Quincunx Grid Sampling 229
        8.1.2.3 Color Interpolation 229
    8.1.3 Color Correction 232
        8.1.3.1 RGB 232
        8.1.3.2 YCbCr 232
    8.1.4 Tone Curve/Gamma Curve 233
    8.1.5 Filter Operation 235
        8.1.5.1 FIR and IIR Filters 236
        8.1.5.2 Unsharp Mask Filter 237
8.2 Camera Control Algorithm 238
    8.2.1 Auto Exposure, Auto White Balance 238
    8.2.2 Auto Focus 239
        8.2.2.1 Principles of Focus Measurement Methods 239
        8.2.2.2 Digital Integration 240
    8.2.3 Viewfinder and Video Mode 241
    8.2.4 Data Compression 242
    8.2.5 Data Storage 243
    8.2.6 Zoom, Resize, Clipping of Image 243
        8.2.6.1 Electronic Zoom 243
        8.2.6.2 Resize 246
        8.2.6.3 Clipping, Cropping 247
8.3 Advanced Image Processing; How to Obtain Improved Image Quality 248
    8.3.1 Chroma Clipping 248
    8.3.2 Advanced Color Interpolation 248

8.3.3 Lens Distortion Correction 251
8.3.4 Lens Shading Correction 251
References 252

In this chapter, we will describe image-processing algorithms used in digital still cameras (DSC) from a theoretical point of view, as well as the many peripheral functions needed to work in a real camera, such as viewfinder, focus and exposure control, JPEG compression, and storage media. This chapter includes simple descriptions to give the reader an outline of these items. The physical aspects of CCD (charge coupled device) and CMOS (complementary metal-oxide semiconductor) sensors and color space are not included here; they are described in Chapter 3 through Chapter 6. Because digital video cameras (DVCs) use similar image-processing technologies and functions, the majority of this chapter is relevant to them also. However, DVCs use other image-processing technology based on the time axis, such as noise reduction in time domain processing. This chapter does not include DVC-specific technologies that are out of the scope of this book.

8.1 BASIC IMAGE-PROCESSING ALGORITHMS

DSCs or DVCs acquire data using a CCD or CMOS sensor, reconstruct the image, compress it, and store it on a storage medium (such as a compact flash) in less than a second. In recent years, the number of pixels of the image sensor for DSCs has become large. Also, many kinds of image-processing algorithms are used in a DSC, such as color interpolation; white balance; tone (gamma) conversion; false color suppression; color noise reduction; electrical zoom; and image compression. Figure 8.1 shows a configuration of a DSC from the viewpoint of the image-processing technology used in it. The figure is a simplified view intended to show how image-processing algorithms are applied in the camera. The details of the actual image-processing hardware/software are described in Chapter 9.

The lens focuses incident light on a CCD sensor to form an image, which is converted to an analog signal. This signal is read out, digitized, and sent to the digital signal-processing block. A CMOS sensor may output an analog signal, in which case it is treated the same way as the output from a CCD sensor. Alternatively, a CMOS sensor may output a digital signal that can be fed to the digital signal-

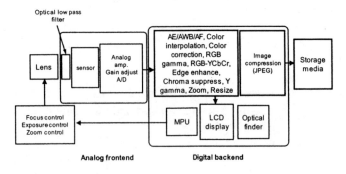

FIGURE 8.1 Digital still camera and its typical configurations.

Image-Processing Algorithms

processing block directly. There is no difference between the two types of sensor from the point view of the image-processing algorithms. A complete working camera has several other important blocks, including: focus; iris and optical zoom controls; optical finder; LCD display; and storage media. Usually, these are controlled by a general-purpose MPU in the camera.

The digital signal-processing block of a DSC must execute all the image-processing algorithms on over 5 Mpixels of data and construct the final image. Recent CCD sensors generate three to five full frames of data within a second. Also, recent high-speed CMOS sensors can generate more than ten frames per second. If images are to be generated with same output speed of sensor signal data, the required processing speed for the image processor becomes over 50 Mpixels/sec. The two possible approaches for the signal processor for the DSC to achieve the speed are: (1) to base the design on a general-purpose DSP (digital signal processor) with image-processing software; or (2) to construct hard-wired logic to achieve maximum performance. Each approach has certain advantages and disadvantages, as discussed in Chapter 9.

8.1.1 NOISE REDUCTION

The number of photons striking the sensor is linearly proportional to the incident light intensity. The number of electrons displaced, and thus the output signal, is proportional to the number of photons. The output signal level from a sensor is in the order of 1 V. It is very important to provide the digital image-processing hardware with the best quality input signal to get the best processed image from measured data. Thus, it is necessary to pay attention to keeping signal accuracy and quality of the input data.

8.1.1.1 Offset Noise

It is best to digitize the input data value between black and maximum levels so as to utilize the dynamic range of a sensor fully. Most CCDs generate some noise signal like thermal noise even when no input light signal is present. Also, the dark noise level of a CCD drifts gradually from beginning to end of readout. Figure 8.2 is a raw data of dark signal acquired with no incident light to a lens. As can be seen on

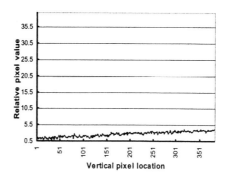

FIGURE 8.2 Vertical offset drift of CCD sensor.

the figure, there is vertical brightness drift from top to bottom. A noise drift between start and end of signal readout is characteristic of this type of noise, so it is called offset drift noise of a sensor. Subtracting an offset that depends on the vertical position of a pixel can compensate for this noise. The compensation is performed at the early stage of the signal-processing path in a camera.

8.1.1.2 Pattern Noise

"Pattern noise" is caused by an imbalance of pixel readout from the sensor. The human eye has very high sensitivity for patterns in a picture. When some feature in a picture is recognized as a pattern or pattern noise, the pattern will stand out very strongly. For example, if a circular pattern (ring) of noise is recognized in a picture, then a solid line circle will be seen or imagined, even if it is only very short broken arcs. To avoid this kind of effect, the pattern noise should be reduced as much as possible. Insufficient accuracy for the data calculation process may generate pattern noise.

8.1.1.3 Aliasing Noise

Each cell spatially samples the image projected onto the sensor. The sampling pitch governs this sampling characteristic. It is defined by cell pitch, the spatial arrangement of the entire cell, and the aperture function of each cell. In many cases, the MTF (modulation transfer function) of a lens has higher response than a sensor, so the output of the sensor will contain alias signals over the Nyquist frequency limit defined by the cell pitch. Once this type of alias signal is mixed in the raw data, it cannot be eliminated. Many cameras use an optical low-pass filter (OLF) between the lens and sensor to limit the spatial frequency of the input image focused on a sensor. Table 8.1 summarizes the types of noise that are generated inside a DSC. It is important to reduce these noise types at an early stage of data acquisition; nevertheless, they will remain in the raw data. It is necessary to eliminate or reduce this noise in the image processing using adaptive image-processing technology.

8.1.2 COLOR INTERPOLATION

The full-color digital image is displayed on a soft copy device such as a CRT monitor so that each pixel has a digital information of red (R), green (G), and blue (B) on a rectangular coordinate grid. In the case of a DSC with three sensors, each sensor has different color filters and does not need color interpolation operation because R, G, and B values for each position are measured by each sensor. However, in cameras with three CCD sensors, there is a method called half-pitch shift of G in reference with R and B sensors in horizontal. This is not used widely for DSC because highly accurate alignment of three sensors is very costly. Some professional DVCs use this type of three-sensor design. The half-pitch shift requires interpolation in the middle of each position of R, G, and B in the horizontal direction to double the horizontal resolution. After operation of color correction and other image-processing operations, the horizontal pixel size is reduced by half to get the correct aspect ratio. This chapter will focus on single-sensor systems because the major

Image-Processing Algorithms

TABLE 8.1
Noise and its Causes

Noise type	Root cause	Solutions
Cross color noise	Caused by mixing of adjacent color pixels signals; typical for single CCD sensor	Calibrate for each pixel
False color noise	Generated when using a digital white balance, which provides insufficient resolution in shadow areas	Use more bit length or analog white balance
Color phase noise	Shows up as color blotches in dark gray areas where there was no color; can also show up as a shift in the original color	Higher resolution (e.g., 16-b) ADC and analog white balance so as to provide high resolution in dark areas of the picture
Digital artifacts	Multiple causes; in an AFE, can be caused by insufficient A/D resolution, which causes step to be seen in smooth tonal areas	Higher resolution (e.g., 16-b) ADC to provide smooth tonal response and provide an effective 42 to 48 b of color depth
Temporal noise	Caused by shot noise in the CCD or insufficient SNR in the AFE	Use a low-noise CDS, PGA, ADC, and black level calibration so as not to add any noise
Fixed pattern noise	Caused by a measurement mismatch in the sampling of alternate pixels	Balanced CDS
Line noise	Shows up as horizontal streaks in dark areas of the image; caused by the black level calibration using adjustment steps that are too large and, therefore, visible	Use an ultraprecision (e.g., 16-b) digital black level calibration circuit

difference is the color interpolation block only and the rest of the processing is common for single- and three-sensor systems.

In a single-sensor system, each cell on the sensor has a specific color filter and microlens positioned above it. The array of color filters is referred to as a "color filter array" (CFA). Each cell has only information for the single color of its color filter. The raw data obtained from this sensor do not have full R/G/B information at each cell position. Color interpolation is required when a full-color image is wanted from a single-sensor system.

There are several color filter sets and filter arrays for DSC and DVC sensors. Figure 8.3 shows three sets and their spatial arrangements. Figure 8.3(a) is called "Bayer pattern color filter array" and widely used for DSCs. The ratio among R, B, and G is 1:1:2.

Figure 8.3(b) is called "complementary color filter pattern" and is also used in DSCs. The advantage of this color filter is better light sensitivity than an R/G/B color filter. Figure 8.3(c) is another complementary color filter pattern and is mostly used in DVCs. It is similar to the layout (b) except that the G and Mg are changed at each line. More details for sensor, color filter array, and microlens can be found in Chapter 3.

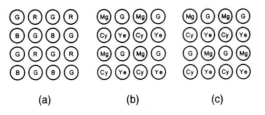

FIGURE 8.3 Sensor color filter array for DSC and DVC.

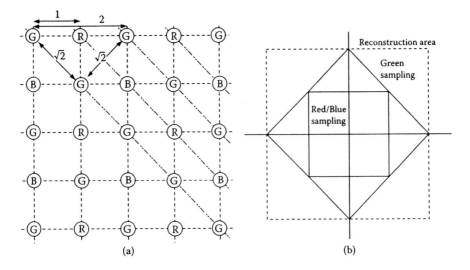

FIGURE 8.4 Rectangular grid sampling and frequency coverage for Bayer CFA.

From the viewpoint of spatial sampling geometry used for actual sensors, two types of cell arrangements are realized in CCD sensors: rectangular and quincunx or diamond-shaped sampling. The basic concepts of color interpolation theory are the same for all the color filter arrays in Figure 8.3, although actual color interpolations for the various color filter spatial and sampling cell layouts differ. In this book, rectangular sampling patterns are mainly considered when describing color interpolation algorithms for the Bayer pattern color filter array.

8.1.2.1 Rectangular Grid Sampling

Figure 8.4 shows the normalized geometry and dimensions of the Bayer rectangular color filter array and its frequency coverage. There are several ways to determine which color starts from the upper-left corner position, but the grid dimensions and color pattern cycle are the same. In part (a), the distance among vertical, horizontal, and diagonal cells is normalized with minimum distance as 1.0.

In the rectangular grid, the G sampling distances for horizontal and vertical directions have minimum value so that the highest Nyquist frequency is in the horizontal and vertical directions. Compared with the G signal, R and B sampling distances are highest in the diagonal directions, but their sampling distances are $\sqrt{2} = 1.4$. This means that the Nyquist frequency for R and B sampling is $1/\sqrt{2} \approx 0.7$ times less than G. R and B cover the same frequency range. Figure 8.4(b) shows frequency coverage of R, G, and B, respectively. Also, the maximum frequency range that can be reconstructed from Bayer filter with rectangular sampling is indicated by a broken line that surrounds the G and R/B solid lines.

8.1.2.2 Quincunx Grid Sampling

Figure 8.5 shows the geometry and dimensions of the alternative spatial sampling that uses quincunx or diamond grid location and frequency coverage. The geometry is the 45° rotated layout of a rectangular grid. As before, the dimensions are expressed with respect to the minimum sampling distance of the G signal. In the quincunx grid sampling, sampling frequency is highest in the diagonal direction (45 and 135°). R and B signals have the same sampling interval with a distance of $\sqrt{2} \approx 1.4$ times larger than the G sampling distance. Based on this geometry, Figure 8.5(b) shows frequency coverage of R, B, and G and the reconstruction area for quincunx grid sampling. It is possible to reconstruct a larger reconstruction area than those described in Figure 8.4 and Figure 8.5; however, the resulting image has less information. This type of expansion can be considered as a magnification operation on the raw data.

8.1.2.3 Color Interpolation

Color interpolation is the process of estimating a value at a location that does not have a measured value by using other actual measured values. It is clear that better estimation can be obtained if more measured pixels are used for the interpolation calculation. However, in an actual implementation for a DSC, the trade-off among hardware cost, calculation speed, and image quality must be considered.

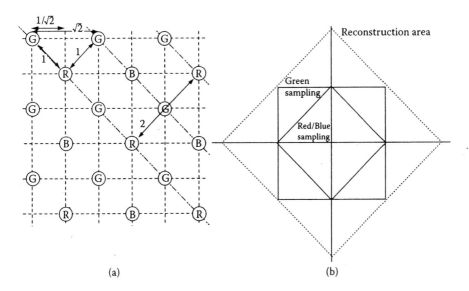

FIGURE 8.5 Quincunx grid sampling and frequency coverage for Bayer CFA.

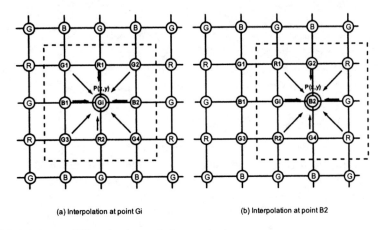

(a) Interpolation at point Gi (b) Interpolation at point B2

FIGURE 8.6 Bayer CFA color interpolation method.

Figure 8.6 shows the basic concept of color interpolation for a rectangular grid having Bayer CFA. The interpolation operation is the equivalent of inserting zero points between each two actual existing points and then applying a low-pass filter. In Figure 8.6(a), $P(x,y) = G_i$ has a measured value of green color only, so it is necessary to calculate the nonexisting values, R and B, from the measured values of R and B points that surround $P(x,y)$. Better estimation can be obtained if more points are used for calculation. An area shown by the broken line that has a center at the interpolation point in Figure 8.6 is defined. The number of points contained within this area affects the quality of the reconstructed image. Figure 8.6(b) is the color interpolation point that has a measured value for B but no values for G and R; thus they must be interpolated from other G and R values.

The simplest interpolation method is a nearest neighbor or zero order interpolation. This is based on a very simple algorithm that the value of an interpolated point will take the same value as the nearest measured point. There are no multiplication/addition arithmetic operations for pixel data that are used for interpolation, but the image quality is not good enough for a commercial camera.

The next simplest interpolation algorithm is a linear interpolation (in one dimension) or bilinear interpolation (in two dimensions). In the linear interpolation algorithm, the middle point takes the arithmetic mean of the two values of the adjacent points. Bilinear interpolation is expressed as a serial calculation of linear interpolation for vertical and horizontal directions. It does not depend on the order of calculation.

In the bilinear color interpolation shown in Figure 8.6(a), R and B values at the location of $P(x,y)$ are calculated with the equation $R = (R1 + R2)/2$ and $B = (B1 + B2)/2$. This is equivalent to one-dimensional interpolation (linear interpolation). The frequency response of the linear interpolation of this equation has a response as shown in Figure 8.7. This is the vertical response for R pixel value and horizontal response for B pixel value interpolation. There is no low-pass effect for the orthogonal direction of each pixel.

If the interpolation location $P(x,y)$ is moved to the right one position to $B2$ as shown in Figure 8.6(b), the geometric relationship of the colors changes and the

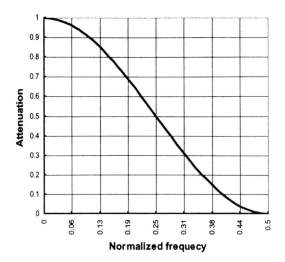

FIGURE 8.7 Frequency response of linear interpolation.

frequency response is interchanged between R and B. If focus is on the G value, $P(x,y)$ in (b) has no measured value for G, so the G value must be interpolated using the four G values located at the diagonal positions.

The bilinear interpolation operation described here is simple and gives better image quality than the nearest neighbor method. However, it still generates cyclic pattern noise because of the cyclic change of direction of the interpolation filter and also the change in its frequency response. To avoid pattern noise caused by them, it is better to use more data points for the interpolation calculation. Also, the adaptive interpolation (described in a later section) gives better image quality. The interpolation algorithm is explained in more detail in Section 8.2.6.1 and Section 8.3.2.

The color interpolation method, which reconstructs a full color image from the raw data, has been briefly outlined in this section. In the color interpolation calculation, it is assumed that no aliasing noise is present in the raw data. As mentioned in Section 8.1.1.3 on aliasing noise, many DSCs have an optical low-pass filter (OLF) between lens and sensor so as to limit the spatial frequency of a real scene to match the Nyquist limit defined by the sensor cell pitch. Figure 8.8 shows the OLF, which is composed of a single or multiple layer of a thin crystal plate. The plate has a characteristic to split input light into two different paths: normal (n_o) and abnormal light (n_e). The distance, S, between two separated light paths is given in Equation 8.1, where d is a thickness of the thin plate and n_o and n_e are diffraction index for normal and abnormal light:

$$S = d \cdot \frac{n_e^2 - n_o^2}{2 \cdot n_e \cdot n_o} \tag{8.1}$$

A single plate can split the light in one direction only, so four slices are required to split the light in all four directions. How many OLFs will be used in a DSC depends on a lens design and lens/sensor combination.

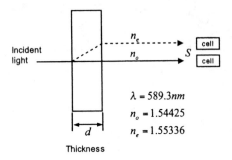

FIGURE 8.8 Optical low-pass filter.

8.1.3 COLOR CORRECTION

8.1.3.1 RGB

RGB color correction is used to compensate for cross-color bleeding in the color filter used with the image sensor. It uses a set of matrix coefficients (a_i, b_i, c_i) in the matrix calculation shown in Equation 8.2. However, with only nine elements, the matrix shown is not enough to correct the entire color space. The matrix parameters must satisfy

$$\sum_{i=1}^{3} a_i = \sum_{i=1}^{3} b_i = \sum_{i=1}^{3} c_i = 1.$$

Increasing the main diagonal values results in richer color in the corrected image. The RGB CFA has better color representation characteristics than the complementary color filter array. The latter has the advantage of better light sensitivity (usability) because of less transmitted light loss than RGB. Many DSCs use RGB CFA, but DVC uses complementary CFA.

$$\begin{bmatrix} R' \\ G' \\ B' \end{bmatrix} = \begin{bmatrix} a_1 & a_2 & a_3 \\ b_1 & b_2 & b_3 \\ c_1 & c_2 & c_3 \end{bmatrix} \begin{bmatrix} R \\ G \\ B \end{bmatrix} \qquad (8.2)$$

8.1.3.2 YCbCr

YCbCr color space is also used in DSCs. The YCbCr and RGB color spaces have a linear relation with transformation Equation 8.3 and Equation 8.4, known as ITU standard D65. Equation 8.5 is a fixed-point number representation of the transformation matrix in Equation 8.3 with 10-b resolution. By converting an image from RGB to YCbCr color space, it becomes possible to separate Y, C_b, and C_r information. Y is luminance data that do not include color information; C_b and C_r are chrominance data that contain color information only. From Equation 8.3, it can be seen that G (green) is the biggest contributor to luminance. This is as a result of the human eye's

Image-Processing Algorithms

higher sensitivity to green than to the other colors, red and blue. Blue is the biggest contributor for C_b and red contributes most to the C_r.

$$\begin{bmatrix} Y \\ C_b \\ C_r \end{bmatrix} = \begin{bmatrix} 0.2988 & 0.5869 & 0.1143 \\ -0.1689 & -0.3311 & 0.5000 \\ 0.5000 & -0.4189 & -0.0811 \end{bmatrix} \begin{bmatrix} R \\ G \\ B \end{bmatrix} \qquad (8.3)$$

$$\begin{bmatrix} R \\ G \\ B \end{bmatrix} = \begin{bmatrix} 1 & 0 & 1.402 \\ 1 & -0.3441 & -0.7141 \\ 1 & 1.772 & 0.00015 \end{bmatrix} \begin{bmatrix} Y \\ C_b \\ C_r \end{bmatrix} \qquad (8.4)$$

$$\begin{bmatrix} 306 & 601 & 117 \\ -173 & -339 & 512 \\ 512 & -429 & -83 \end{bmatrix} \qquad (8.5)$$

The spatial response of the human eye for chrominance is very low compared to that for luminance signal. From this characteristic, the data rate can be reduced when the chrominance signals are processed. The image compression that will be described later in this chapter uses this characteristic and reduces spatial resolution of chrominance to one half or one fourth of the original data without severe degradation of image quality. The reduced bandwidth signal format is known as 4:2:2 and 4:1:1 for the JPEG standard.

The RGB and YCbCr color spaces are used for DSC and DVC, although some other color spaces exist. YCbCr, RGB, and other color spaces are described in detail in Chapter 7.

8.1.4 TONE CURVE/GAMMA CURVE

Many imaging devices have nonlinear characteristics when it comes to the process of capturing the light of a scene and transforming it into electronic signals. Many displays, almost all photographic films, and paper printing have nonlinear characteristics. Fortunately, all of these nonlinear devices have a transfer function that is approximated fairly well by a simple power function of the input, of the general form given in Equation 8.6:

$$y = x^\gamma \qquad (8.6)$$

This is called the tone or gamma curve. The conversion of an image from one gamma curve function to another is called "gamma correction." In most DSCs, gamma correction is done in the image acquisition part of the signal-processing chain: the intensity of each of the linear R, G, and B components is transformed to a nonlinear signal by an RGB gamma correction function.

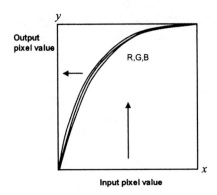

FIGURE 8.9 RGB gamma with table look-up method.

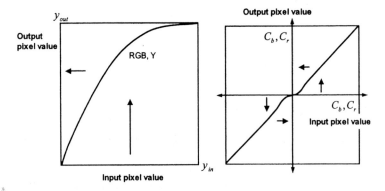

FIGURE 8.10 Y and chroma gamma with table look-up method.

The gamma curve converts each input pixel value to other value with a nonlinear function that modifies the histogram distribution of the source image. The value γ (gamma) in the Equation 8.6 lies around 0.45 for most CRT display systems. However, it is not constant for a DSC.

Gamma conversion in the DSC is often used in conjunction with bit depth compression such as 12- to 8-b compression of bit depth resolution. Gamma correction also adjusts the noise floor and contrast of the original data near the black level. For this purpose, many DSCs use a table lookup method to get high flexibility and better accuracy of bit depth conversion.

There are two implementations of tone curves for DSCs, one in RGB and the other in YCbCr space. Usually, the former is used in DSCs that use an RGB CFA sensor and the latter in DSCs/DVCs that use a complementary CFA sensor. In some applications, RGB gamma uses three channels of gamma curves for each of R, G, and B. Also, YCbCr uses Y, Cb, and Cr gamma curves for the three components. Figure 8.9 shows an example of RGB gamma curves and Figure 8.10 shows Y (luminance) and CbCr (chrominance) gamma curves. RGB and Y curves affect the contrast of the image. The curvature near the origin controls the dark level of the image. Granularity at low-level input changes the contrast and black level of an image greatly.

8.1.5 FILTER OPERATION

Filtering operations on images can be viewed as modifications in the two-dimensional spatial frequency domain. The low-pass filter functions used in image manipulation should have a smooth impulse response to avoid excess ringing or artifacts on edge boundaries within an image. Actual filter calculations can be performed using arithmetic multiplication in the spatial frequency domain of Fourier transformed filter function and image. Actual Fourier transform uses the FFT (fast Fourier transform) algorithm that needs only of the order of $M \cdot N \cdot (\log_2 M + \log_2 N)$ multiplications and additions for a two-dimensional array with the image size of $M \times N$. However, it requires complex hardware and large amounts of memory to store temporary results. The mathematically equivalent operation in the spatial domain is a circular convolution. Using a $K_1 \times K_2$ matrix of coefficients, called the filter kernel or convolution kernel, the convolution operation takes $K_1 \times K_2 \times M \times N$ multiplications and accumulations (MAC), where $M \times N$ is the image size.

Most DSCs use convolution operations because the hardware implementation is easier and a relatively small kernel size can be used. The number of multiplies is directly proportional to $K_1 \times K_2 \times M \times N$, which further restricts the choice of kernel size. When a filter is to be designed, it is necessary to specify its features, such as cut-off frequency, roll-off characteristics, stop-band attenuation, and filter taps. Usually, the filter synthesis starts from a one-dimensional filter design, which satisfies a desired response; it is then converted to a two-dimensional filter. The time domain response of a unit impulse input is called its impulse response and its Fourier transform shows the frequency response of a filter. In other words, the impulse response is equivalent to a filter kernel. There are good books that describe the theoretical background and actual synthesis of one- and two-dimensional filters.[1,3,4,8,10]

Two-dimensional filter synthesis from a one-dimensional filter is classified into two types: separable and nonseparable. Figure 8.11 shows how to construct a separable two-dimensional filter from a one-dimensional one. The two-dimensional filter kernel is constructed from two one-dimensional filters, $f_1(x,y)$ and $f_2(x,y)$, that have different impulse responses. Equation 8.7 is a serial synthesis that is expressed as $f_1(x,y) \otimes f_2(x,y)$, where $\otimes$ denotes convolution operation of two filters, f_1 and f_2. Equation 8.7 means filtered output from the f_2 filter is then convoluted by f_1 to get the final result, $g(x,y)$:

$$g(x,y) = f_1(x,y) \otimes f_2(x,y) \otimes f_{org}(x,y) \qquad (8.7)$$

Equation 8.8 is a nonseparable filter structure:

$$g(x,y) = f(x,y) \otimes f_{org} \qquad (8.8)$$

In this case, the filter, f, cannot separate into two filters, f_1 and f_2; therefore, it is necessary to calculate a two-dimensional convolution with the source image and a two-dimensional filter kernel. Note that, in Equation 8.7, the two filters, f_1 and

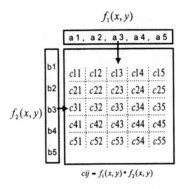

FIGURE 8.11 Construction of two-dimensional filter from one-dimensional filter.

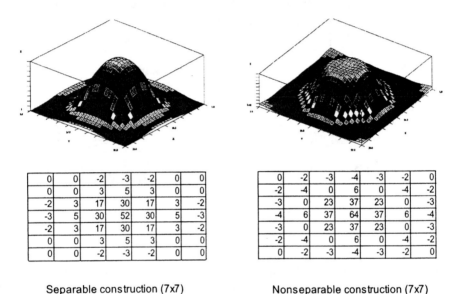

FIGURE 8.12 Separable and nonseparable low-pass filters.

f_2, can be combined into a two-dimensional filter, f_3, by convolving them together: $f_3 = f_1 \otimes f_2$. The resulting filter, f_3, is then applied to the input data.

Figure 8.12 shows the two-dimensional frequency response of the two different kinds of filter implementation. These two configurations use the same one-dimensional impulse response. Figure 8.12(a) is a serial or separable construction as described in Figure 8.11. Figure 8.12(b) is a nonseparable configuration that was designed directly in the two-dimensional Fourier domain.

8.1.5.1 FIR and IIR Filters

Digital filters are classified by their structure as finite impulse response (FIR or nonrecursive) filters and infinite impulse response (IIR or recursive) filters. Figure 8.13 shows the basic structures of these two kinds of filters with time domain representation and also Z-transform notation.[3,4,12] The FIR filter is always stable for

Image-Processing Algorithms

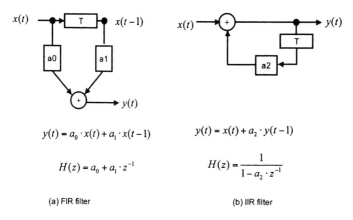

FIGURE 8.13 FIR and IIR filters

any input data stream and filter parameters because of its nonrecursive structure; however, it needs much longer filter coefficients (taps) than an IIR to achieve the same specification. On the other hand, the IIR has much shorter filter structures as a result of its recursive structure. In image-processing applications, care must be taken not to introduce phase distortion and instability when IIR is used. FIR structure is widely used in the image-processing field; IIR is only used for the image application in which phase distortion can be neglected.[3,4,9]

8.1.5.2 Unsharp Mask Filter

Unsharp mask filter operation is a kind of high-pass filter. It is used widely in DSCs to boost the high-frequency component of the image. Equation 8.9 describes the theoretical background. The basic algorithm is that filtered output, $g(x,y)$, is obtained by multiplying a blurred image by a coefficient, α, and subtracting the result from the original image, f_{org}. Blurring is equivalent to a low-pass filter operation. The blurred image is generated by convolution with a rectangular filter kernel such as 5×5, 9×9, which has a constant coefficient. A nonrectangular kernel may also be used. The convolution operation is simply a summation of pixels within an area that corresponds to the kernel size. The h in Equation 8.9 is called the "mask" and is equivalent to a simple arithmetic mean. Parameter α controls amplitude of high-frequency emphasis of $g(x,y)$; α is set between 0 and 1. A larger value of α increases the high-frequency content of the image.

$$g(x,y) = \{f_{org} - \alpha \cdot (h \otimes f_{org})\} / (1-\alpha) \qquad (8.9)$$

The Fourier transform is applied to both sides of Equation 8.9, yielding:

$$G(\omega_x, \omega_y) = \{F_{org} - \alpha \cdot H \cdot F_{org}\} / (1-\alpha) = \{(1-\alpha \cdot H) \cdot F_{org}\} / (1-\alpha) \qquad (8.10)$$

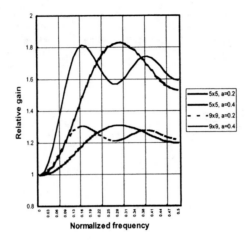

FIGURE 8.14 Frequency response of 5 × 5 and 9 × 9 unsharp mask filters.

The H in Equation 8.10 is the Fourier transform of the mask h. Figure 8.14 shows frequency response with 5 × 5 and 9 × 9 mask sizes and α set to 0.2 and 0.5. There is a very low-frequency bump in the response curve in the figure, but it does not affect the filtered results, $g(x,y)$. It is possible to use other masks that have different values in the kernel than a simple constant valued mask. To do this requires a convolution instead of simple summation to generate the blurred image and consequently increased calculation time. A trade-off must be made among processing time, hardware cost, and required filter response.

8.2 CAMERA CONTROL ALGORITHM

8.2.1 Auto Exposure, Auto White Balance

In a DSC, an auto exposure (AE) block adjusts the amount of incident light on the sensor so as to utilize its full dynamic range. Exposure is typically handled by electronic devices in a camera that simply record the amount of light that will fall onto the image sensor while the shutter is open. This amount of light is then used to calculate the correct combination of aperture and shutter speed at a given sensitivity.

An analog or digital gain amplifier that is set before nonlinear operation and color control executes the AE information sensing and control. The luminance value, Y, is used as an index to control exposure time. The entire image-sensing area is divided into nonoverlapped sub-blocks, called AE windows. Figure 8.15 shows a typical layout of AE window within an image. This window layout is also used for calculation of auto white balance (AWB). However, the index of AWB control is different from AE. The AWB uses mean value of each R, G, and B pixel instead of Y for AE.

For an RGB CFA sensor, color information is converted to Y value to measure incoming light intensity. The AE signal-processing block calculates mean and peak values for each AE window and passes them to the MPU (as shown in Figure 8.1).

Image-Processing Algorithms

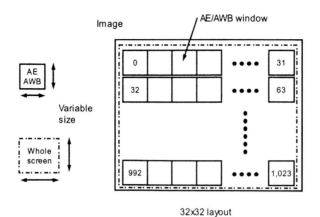

FIGURE 8.15 AE/AWB window layout.

TABLE 8.2
Classification of Automatic Focus Control Algorithms

Measurement method	Detection algorithm	Actual implementation
Range measurement	Active method	IR(Infrared), ultrasonic
	Passive method	Image matching
Focus detection	External data	Phase detection
	Internal data	Digital integration

The AE control application program running on the MPU will evaluate each value and decide the best exposure time. There are many kinds of evaluation algorithms; some use the variance of each window.

Another important camera control parameter is the white balance. The purpose of the white balance is to give the camera a reference to white for captured image data. The white balance adjusts average pixel luminance among color bands (RGB). When the camera sets its white balance automatically, it is referred to as auto white balance (AWB). For an RGB CFA sensor, an image processor in the DSC controls each color gain for incident image light at an early stage of the signal flow path. Details of white balance are described in Chapter 7. The AWB gain correction is adjusted in the raw data domain or just after the color interpolation.

8.2.2 Auto Focus

8.2.2.1 Principles of Focus Measurement Methods

Auto focus (AF) control is one of the principal functions of a DSC. Table 8.2 summarizes several kinds of focus control algorithms and their actual implementations in DSCs. In addition, Figure 8.16 illustrates the principles of three focus measurement methods other than the digital integration that is described next.

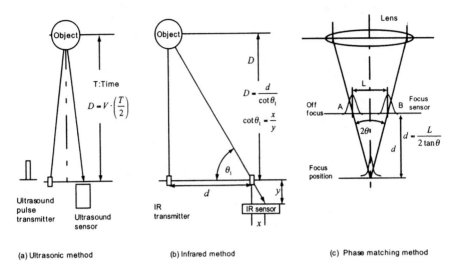

FIGURE 8.16 Focus measurement.

8.2.2.2 Digital Integration

Many DSCs use a digital integration method for auto focus that uses the acquired image from the sensor and a digital band-pass filter because the algorithm is simple and easy to implement in a digital signal-processing block. The digital integration focus method works on the assumption that high-frequency (HF) components of the target image will increase when in focus. The focus calculation block is composed of a band-pass filter and an absolute value integration of the output from the filter. Digital integration AF uses the acquired image data only so that it does not use any other external active or passive sensors. The host processor of the DSC uses the AF output value and adjusts the lens to get peak output from the AF block. Usually, the MPU analyzes multiple AF windows within an image.

Figure 8.17 shows a typical AF window layout within an image. The layout and size of each AF window depends on the type of scene to focus. Five windows and 3×3 layout are common for many scenes. The single-window layout is useful to get easy focus calculation for moving objects. The AF output is calculated for each window every frame. Each window generates AF data every 1/30 sec using draft mode frames of the CCD sensor and sends them to the host processor of the DSC. This method does not give absolute range distance value from the DSC to the object as an active sensor does, so the host processor will search for the peak of the AF data by using lens movement. Also, the host processor must judge which direction of lens movement will improve the focus based on the past history of lens movement and AF output data.

Figure 8.18 shows how to calculate AF data within a window and the AF filter response. In Figure 8.18(a), the AF block receives horizontal pixel data and calculates the filter output. Then it accumulates the absolute values of the AF filter outputs. It is desirable to calculate in vertical and horizontal directions in each window. From the viewpoint of hardware implementation, horizontal line scan is easy but vertical

Image-Processing Algorithms

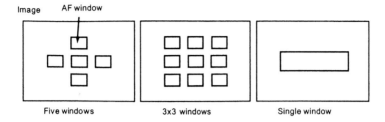

FIGURE 8.17 Multiple window layout for AF operation.

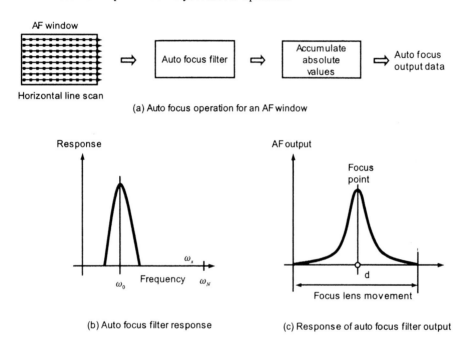

FIGURE 8.18 Auto-focus filter and output.

line scan requires some amount of line buffer memory. Figure 8.18(b) shows a typical band-pass filter for AF use. The resonant frequency depends on the characteristics of the lens system in the DSC. Usually, a low resonant frequency gives better response at the initial stage when a scene is out of focus. Figure 8.18(c) shows how the output from the AF block will vary with lens movement around the point of focus.

8.2.3 Viewfinder and Video Mode

Most DSCs have an electronic viewfinder composed of thin film transistor (TFT) liquid crystal display as well as a conventional optical viewfinder. The electronic viewfinder image must refresh very quickly for the photographer to find and focus on the subject. For this purpose, the DSC uses a draft mode output of a CCD sensor. Details of a draft mode readout mechanism are described in Section 4.3.5 and Section 5.4.3. The draft mode data are full size in the horizontal direction but decimated in the vertical direction; the output speed is 30 frames per second. The DSC uses draft mode data for auto-focus operation and for the viewfinder by resizing it in the horizontal direction using a decimation low-pass filter.

8.2.4 Data Compression

In a single-sensor DSC, the raw data are spatially compressed by the CFA on the sensor by one third. The amount of raw data is given by $X \times Y \times (12 - 14)$ b/pixel, where X and Y are number of pixels for horizontal and vertical directions. However, once the full-color image is reconstructed by color interpolation, it becomes $X \times Y \times 3$ bytes. For example, in the case of a 5-Mpixel sensor having 12 b/pixel depth, the raw data have 7.5 Mbytes; however, the reconstructed full-color image has 15 Mbyte, which uses 15 Mb of storage area for every single image.

The image compression technology plays an important role. There are two data compression categories; one is reversible or nondestructive and the other is irreversible or destructive compression. The former recovers 100% of the original image and the latter loses a certain amount of information of original data. In the ideal case, the former can achieve 30 to 40% compression from original data (reduction to 0.7 to 0.6 from 1.0), but this is not enough for most DSC systems. Reversible compression is used for raw data saving and offline image processing on a personal computer. Irreversible compression can achieve a large amount of size reduction that depends on how much image degradation is acceptable. Usually, the image has a large amount of redundancy, so irreversible compression techniques have been used to reduce its redundancy.

Many image compression algorithms exist today, but the need for performance and a widely used standard favors the choice of JPEG (joint photographic expert group) compression. Figure 8.19 shows the DCT-based JPEG image compression flow. JPEG technology stands on many kinds of technology. Because the human eye has highest spatial resolution for luminance but very low for chrominance, the JPEG handles YCbCr image data. It also uses discrete cosine transform (DCT) and Huff-

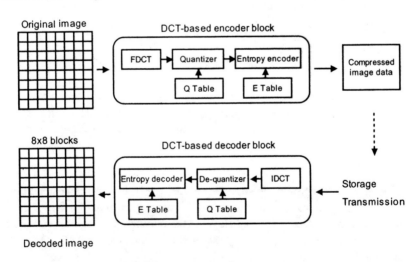

FIGURE 8.19 DCT-based JPEG image compression.

Image-Processing Algorithms

TABLE 8.3
Removable Data Storage

Name	Media size	Memory
HDD card	20 g, 42.8 × 36.4 × 5.0 mm	Hard disk
Compact flash	15 g, 36 × 43 × 3.3 mm	Flash memory
Smart media (SSFDC)	2 g, 45 × 37× 0.76 mm	Flash memory
SD card	1.5 g, 32 × 24 × 2.1 mm	Flash memory
Mini SD card	21.5 × 20 ×1.4 mm	
Memory stick	4g, 50 × 21.5 × 2.8 mm	Flash memory
Memory stick duo	31 × 20 ×1.6 mm	
XD-picture card	24.5 ×20 ×1.8 mm	Flash memory

man coding. The standard and related technologies are described in Pennebaker and Mitchell.[2] The image file format for DSCs is standardized as the digital still camera image file format standard (exchange image file format for digital still camera: exif) by JEIDA (Japan Electronics Industry Development Association Standard).[6,7]

8.2.5 DATA STORAGE

Thanks to the recent spread of personal data assist (PDA) equipment and the demand for large-capacity compact storage media in many fields, a wide variety of small removable media has been developed and introduced to the DSC-related market. Table 8.3 summarizes current removable data storage media used in DSC systems. Some media have storage capacity over the GB (10^9 byte) boundary. All of them have a capability to transfer data to a personal computer (PC) using an adapter. Read/write speed is also important for application in future DSCs. All these media achieve at least 2 Mb/sec and some over 10 Mb/sec R/W speed.

8.2.6 ZOOM, RESIZE, CLIPPING OF IMAGE

8.2.6.1 Electronic Zoom

Many DSCs have an optical zoom that uses a zoom lens system. However, the zoom described in this section is based on the digital signal processing in the DSC, which is called "electronic zoom" or "digital zoom" in some books. This section describes the principles of the electronic zoom algorithm and the underlying two-dimensional interpolation algorithm in detail. Unlike optical zoom, electronic zoom magnifies the image with digital calculation so that it does not expand the Nyquist limit of the original image. The electronic zoom operation inserts pixels between the original pixels, which are calculated by interpolation. Discussion will start with one-dimensional interpolation theory and then expand to two dimensions.

A basic, special case of interpolation was explained in Section 8.1.2 as color interpolation in which a value at nonexisting location is estimated using values of surrounding existing locations. In this section, a more general form will be explained. Equation 8.11 describes the general interpolation calculation using convolution.

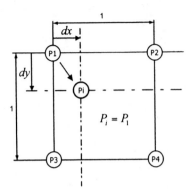

FIGURE 8.20 Nearest neighbor interpolation.

$$p(x) = \sum_i f(x - x_i) \cdot g(x_i) \qquad (8.11)$$

The function, $f(x)$, of Equation 8.11 is an interpolation function and describes the entire characteristics of the interpolation results. It is a kind of low-pass filter and affects the image quality of zoomed results. A number of interpolation functions have been reported by several authors[5]; three well-known interpolation functions will be introduced here: nearest neighbor, linear, and cubic spline interpolation functions.

8.2.6.1.1 Nearest Neighbor Interpolation Function

In the nearest neighbor interpolation, each interpolated pixel is assigned the value of the pixel value that is the nearest from the original data. Figure 8.20 shows nearest neighbor interpolation in two-dimensional space. In the figure, the interpolation result, P_i, has the same value as the original image pixel, P_1. The interpolation function is shown in Equation 8.12. Nearest neighbor interpolation is the simplest method and requires little computational resource. The drawback is the poor quality of the interpolated image.

$$\begin{aligned} f(x) &= 1 & 0 \leq |x| < 0.5 \\ f(x) &= 0 & 0.5 \leq |x| < 1 \\ f(x) &= 0 & 1 \leq |x| \end{aligned} \qquad (8.12)$$

8.2.6.1.2 Linear Interpolation Function

Linear interpolation uses two adjacent pixels to obtain the interpolated pixel value. The interpolation function is expressed in Equation 8.13. When linear interpolation is applied to an image, it is called bilinear interpolation (Figure 8.21). Four surrounding points are used to estimate a pixel. Bilinear interpolation is a relatively simple interpolation method, and the resulting image quality is good. The calculation

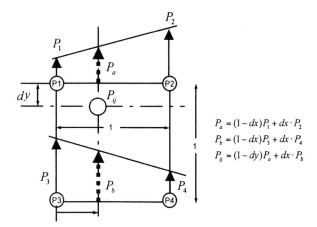

FIGURE 8.21 Bilinear interpolation.

can be divided into each direction (x and y) and the calculation order of x or y makes no difference.

$$f(x) = 1 - x \quad 0 \le |x| < 1$$
$$f(x) = 0 \quad 1 \le |x| \tag{8.13}$$

8.2.6.1.3 Cubic Interpolation Function

Cubic interpolation functions are a family of third-order interpolation functions. When used for two-dimensional interpolation, this is called bicubic interpolation. Figure 8.22 shows a bicubic interpolation algorithm that uses 16 surrounding points to estimate the value of pixel Q. Equation 8.14 shows one of the cubic interpolation functions called a cubic spline. The bicubic interpolation function gives better image quality than the two methods described earlier, but at the cost of much higher computational resource requirements than nearest and linear interpolations.

$$f(x) = (1-x)(1+x-x^2) \quad 0 \le |x| < 1$$
$$f(x) = (1-x)(2-x)^2 \quad 1 \le |x| < 2 \tag{8.14}$$
$$f(x) = 0 \quad 2 \le |x|$$

Figure 8.23 summarizes the three interpolation functions in one dimension. The horizontal axis of the graph in this figure is the normalized distance between original pixels; each location of 0, ±1, and ±2 has a value of the original pixel.

Figure 8.24 shows the frequency response of the three functions. The frequency response of the interpolation functions is obtained by taking the Fourier transform of the functions, as shown in Figure 8.24; the nearest neighbor has a wide frequency response over the value 0.25, which causes a strong aliasing error for the zoomed

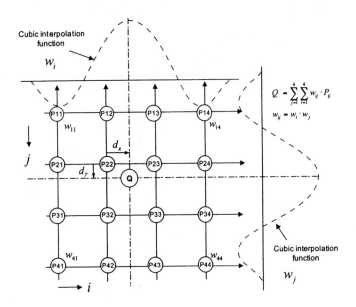

FIGURE 8.22 Bicubic interpolation.

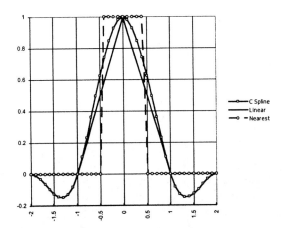

FIGURE 8.23 Three different kinds of interpolation functions.

image. Linear interpolation gives better attenuation characteristics over the value 0.25 than the nearest neighbor. However, linear interpolation gives lower response than the cubic spline between 0 and 0.25. Cubic spline has the best cut-off frequency of the three interpolations, giving the best image quality. Table 8.4 shows the total number of multiply/add operations for each of these interpolations.

8.2.6.2 Resize

Resize is the opposite operation to zoom: it reduces the size of the image. In some cases, it is desirable to reduce image size smaller than the original pixel. Resize reduces the number of pixels in the image while keeping the entire scene. It is widely used for generating thumbnails, reducing image size, placing multiple images within

Image-Processing Algorithms

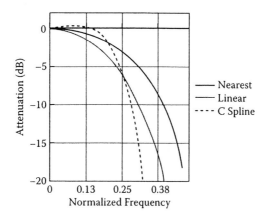

FIGURE 8.24 Frequency response of three interpolation functions.

TABLE 8.4
Number of Multiply/Addition of Interpolation Functions

Interpolation function	Order	Multiply/add (one dimension)	Multiply/add (two dimensions)
Nearest neighbor	0	0	0
Linear	1	2	4
Cubic spline	3	4	16

a frame, and movie display. The simplest way to reduce pixel size is to decimate points from the original data without any low-pass filter operation. The simplest decimation is just to extract pixels that fit at a location in the resized image or to use nearest neighbor interpolation like that described previously. However, this does not give a good-quality resized image. A low-pass filter should be applied before the resize operation, depending on the resize ratio between original data and result. If the resize is a noninteger ratio like 0.3, an interpolation operation must be used to get the resized image. The zoom filter can be applied to the resize operation up to 1/2 because resize has little effect on image quality. However, if a large resize ratio is needed, it is necessary to apply a low-pass filter with low cut-off frequency.

8.2.6.3 Clipping, Cropping

Clipping or cropping is the operation of cutting out a region of interest from an original image. The clipping process is different from zoom and resize operations in that it only cuts out a part of the original image. If a clipped pixel location is not the same as an original pixel location, it must use interpolation to calculate the pixel value. In addition, if it uses subpixel offset in a clipping, the interpolation operation must also be used when a clipped region is offset by a noninteger number of pixels. In this case, the interpolation operation is the same across the entire image. The interpolation will cause a very small amount of degradation in the frequency response near Nyquist. The amount of degradation depends on the interpolation function used.

8.3 ADVANCED IMAGE PROCESSING; HOW TO OBTAIN IMPROVED IMAGE QUALITY

Previous sections have described many linear image-processing operations used in DSCs. These are very useful for constructing better quality images from a sensor and also for modifying the images. However, DSCs have several noise sources, as described in the Section 8.1.1 and Table 8.1. In addition, the CCD/CMOS sensor is not a perfect detector for real-world scenery that has very wide dynamic range and high spatial resolution. The grain of silver halide film has a much smaller cell than current CCD/CMOS sensors. The limitations of CCD/CMOS cell size and dynamic range cause aliasing noise and also many types of color noise at severe light conditions.

This section will describe some of advanced image-processing techniques that use nonlinear image-processing. The word "nonlinear" means that the local processing parameter for each pixel depends on a feature or index of the location of the pixel so that the image-processing operation is no longer constant for the entire image. In addition, a class of nonlinear processing depends on the value of each pixel. Some of the technologies described here are used in DSCs; others are not used because of their complex implementation.

8.3.1 CHROMA CLIPPING

Image reconstruction from raw data must interpolate pixel values from undersampled output from the sensor. This interpolation will generate excess color noise caused by undersampling of the object space. This color noise can be seen around thin edges that have relatively large RGB values (gray edge). Also, it is visible in high-luminance areas, where R, G, and B have high values so that any imbalance between them is seen as color noise clearer than in a dark area. This type of excess color noise can be suppressed by using a nonlinear chrominance suppression table with a luminance-dependent index value. This operation is called chroma clipping or chroma suppression.

Figure 8.25 shows a typical chroma clipping operation. The horizontal axis is a luminance value and the vertical axis is a chrominance gain control value for each pixel. In the figure, between luminance values Y_1 and Y_2. chrominance gain is set to 1.0. The chrominance gain between Y_2 and Y_{max} gradually decreases to zero and the transition zone gradually clips the chrominance noise of an image. This clipping operation can be realized by using a table lookup method, which has good flexibility for many situations. In some cases, the dark area between zero and Y_1 is suppressed because the color noise in a dark area should contain less color information.

8.3.2 ADVANCED COLOR INTERPOLATION

Section 8.2.1.3 described that the sampling interval of single sensor such as an RGB CFA is 1.4 to 2 times lower than the reconstructed image. Thus, pixel values must be estimated using insufficient spatially sampled data. The basic method of color interpolation for the RGB CFA is that the interpolation is executed for R, G, and B with a constant interpolation area in the entire raw image. This means that recon-

Image-Processing Algorithms

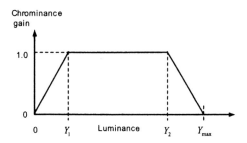

FIGURE 8.25 Chroma clipping.

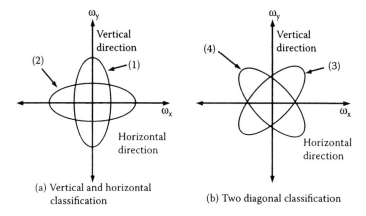

FIGURE 8.26 Adaptive color interpolation with directional interpolation filters.

struction is a linear operation for each interpolation point with respect to spatial operation. However, another idea, called adaptive color interpolation or nonlinear color interpolation method, gives good results compared with the linear method. Adaptive color interpolation has several variations and also depends on how many pixels are used for an interpolation calculation. The outline of adaptive interpolation will be described here.

Some DSC processors use a similar algorithm and get better image quality than with a linear reconstruction method. One of the underlying ideas of adaptive interpolation is to use different interpolation filters according to a local feature index of a raw image. A widely used index is edge information from a small area surrounding each interpolation point. Figure 8.26 shows the response of two-dimensional frequency characteristics of the filters that are applied for the interpolation calculation. Figure 8.26(a) and (b) shows two different filters, each with narrow- and wide-frequency bands for each orthogonal direction. Response (1) in Figure 8.26(a) has a wide band for vertical direction and narrow for horizontal direction. Response (2) has the opposite characteristics.

The application strategy of these two filters is described as follows. In Figure 8.26(a), if the local edge information of an interpolation point has a tendency to have horizontal direction, then the interpolation filter should keep the horizontal edge from the raw image. It should apply a wider response in the horizontal direction than in the vertical. Therefore, a filter (2) should be applied that has wider frequency response in a horizontal direction. In the opposite case, a filter (1) is applied.

This algorithm is simple in the manner of its interpolation strategy but gives better image quality than the basic algorithm. Several variations can be obtained. In the preceding case, two directions have been classified: vertical and horizontal. However, this can be expanded to a more complex class, such as diagonal, in addition to V and H, as shown in Figure 8.26(b). This method has many design parameters, such as filter response and number of edge directions to classify.

Usually edge information of the G signal in a raw image is used for the filter selection index. The raw image has R, G, and B; however, G has highest spatial frequency and light sensitivity. In addition, the simplest differential operator or Laplacian operator can be used for the edge detection filter. Based on the output value from these operators, color interpolation logic will select interpolation filters as described previously. Figure 8.27 shows the frequency response of two kinds of edge detection filters: (a) differentiator; and (b) Laplacian. The granularity of edge direction classification can be increased to $\pi/8(22.5°)$ or $\pi/16(11.25°)$, but consideration must be given to the reliability of the direction index when there is noise in the raw image. Actual hardware implementation in a DSC limits algorithm complexity.

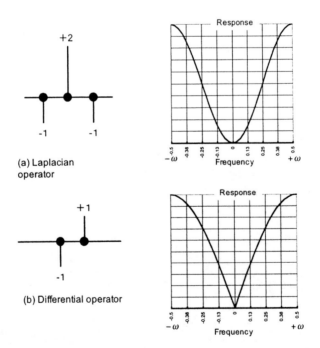

FIGURE 8.27 Two edge detectors.

Image-Processing Algorithms

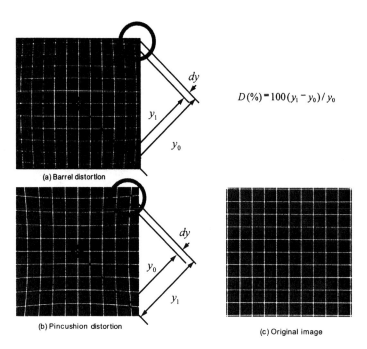

FIGURE 8.28 Lens spatial distortion.

8.3.3 LENS DISTORTION CORRECTION

There is always some degree of distortion or aberration introduced by the lens that causes the image to be an imperfect projection of the object. This section discusses only two kinds of lens distortions: spatial distortion and shading distortion that can be compensated by digital image-processing technology. Spatial distortion is a kind of geometrical distortion, as shown in Figure 8.28. The two types are called (a) barrel and (b) pincushion distortions. The amount of distortion is defined by $D(\%) = 100 \times (y_1 - y_0)/y_0$ in the figure. The distortion amount, D, is a function of the image height from the lens center.

8.3.4 LENS SHADING CORRECTION

Many lens systems have a tendency that light sensitivity decreases towards the edge of the field of view. Figure 8.29 shows sensitivity fall-off with lens-sensor geometry. Figure 8.29(a) is shading of a flat image and (b) is a section profile of it. The profile of the sensitivity decrease is given by $\cos^n \theta$. The degradation characteristics depend on a lens optical design so that the value of n varies between 3 and 5. It is often around 4, so the intensity depression is known as the $\cos^4$ law. The compensation algorithm is relatively simple: to amplify each pixel value depending on the distance from the lens center to the pixel position. The compensation factor is obtained by $1/\cos^n \theta$. A table lookup method or a direct calculation using a simplified $\cos^4$ equation can be used. The shading amount is defined as the ratio of light loss at the diagonal corner of the sensor and the lens center (image center).

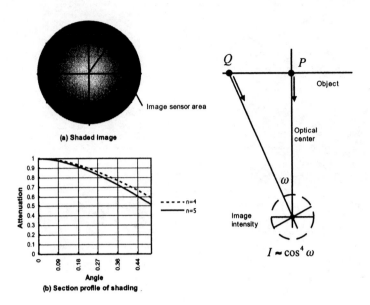

FIGURE 8.29 Lens shading distortion.

In this chapter, we described some basic algorithms and actual implementations for high-speed processing in DSCs. In recent years, the speed of CCD and CMOS sensors has been increasing, and the trend is towards higher output rates. Owing to progress in the design of peripheral devices, the boundary between DSCs and DVCs is merging and the so-called hybrid camera is becoming a reality. Standard movie and high-quality still images can be captured with a single camera. Consequently, when new image processing is implemented for a camera, it is important to investigate high-speed algorithms.

REFERENCES

1. K.R. Castleman, *Digital Image Processing*, Prentice Hall Signal Processing series, Prentice Hall, Inc., Englewood Cliffs, NJ, 1979.
2. W.B. Pennebaker and J.L. Mitchell, *JPEG Still Image Data Compression Standard*, International Thomson Publishing, Chapman & Hall, NY, 1992.
3. L.R. Rabiner and B. Gold, *Theory and Application of Digital Signal Processing*, Prentice Hall, Inc., Englewood Cliffs, NJ, 1975.
4. A.V. Oppenheim and R.W. Shafer, *Digital Signal Processing*, Prentice Hall, Englewood Cliffs, NJ, 1975.
5. P. Thevenaz, T. Blu, and M. Unser, Interpolation revisited, *IEEE Trans. Med. Imaging*, 19(7), 739–758, 2000.
6. Exchangeable image file format for digital still cameras: exif version 2.2, JEITA CP-3451, standard of Japan Electronics and Information Technology Industries Association, 2004.
7. Digital still camera image file format standard (exchangeable image file format for digital still camera: exif) version 2.1, JEIDA-49-1998, Japan Electronic Industry Development Association Standard, 1998.

8. W.S. Kyabd and A. Abtibuiy, *Two-Dimensional Digital Filters*, Marcel Dekker, New York, 1992.
9. R.E. Bogner and A.G. Constantinides, *Introduction to Digital Signal Processing*, John Wiley & Sons, New York, 1975.
10. A. Apoulis, *The Fourier Integral and Its Applications*, McGraw-Hill, New York, 1962.
11. N. Ahmed and K.R. Rao, *Orthogonal Transforms for Digital Signal Processing*, Springer-Verlag, New York, 1975.
12. S.D. Stearns, *Digital Signal Analysis*, Hayden, Rochelle Park, NJ, 1975.
13. H.C. Andrews and B.R. Hunt, *Digital Image Restoration*, Prentice Hall, Englewood Cliffs, NJ, 1977.

9 Image-Processing Engines

Seiichiro Watanabe

CONTENTS

- 9.1 Key Characteristics of an Image-Processing Engine 257
  - 9.1.1 Imaging Functions 257
  - 9.1.2 Feature Flexibility 257
  - 9.1.3 Imaging Performance 258
  - 9.1.4 Frame Rate 258
  - 9.1.5 Semiconductor Costs 261
  - 9.1.6 Power Consumption 261
  - 9.1.7 Time-to-Market Considerations 261
- 9.2 Imaging Engine Architecture Comparison 262
  - 9.2.1 Image-Processing Engine Architecture 262
  - 9.2.2 General-Purpose DSP vs. Hardwired ASIC 263
  - 9.2.3 Feature Flexibility 263
  - 9.2.4 Frame Rate 264
  - 9.2.5 Power Consumption 265
  - 9.2.6 Time-to-Market Considerations 265
  - 9.2.7 Conclusion 266
- 9.3 Analog Front End (AFE) 266
  - 9.3.1 Correlated Double Sampling (CDS) 266
  - 9.3.2 Optical Black Level Clamp 267
  - 9.3.3 Analog-to-Digital Conversion (ADC) 267
  - 9.3.4 AFE Device Example 269
- 9.4 Digital Back End (DBE) 271
  - 9.4.1 Features 272
  - 9.4.2 System Components 272
- 9.5 Future Design Directions 272
  - 9.5.1 Trends of Digital Cameras 272
  - 9.5.2 Analog Front End 274
  - 9.5.3 Digital Back End 275
- References 276

The focus for this chapter is to discuss digital photography functional and performance requirements; to describe two different semiconductor approaches for the image-processing engine; and to suggest future trends in digital cameras.

The recording part of a digital camera is made up of semiconductors, including an image sensor, an image-processing engine, and a storage device. This is analogous to what occurs on film; that is, the photons from the subject cause a complex chemical process in the film in which the image is instantly stored. Modern silver halide films provide large dynamic range performance, allowing them to adapt to suboptimal exposures and uneven lighting conditions.

The image sensor and electronic storage device perform functions similar to silver halide film. The semiconductor devices capture light photons, process the image elements, and store them as digital information. The image-processing engine of a digital camera typically receives an analog signal from the image sensor and converts it to a digital format. Then it performs various pixel functions, compresses the image, and stores, transmits, and/or displays the image. A diagram of a digital imaging system is shown in Figure 9.1.

The chemistry of film is analogous to the art of the image-processing engine. In addition, the film and digital images can go through postprocessing to refine, enhance, and colorize the pictures. For film, photo labs generally perform the postprocessing function. Modern photo labs have moved to digital technology to enhance, manipulate, and handle images.

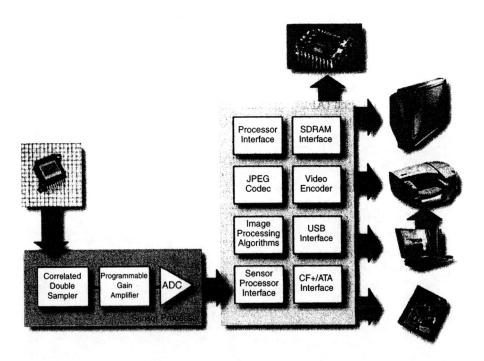

FIGURE 9.1 (See Color Figure 9.1 following page 178.) Camera block diagram.

9.1 KEY CHARACTERISTICS OF AN IMAGE-PROCESSING ENGINE

9.1.1 Imaging Functions

After the image processor converts the analog signal from the image sensor into a digital format, the data go through various pixel and frame operations to obtain the desired results. The order in which the functions are performed, the algorithms inside the functions, and the precision of the operations are driven by the requirements of the camera manufacturer; they are typically based on their color science, performance goals, and cost constraints. Optimal performance depends on throughput and appropriate quantization. The image-processing engine performs many different types of operations. The essential imaging functions for digital still camera (DSC) and digital video camera (DVC) applications are as follows:

- Analog image functions:
    Accurately sample analog signal from image sensor
    Convert sensor signal from analog to digital
    Perform sensor black level calibration
- Digital image functions:
    Color interpolate pixels (single CCD systems)
    Correct image color
    Apply tone and gamma curves
    Apply digital filters
    Provide AE/AF detection for lens control
    Perform image zoom and resize
    Compress image data
    Format and store data

In addition, several image enhancement techniques are used on a case-by-case basis, including two-dimensional, three-dimensional (movie) noise reductions, color clipping, lens compensation, antialias filters for various compression methods, etc.

9.1.2 Feature Flexibility

Feature flexibility is the ability of the system designer to obtain the desired algorithms for the camera functions from a standard set of algorithms. The need for flexibility depends on the maturity of imaging algorithms and the necessity to satisfy regional needs (localization) or make changes to the algorithms for specific camera applications. In terms of imaging algorithms, the relatively long history of developing digital-processing methods starts with the pulse code modulation (PCM) audio systems implemented in the 1960s. Digital-processing methods for imaging had an early start in medical and military imaging.

Basics of digital signal processing are *sampling* and *quantization*. Necessary follow-on technologies are reconstruction methods, filtering methods, and color managements and data compression/decompression methods. Although its maturity can be argued, almost all of the key algorithms are well defined and widely used as proven techniques as described in Chapter 8.

9.1.3 Imaging Performance

CCD analog signal capture and system noise. Like other analog signal processing, a quiet design is key to optimizing the performance of the hardware to maximize the signal-to-noise ratio (SNR) of the system. It is important to reduce noise that comes from each source, including:

- Outside the system (such as photon shot noise in an image sensor)
- Cumulative rounding error in computational units caused by insufficient quantization steps

Computational error. The internal cumulative rounding errors in an imaging engine present a complicated situation due to the nature of unavoidable errors (noise) in quantized, digital-processing systems. Hardware and software designers must choose bit depth carefully at all stages of the imaging engine to avoid inserting excess inaccuracies. A poorly quantized calculation in any part of the imaging engine will introduce excessive rounding errors and potentially dominate the overall SNR because the error propagates to later stages of the imaging engine.

Quantization step. Similar to computation errors, the *quantization* step can have a major impact on picture quality. Less quantization generates more cumulative quantization error as the data is processed. This causes several artifacts in the final image. One design method that allows control over bit depth is a custom, fully pipelined image processor, which can produce optimal picture quality. A conceptual mapping of today's imaging products, expressed in terms of ADC quantization resolution, is shown in Figure 9.2.

In some cases, the imaging calculations need to include over 72 b (24 b × 3) of bit depth to keep the image *clean*. If the bit depth is restricted, significant rounding errors can result. Similar to quantization errors, the rounding errors can produce digital artifacts, which are due to an increase in the color noise to a point of becoming observable. After determining the allocation requirements for the quantization size of each stage in the imaging pipeline, the camera designer can estimate the necessary computational performance requirements for the target application.

9.1.4 Frame Rate

Historically, digital still cameras and digital video cameras have developed from different beginnings. The digital still camera originated with silver halide film technologies; digital video cameras have evolved from TV broadcasting technologies.

Movie evolution. Conventional TV (analog) broadcasting technology addressed the limitations of transmission bandwidth bottlenecks of the economically feasible signal-processing technologies of 30 frames per second (fps) for motion pictures. Several optimization techniques have been widely used — for example, a one-

Image-Processing Engines

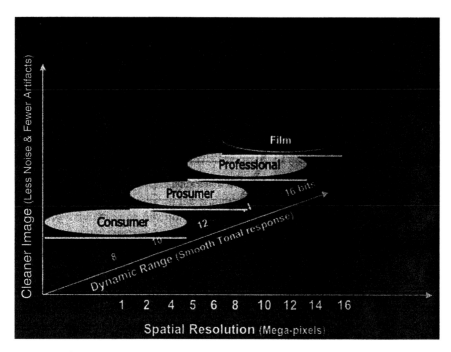

FIGURE 9.2 (See Color Figure 9.2 following page 178.) Conceptual mapping of imaging products.

dimensional horizontal filter for edge enhancement. The *video cameras* or *camcorders* are widely used to produce images using the NTSC, PAL, and SECAM standards found in TV broadcast. Recent digital advances in video technology, such as HDTV and low-cost, high-speed digital video processors, are expanding the number of video camera applications without the constraints of TV broadcasting standards.

DSC and DVC frame rates. The ultimate goal of digital image processing, as a successor technology to film for still and movie applications, is to exceed the quality of traditional silver halide film images with comparable image recording rates. To do this, the requirement for image-processing performance is very high. For example, 35-mm film can capture images at well over 60 frames per second and obtain high-quality images. The image quality of 35-mm film is estimated to be equivalent to over 12 Mpixels with a color depth of 16 b. The processing rates needed for various frame rates and trends over the last few years are shown in Figure 9.3.

As an alternative camera technology, digital imaging will supersede film-based applications in its ability to capture, process, and save images. Silver halide film can capture and store images simultaneously, meaning that if the shutter speed can be 1/30 sec or faster, it is possible to capture a series of images at 30 fps. Similarly, with still pictures, capturing instant action is often one of the important features.

The number of computations required per image as a function of the number of the neighboring pixels for the digital algorithm is shown in Figure 9.4. For example, assume a 3-Mpixel image captured at ten frames per second using an image processor with ten processing stages. If the processing stages require 16 operations on average

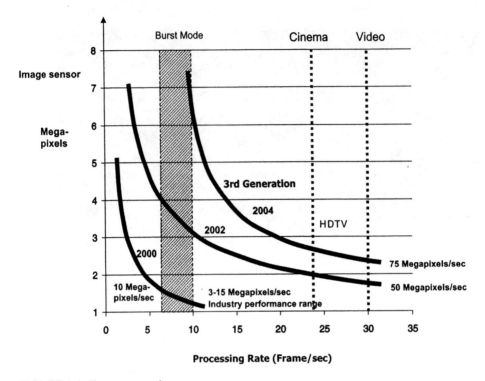

FIGURE 9.3 Frame-processing rates.

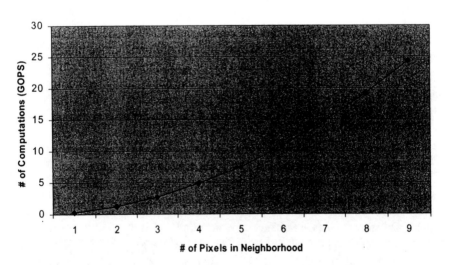

FIGURE 9.4 Computational requirements (in case of 3 Mpixels, 10 fps).

(e.g., 16 multiplication and addition for 4 pixels in neighborhood), the required computation power will be:

$$3 \times 10^6 \times 10_{(fps)} \times 16 \times 10_{(stages)} = 4.8 \text{ billion operations per second or GOPS}$$

Modern DSCs use a 3×3 to 7×7 filter and more complex algorithms are sometimes used. This example shows how heavy the image processing for a processor is.

9.1.5 SEMICONDUCTOR COSTS

The cost for the image-processing engine can be a major portion of the camera's total system cost, depending on the architectural approach and desired performance level. When the imaging functions can be integrated into a denser system-on-a-chip (SOC) device, the system costs can be reduced. For consumer products, the cost reduction generates a stronger demand in the highly competitive camera market.

The choice of packages and the number of packages are important cost factors. The cost savings of integration in the form of fewer packages and smaller boards is more than offset by the increase in die size. As semiconductor technology advances and silicon costs go down, the costs associated with package and PCB assemblies will become relatively larger. Ultimately, the higher integration chips result in a total cost reduction that enables small size applications, like phone cameras, to produce film-quality images and full-size cameras to have inexpensive lens options.

9.1.6 POWER CONSUMPTION

Power consumption in a digital camera is important because consumers want small, lightweight products that have long battery lives. If power consumption is not kept low, the product will have a relatively short battery life (causing the user to change or recharge the battery frequently) or it will require a larger, heavier battery. These situations are not good for consumers. To reduce power consumption, camera designers use a variety of approaches. The most straightforward method is to choose lower power components. Beyond that, power can be dynamically reduced by slowing down clock signals or shutting down camera functions after a period of nonuse.

In 2004, the typical consumer DSC battery provided 2 to 4 watt-hours (Wh) of power, usually with a current output in the range of 700 to 2500 mAh at 2.4 to 4.8 V. The requirements for battery life depend on the camera and the usage. A typical demand for 2 h of use provides 20 to 100 still image pictures.

The power consumption of the major functional blocks in a typical DSC, excluding the flash unit and other camera features, is shown in Table 9.1. The continuous viewing time for this camera using a 3.6-V, 700-mAh battery (available energy of 2.52 Wh) is 1.9 h (2.52 Wh/1.3 W). A standard for measuring digital still camera battery life was issued by the CIPA (Camera and Imaging Products Association) in 2003.

9.1.7 TIME-TO-MARKET CONSIDERATIONS

It is important to consider lead time when bringing products to market. Lead time consists of two parts: hardware development time and software development time.

TABLE 9.1
Camera Power Consumption

CCD image sensor	100 mW
AFE	50 mW
DBE	400 mW
Other circuits	200 mW
Lens motors	50 mW
LCD back light	500 mW
Total	1300 mW

9.2 IMAGING ENGINE ARCHITECTURE COMPARISON

Let us examine two common image-processing engine architectures and contrast their impact on the key characteristics discussed in Section 9.1. These characteristics include: imaging functions; feature flexibility; imaging performance; frame rates; semiconductor costs; power consumption; and time-to-market considerations.

9.2.1 IMAGE-PROCESSING ENGINE ARCHITECTURE

We will consider two distinct architectures for the image-processing engine:

- General-purpose DSP
- Hardwired ASIC

A general-purpose DSP architecture is shown in Figure 9.5. The components typically include a microprocessor and a DSP with SOC peripherals. The image-processing input comes from the analog front end (shown as AFE in the block diagram), which conditions and digitizes the analog pixel information from the image sensor.

The hardwired ASIC includes a custom imaging pipeline, RISC processor, and SOC peripherals as shown in Figure 9.6. The camera custom-hardwired ASIC provides specially tailored logic functions that are optimized for digital camera image processing.

The DSP approach uses a software programmable microprocessor with signal-processing hardware to perform the image processing and, often, a separate image compression device. Traditionally, the DSP architecture dominates the early stages of almost all embedded systems because it requires the shortest hardware development time and provides room for change and improvements in the imaging algorithms. When the imaging requirements are well understood and have matured, the optimal ASIC approach is more attractive because it implements the well-tuned imaging functions efficiently. In addition to providing optimal functionality, the hardwired ASIC often has features to allow the camera manufacturer to configure the key image algorithm parameters and includes an RISC processor for general imaging operations, if needed.

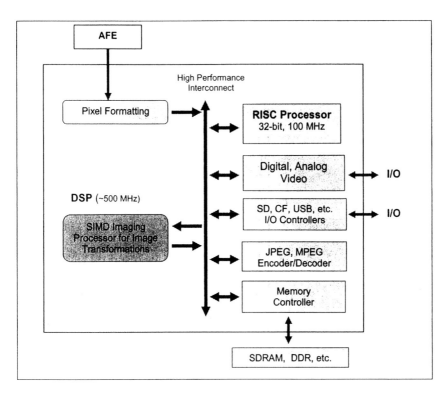

FIGURE 9.5 General-purpose DSP-based image-processing engine.

9.2.2 GENERAL-PURPOSE DSP VS. HARDWIRED ASIC

Let us dig a little deeper to understand how general-purpose DSP and hardwired ASIC are compared for digital camera applications. We will discuss each of the key characteristics described in Section 9.1. The following list shows several primary trade-offs that are part of determining the camera design architecture:

- Integration level vs. die size
- Pin count vs. integration level
- Power consumption vs. integration level
- Bit depth (SNR) vs. die size/target market
- Package selection vs. die size
- Noise immunity vs. integration level

9.2.3 FEATURE FLEXIBILITY

General-purpose DSP — high flexibility. The DSP is a programmable processor and thus can utilize software routines to achieve various functions and features, thus saving hardware development time when compared to a similarly hardwired ASCI implementation. However, with all software-based systems, real-time needs are often better served by hardwired solutions.

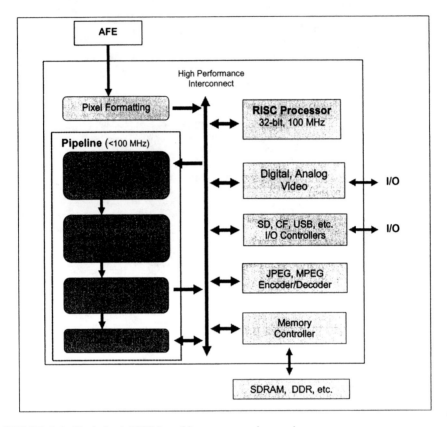

FIGURE 9.6 Hardwired ASIC-based image-processing engine.

Hardwired ASIC — depends on the implementation. With an ASIC, once logic is committed for the application, if a change is needed, it is necessary to start over and redesign to add features. Trying to achieve differentiated features may slow down development turn-around time. Specific focus and optimization may trade off flexibility. If the design uses a hardwired ASIC approach for fundamental image-processing blocks such as a compression engine, two-dimensional filter bank, interpolator, etc. and uses a DSP/MPU/reconfigurable method for additional feature flexibility, a good compromise between performance and flexibility may be achieved for the target application.

9.2.4 Frame Rate

General-purpose DSP — depends on processor and system architecture. Performance of DSP-based cameras depends on processor/system parallelism and the clock frequency available. DSPs can provide VLIW (very long instruction word) architectures and/or multiprocessing units for the image-processing engine. However, because image processing is driven by data flow, if multiple processing units (PUs) are implemented in order to meet the required performance, data passes may end up changing many times during processing of a frame data. Changing the data passes requires overhead. Thus, it is usually difficult to utilize fully the parallel processing capabilities of PUs for image-processing applications. In addition, the instruction code fetch overhead is unavoidable with the DSP/CPU approach. To compensate for the overhead and inefficient use of the CPU, it is necessary to have higher clock frequencies, which could range from 500 MHz to the gigahertz range. The trade-off with higher frequency operation is that as clock frequency goes up, so does power consumption.

Hardwired ASIC — excellent frame rates. Image-processing engine architectures based on hardwired ASICs depend less on clock frequency. In most cases, the ASIC approach provides hardwired pipeline processing. Because this approach enables configurations with massively parallel image processing and distributed memories inside the pipeline, several advantages result. If the ASIC is designed with full parallel image processing, the performance will be maximized and the required clock rate will be minimized. The quantization step size for image-processing performance can be tailored in a custom image pipeline. The designer can choose the best quantization step at each stage of the pipeline to optimize picture quality.

9.2.5 POWER CONSUMPTION

General-purpose DSP — implementation dependent. A wide range of choices is available for a general-purpose DSP, from high-power, high-performance to low-power, compromised-performance, off-the-shelf products. If the design requires very low power at the expense of image performance, a slower, ultralow-power DSP can be chosen. For higher imaging performance in a DSP architecture, the clock rate must go up significantly to provide the desired imaging performance. This leads directly to higher power consumption.

Hardwired ASIC — efficient use of power. Many hardwired ASICs designed for the mid- to high-end camera market segments are designed for high-performance image processing. The complexity of logic functions can be high, but because of its inherently massive parallel architecture, the clock frequency is relatively low — typically at the pixel rate of 27 to 100 MHz. In many cases, power consumption (energy per picture) is lower in a hardwired ASIC design than with a DSP system with equivalent imaging performance.

9.2.6 TIME-TO-MARKET CONSIDERATIONS

General-purpose DSP — good for hardware development, depends on software implementation. The general-purpose DSP architecture provides programmable processors and, thus, can utilize software programmability to achieve different functions and features; this saves hardware development time that would be required to implement a hardwired ASIC. However, for image-tuning activities (in addition to completing a hardware platform, which is relatively easy in many cases), the firmware architecture needs to be designed carefully to keep it fully independent of the tuning requirements in order to ensure design convergence. The software architecture should be designed or directed by an imaging expert to ensure that efficient image processing is performed.

Hardwired ASIC — slower hardware development, faster software development. The effort required for creating configurable processors and ASIC devices is similar because long design cycle times are required to complete and test the device design, fabricate the wafers unique to the design, and validate the solution, which can take many months. On the other hand, once a hardwired ASIC design is validated, the overall system development burden is greatly reduced because of the reduced demand for software development time.

9.2.7 CONCLUSION

The DSP-based system approach is suitable for early stages of developing emerging applications and for applications in which localized requirements in different geographical locations need to be adopted, such as cell phones, radios, etc. The differences between designs may result in changes in performance allocation; real-time issues may necessitate changes in image-processing requirements. It can be very difficult to retune and iterate the software design during the debugging stage; this can slow the development schedule. Also, the inherent requirement to speed up the processor, especially for high-end applications, can result in higher (and less competitive) power consumption.

The hardwired ASIC approach is suitable for well-defined applications and nonlocalized systems. Also, this approach is recommended when there is a strong desire to minimize power consumption and maximize performance for demanding imaging applications such as movie capture. The drawbacks of the hardwired ASIC include the cost of chip development and the additional time required to bring the chips to market.

9.3 ANALOG FRONT END (AFE)

The main function of the AFE is to convert the analog signal from a CCD or CMOS image sensor to a digital data format. The AFE prepares the data for digital image processing in the digital back end (DBE) processor. The AFE may also generate timing pulses to control an image sensor operation as well as operations of circuitry implemented in the AFE. The major sections of an AFE are shown in Figure 9.7.

9.3.1 CORRELATED DOUBLE SAMPLING (CDS)

In CCD-based cameras, it is necessary to have a special analog circuit to sample the sensor signal. In a raw output signal of a CCD image sensor, temporal noise called reset noise or kTC noise appears associated with the reset of the charge sense node of the output amplifier of the CCD image sensor. Also, the CCD output signal contains a low-frequency noise (i.e., 1/f noise; see Section 3.3.3 in Chapter 3) from the output amplifier of the CCD image sensor. To suppress the reset noise and 1/f noise, an analog circuit — the correlated double sampling (CDS) circuit — is commonly employed. As shown in Figure 9.7, the CDS circuit takes two samples: one at R_LAT sample point and the other at D_LAT sample point. Because the reset noise components on these two samples are identical, reset noise can be suppressed

Image-Processing Engines

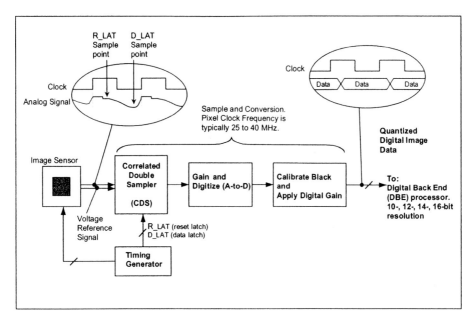

FIGURE 9.7 Simplified AFE block diagram.

by subtracting one sample from the other. The difference between the two samples, thus, represents the signal without the reset noise and low-frequency noise. Details of the reset noise and the CDS are provided in Section 3.3.3, and Section 4.1.5 and Section 5.2.1.4, respectively.

9.3.2 OPTICAL BLACK LEVEL CLAMP

As described in Section 3.5.3 in Chapter 3, an AFE should generate the reference black level of a reproduced image using signals from optical black (OB) pixels in an image sensor. Figure 9.8 shows an example of the black level adjust circuit together with a CDS circuit. A signal from an OB pixel that contains a temperature-dependent dark current component is stored on a capacitor, which is then subtracted from a signal from a photosensitive pixel.

9.3.3 ANALOG-TO-DIGITAL CONVERSION (ADC)

How many pixel bits are required for the ADC? Currently, most consumer level cameras use 10 or 12 b of pixel resolution and cameras for professional applications use 12 or 14 b. In order to achieve silver halide film quality, 14 or 16 b of pixel resolution is desirable.

In recent DSC's, the gamma correction (see Section 8.1.4 in Chapter 8) is applied digitally at DBE. In many cases, the digital gain for the low-light regions of the image is greater than ten times that for the high-light region of the image. If the ADC bit resolution is not enough for such high gains, visible contour lines and color distortion after the gamma correction can occur.

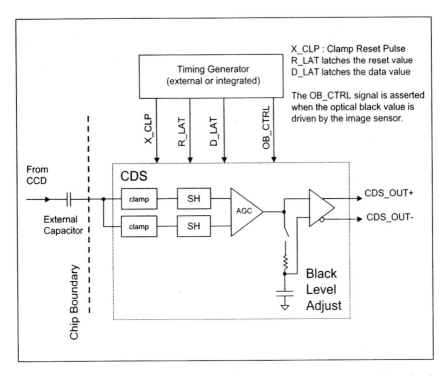

FIGURE 9.8 Block diagram of correlated double sampling and black level adjust circuits.

This occurs due to quantization noise (error). Quantization error is caused by the nature of the analog-to-digital conversion process in which a range of analog inputs can produce the same discrete digital output. This uncertainty, or error, produces an effective noise given by

$$<v_n> = \frac{V_{LSB}}{\sqrt{12}} \qquad (9.1)$$

where V_{LSB} is the voltage associated with the least significant bit.[1] Obviously, this error decreases as the resolution increases. The bit resolution and ADC design need to be chosen carefully in order to avoid such unexpected digital image artifacts.

How fast an ADC conversion rate is required? For low-resolution images, a conversion rate of approximately 20 Mpixels/sec is used. For modern, high-resolution cameras with the capability to perform motion picture recording and continuous burst shooting, over 50 Mpixels/sec is needed. For HDTV, over 75 Mpixels/sec is necessary. In order to achieve such a high performance, parallel processing is often used instead of adopting higher clock frequencies. It is always necessary to take care to overcome complications (e.g., stitching boundary errors or a channel-to-channel performance mismatch) that occur to reconstruct the full image. The design choices are a trade-off between a slower, multichannel processing design and a high-speed, single-channel design.

Nonlinearity. In imaging applications, special care is needed to keep the signal monotonic over the dynamic range of the image data. Differential nonlinearity (DNL) is the indicator to evaluate the monotonicity of the conversion characteristic (±1/2 LSB is needed to guarantee no missing codes). Integral nonlinearity (INL) should also be addressed because the smoothness of an INL curve is quite important to obtain a good monotonic signal (±1 LSB is desirable to avoid affecting gamma compensation).

9.3.4 AFE Device Example

The NuCORE NDX-1260 AFE device[2] is used in high-end still and video consumer/professional camera applications. A block diagram of the AFE is shown in Figure 9.9. It is designed to minimize color noise and digital artifacts to allow digital cameras to capture a truer representation of the actual subject being photographed. Also, it features the industry's highest throughput of 50 megasamples (pixels) per second. The CDS circuit employs a differential amplifier design, which ensures power-supply rejection. It also converts the inherently single-ended input from the CCD image sensor to a differential output signal to improve the signal-processing performance of the downstream blocks in the AFE. A programmable gain amplifier (PGA), prior to a precision ADC, takes the CDS's differential output and amplifies it under the control of an 8-b programmable word from 0 to 32 dB in 0.125-dB steps.

The PGA can have four different gain settings to accommodate four colors, and these gains can be switched at the pixel rate. This pixel-by-pixel gain setting permits white balancing to perform in analog domain at the prequantization stage. Thus, it is possible for each color pixel (RGB or CYMG) to have the same quantization step (error) at ADC, which in turn avoids generating color noise (false color noise) that could appear if the white balancing is performed by digital scaling circuits in the DBE/ASIC/DSP.

Figure 9.10 shows an example of the false color generation that can occur. The same CCD device is used for both photographs. The upper image is reproduced by the NDX-1260 device, and the lower image is obtained by another AFE that does not implement different analog gain settings for the different color channels. The improvement provided by having separate analog gain settings for each color channel can be easily seen: the lower picture shows a significant amount of false color that is not visible in the upper picture.

The ADC is a 16-b, high-speed, dual-pipeline ADC. A wide input range of 2 V realizes maximized SNR. The 16-b ADC output data are truncated to 12-b just prior to the DBE interface. If only lower bits of data are needed for an ASIC or DSP, the chip has a programmable 8-, 10-, or 12-b wide output data bus for 8-, 10-, and 12-b image-processing applications.

Black level calibration is performed in the analog domain prior to the PGA stage for each color separately through 10-b offset DACs to which digital calibration values are fed. A 16-b vertical drift compensation block suppresses the offset noise (see Section 8.1.1.1 in Chapter 8). Descriptions of a predecessor AFE to the NDX-1260, NC1250,[3] are found in Opris et al.[3] and Opris and Watanabe.[4]

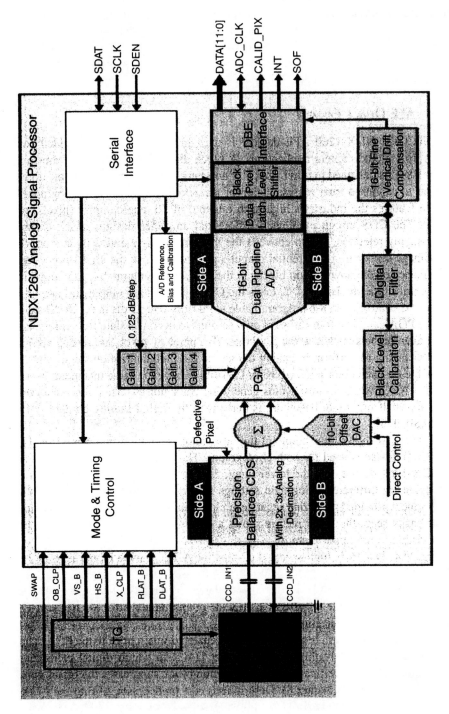

FIGURE 9.9 Block diagram of NDX-1260 analog front end device.

Image-Processing Engines

FIGURE 9.10 (See **Color Figure 9.10 following page 178.**) Reproduced image comparison. The upper image is reproduced by the NDX-1260 device and the lower image is obtained by other AFE.

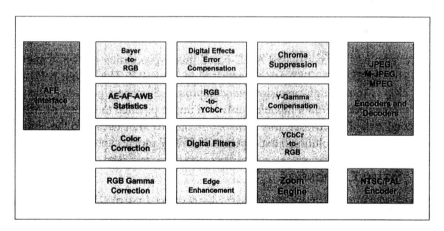

FIGURE 9.11 Image-processing engine components.

9.4 DIGITAL BACK END (DBE)

An example of the image-processing engine components for still and movie applications is shown in Figure 9.11. This design covers mid- and high-end digital still and video cameras. Image-processing algorithms for each component are described in Chapter 8.

TABLE 9.2
NuCORE DBE Imaging Features

Feature	Description
Input pixel data width	14 b
Maximum still image size	24 Mpixels
Burst capture speed	3.3 fps @ 6-Mpixel resolution
Video/movie capture speed	Motion JPEG and MPEG2: 1280×960 @ up to 30 fps; 640×480 @ up to 60 fps
JPEG format	Exif 2.2
Digital zoom	0.5× to 256× scaling
On-screen display (OSD)	Full or partial image bitmap
Video encoder	NTSC/PAL
AF	15-Area statistical data
Auto-exposure (AE)	Four-channel exposure meter Up to 1000 spots, programmable
Auto-white balance (AWB)	Software selected
Storage device interfaces	CF/ATA, SDIO, USB
CPU	Embedded ARM 922T 32-b, 200 MHz
Power consumption	400 mW without DACs
Power-up speed	100 ms (DBE only)
Color effects	Black and white, sepia, negative
Digital video ports	ITU-R656 with 8-b interface and 16-b ITU-R601 interface

9.4.1 FEATURES

As an example of the hardwired ASIC-based image-processing engine, features of NuCORE's DBE (SiP-1280) are summarized in Table 9.2. The design philosophy of this chip is to use high-precision, noncompromised image algorithms. The algorithms are embedded into a flexible hardwired image pipeline to deliver 81 Mpixels per second. For the end-user, this architecture directly translates into the ability to perform very high-resolution burst capture and HDTV movie capture and play-back.

9.4.2 SYSTEM COMPONENTS

A system design example is shown in Figure 9.12. The DBE controls and communicates with peripheral devices in addition to generating images.

9.5 FUTURE DESIGN DIRECTIONS

9.5.1 TRENDS OF DIGITAL CAMERAS

The move from traditional silver halide film to electronic systems has accelerated in recent years. Film technology has matured greatly over the last century, but the semiconductor technology promises to provide greater room to improve image quality and to do it faster except, perhaps, in some esoteric applications. The driving

Image-Processing Engines

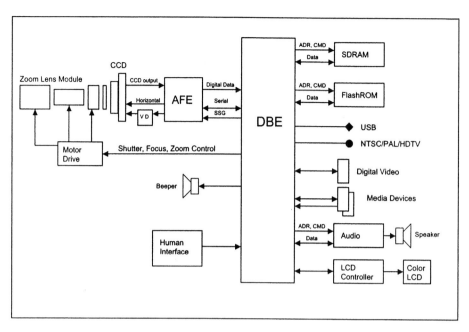

FIGURE 9.12 System wiring diagram.

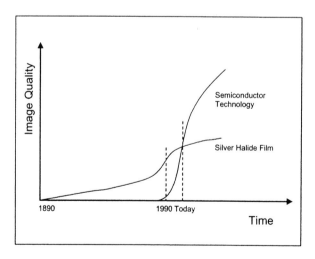

FIGURE 9.13 Film and digital camera performance trends.

force is the ability of semiconductor technology to progress well past film as shown in Figure 9.13. With equal lens and mechanical control systems, the images from today's digital technology are comparable to film cameras.

One major attraction of digital cameras that should not be overlooked is their ability to have an image immediately available that can be quickly e-mailed to others, easily manipulated by an imaging application program on a personal computer, and sent to a photo lab over the Internet to make prints. This is a profound shift in camera usability as shown in Figure 9.14.

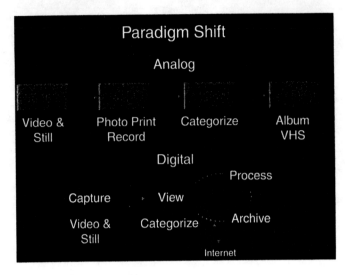

FIGURE 9.14 Camera paradigm shift.

The cell phone with a digital camera is a good example of ubiquitous camera usage connected to the Internet. The business model for cell phones is quite different and open ended compared to the traditional camera business. However, as technology marches forward, the requirements for *pure* DSCs and cell phone digital camera modules will become similar because of the available higher image quality thanks to advancements in silicon technology.

Another key trend is toward high-definition (HD) image capture. The big transition period of switching from traditional TV broadcast technologies to the HDTV era is progressing after a delayed start. HDTV broadcasting with resolutions of over 2 Mpixels per frame will commonly be seen.

Technology has fueled higher resolution progressive scan PC displays and could be unleashed in game consoles:

- Higher resolutions: VGA → SVGA → SXGA
- Faster frame rates 30 p → 60 p → 120 p ("p" stands for the progressive scan)

Camcorder technologies will be able to deliver images that are increasingly like the electronic-cinema market with robust image quality, with each frame nearly equal to what is expected in a still image. Based on the preceding discussions, the challenges for future ubiquitous digital cameras will be to achieve higher image quality than that of silver halide film, as shown in Figure 9.13, and to achieve high-definition movies.

9.5.2 Analog Front End

To achieve image quality higher than that of film, higher performance AFE should be developed. Because of the inherent nature of digital quantization, image losses

Image-Processing Engines

will be added in the image-processing engine. These losses are unlike those found in silver halide film system.

In recent DSCs, the image sensor output signal has been digitized linearly by a high-resolution ADC. In this case, it seems reasonable that the sensor noise floor is assigned to 1 LSB (least significant bit) of the ADC. For example, if the sensor dynamic range (the ratio of the sensor [linear] saturation signal to the noise floor) is 60 dB, then ADC resolution of 10 b is considered good enough. However, implementing a higher number of quantization steps than the sensor dynamic range has the potential for improving system performance.

Downstream from the ADC, in a later stage in the DBE, gamma compensation (a kind of logarithmic curve; see Section 8.1.4 in Chapter 8) is applied in the digital domain. In this step, ADC's quantization step error in a dark image will tend to be amplified and become visible as *digital artifacts*. These include color phase noise, false contour lines, etc.

Also, inside the Bayer-to-RGB reconstruction engine, digital calculations take place. This digital-processing stage could generate color phase noises (clustering color noises) if the quantization steps are too sparse; however, this processing stage averages random noise components in the raw pixel data if the quantization resolution is sufficiently high. Because the human eye sees an image on a cluster basis (a strong correlation in a noisy image can be recognized) — not on a pixel-by-pixel basis — and the quantization noise given by Equation 9.1 reduces as the quantization resolution increases, details in a dark region of an image could become recognizable. Therefore, raising the quantization resolution of the ADC in the AFE helps to increase the total system SNR and the dynamic range (latitude, tone smoothness) for all camera applications.

Every camera system needs to be equalized for color balance under different light sources, such as fluorescent, tungsten or sun, etc. Thus, there is a need to adjust each color gain (R, G, B, for example) depending on light sources. If the gain setting for each color is adjusted in the digital domain, the ADC's quantization error for each color is now different. This will cause digital artifacts such as color phase noise and false color noise in a processed image. The way to avoid such artifacts is to increase ADC's quantization steps and color balance in the analog domain. An example of addressing these issues is found in NuCORE's NDX1260, as described in Section 9.3.4.

9.5.3 DIGITAL BACK END

Progress in imaging technology is open ended. Digital technology can and will deliver in the future in two major areas. The first one is to merge digital still and digital video capabilities into one as a *hybrid camera* that enables capture of printable quality frames in a movie rate. To realize the hybrid camera, an image sensor with 2 to 6 Mpixels, with high-quality VGA to HDTV video capability, a high-density, high-speed semiconductor storage device, and a high-performance DBE, will be needed. A DBE with higher quantization steps and higher data processing throughput should help in generating wider latitude images. A digital cinema (electronic cinema) camera will be a reasonable outcome in this direction.

The second area that future DBE should address is in advanced image-processing technologies, such as real-time image recognition for improving AF/AE performance of a digital camera; advanced data compression technology that requires an object-adaptive compression algorithm; and an image-mining algorithm for finding images in storage devices in a faster and more user-friendly manner.

The improving performance of semiconductor technology will help continue to reduce the costs of other camera components similar to the way in which modern silver halide film (with all its limitations) has enabled inexpensive point-and-shoot film cameras. The digital camera trend will eventually exceed the performance of film in most camera applications due to advances in the image sensor and image-processing engine technologies.

REFERENCES

1. B. Razavi, *Data Conversion System Design*, chapter 6, IEEE Press, Piscataway, NJ, 1995.
2. http://www.nucoretech.com.
3. I. Opris, J. Kleks, Y. Noguchi, J. Castillo, S. Siou, M. Bhavana, Y. Nakasone, S. Kokudo, and S. Watanabe, A 12-bit 50-Mpixel/s analog front end processor for digital imaging systems, presented at Hot Chips 12, August 2000.
4. I.E. Opris and S. Watanabe, A fast analog front end processor for digital imaging system, *IEEE Micro*, 21(2), 48–55, March/April 2001.

10 Evaluation of Image Quality

Hideaki Yoshida

CONTENTS

10.1 What Is Image Quality?...278
10.2 General Items or Parameters..278
    10.2.1 Resolution...278
    10.2.2 Frequency Response 279
    10.2.3 Noise...280
    10.2.4 Gradation (Tone Curve, Gamma Characteristic)..................282
    10.2.5 Dynamic Range..284
    10.2.6 Color Reproduction..285
    10.2.7 Uniformity (Unevenness, Shading) 286
10.3 Detailed Items or Factors...287
    10.3.1 Image Sensor-Related Matters 287
        10.3.1.1 Aliasing (Moiré) 287
        10.3.1.2 Image Lag, Sticking 289
        10.3.1.3 Dark Current 289
        10.3.1.4 Pixel Defect 289
        10.3.1.5 Blooming/Smear 290
        10.3.1.6 White Clipping/Lack of Color/Toneless Black.......291
        10.3.1.7 Spatial Random Noise/Fixed Pattern Noise............292
        10.3.1.8 Thermal Noise/Shot Noise 292
    10.3.2 Lens-Related Factors..293
        10.3.2.1 Flare, Ghost Image 293
        10.3.2.2 Geometric Distortion 294
        10.3.2.3 Chromatic Aberration 294
        10.3.2.4 Depth of Field..295
        10.3.2.5 Perspective 296
    10.3.3 Signal Processing-Related Factors...297
        10.3.3.1 Quantization Noise 297
        10.3.3.2 Compression Noise 298
        10.3.3.3 Power Line Noise, Clock Noise..............................298
    10.3.4 System Control-Related Factors 299
        10.3.4.1 Focusing Error 299
        10.3.4.2 Exposure Error...299

10.3.4.3 White Balance Error 300
       10.3.4.4 Influence of Flicker Illumination 300
       10.3.4.5 Flash Light Effects (Luminance Nonuniformity/White
                Clipping/Bicolor Illumination) 300
   10.3.5 Other Factors: Hints on Time and Motion 301
       10.3.5.1 Adaptation 301
       10.3.5.2 Camera Shake, Motion Blur 301
       10.3.5.3 Jerkiness Interference 302
       10.3.5.4 Temporal Noise 302
10.4 Standards Relating to Image Quality 302

10.1 WHAT IS IMAGE QUALITY?

There is little doubt that the main concern for technical experts who work with images is image quality. Image devices (or systems) that have poor quality cannot accomplish their purpose. But what is image quality? In a nutshell, the purpose of image devices is "to transfer the subject information visually by reproducing images to present to the viewer." Thus, it can be said that image quality is the estimation of viewer satisfaction.

Nevertheless, it is rather difficult to define clearly. The reason for the difficulty is that an image is an information medium based on a human's visual experience and, therefore, the total evaluation of quality is made as the satisfaction level for the information transfer that depends on the purpose of the image (including the psychological effect). For instance, photographs taken by war photographers sometimes have low image quality in the usual sense because they have been taken in very inferior photographic conditions. However, such photographs never lose acclaim by reason of their low image quality. If anything, it enhances the realism of the subject.

Of course, there is a necessary level of image quality, even in a "low-quality" photograph. A picture would be nonsense if its image quality were too low for the subject to be recognizable. In other words, it is assumed that the minimum information required is transferred to the appreciator still in that case. Thus, "image quality" is a term used to express the level of satisfaction in the expected (or necessary) information transmission from the subject by the image and, in the usual meaning is the requirement to be able to obtain the expected transmission for general-purpose photography that is not limited for a particular usage. See Figure 10.1.

Each of the following characteristics for image evaluation is a particular aspect of the required performance of the images, and higher quality is obtained when every characteristic is improved. However, notice that each of the characteristics is related to the others and that sometimes a trade-off is necessary.

10.2 GENERAL ITEMS OR PARAMETERS

10.2.1 RESOLUTION

In practical terms, image resolution is the measure of the size of details that can be depicted by a given image transmission system. It should be clearly distinguished from

Evaluation of Image Quality

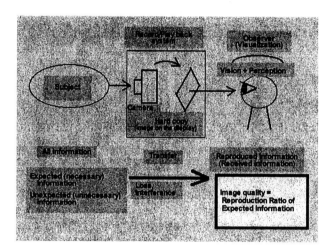

FIGURE 10.1 What is image quality?

the sharpness of the whole image or the total evaluation of image quality. Resolution is just a parameter determining whether the output signal can contain enough information to read detail of a given size. It does not have much relation to whether the picture is sharp or whether it is free from interference (for example, color moiré).

Resolution is usually measured as one-dimensional or line resolution because of convenience of measurement and facility for correspondence to the theory. The camera to be tested photographs a test pattern called the "wedge," which has black and white lines converging gradually, at the appropriate magnification. The estimator visually observes the picture taken as a photo film, a photo print or a reproduced image on an electronic display, then decides the limiting point at which the black and white lines of the wedge can be seen separately without spurious resolution. In other words, resolution is the highest spatial frequency that the target image system can transmit.

The test chart defined in the international standard ISO 12233 is shown in Figure 10.2. It is particularly specified for digital still cameras and has the hyperbolic type resolution wedges optimized for visual resolution measurement as specified in this standard.

A software tool developed by the author of this chapter, which provides an easy and accurate measuring method using estimating software for ISO 12233's visual resolution measurement, is free and downloadable from the Web site of CIPA (the Camera and Imaging Product Association; the URL is http://www.cipa.jp/english/index.html).

10.2.2 Frequency Response

Frequency response is a parameter related to spatial output response like resolution, but it covers the response at all frequencies (as opposed to resolution, which treats only the limiting frequency of transfer). It is usually evaluated as a curve plotted of the amplitude response of the output signal (normalized at the lowest frequency) vs. the frequency, where a frequency sweep pattern is input to the testing system. The

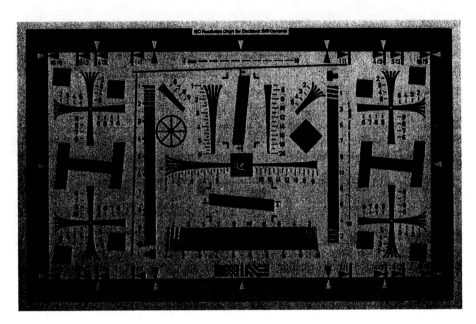

FIGURE 10.2 Appearance of the ISO 12233 test chart.

ISO12233 chart mentioned earlier has a frequency sweep pattern intended for measuring limiting resolution, but it is equally valid for measuring frequency response. In addition, the third method described in ISO12233, called SFR (spatial frequency response), is just a measurement of frequency response, as indicated by its name.

Frequency response is almost equivalent to the MTF (modulation transfer function) of the optical system. Various image quality-related information can be read from the response curve. For example, a higher output level at every frequency gives greater sharpness. The resolution value nearly corresponds to the highest frequency of the response curve. Therefore, a system with a response in the middle frequency range that is greater than that of another system has higher sharpness, even when they have almost the same resolution. For instance, Figure 10.3 shows that the left image is apparently sharper than the right one even though they have almost the same visual resolution.

However, the frequency response curve holds much more information than just the sharpness and resolution of the image, so it is difficult to say what curve is good or bad in general. There is no standard interpretation for the overall response curve, and thus evaluation or use of the frequency response may be arbitrary.

10.2.3 Noise

Broadly speaking, any signal element mixed in the output that causes differences from the expected signal is called "noise." The term is a relic of audio technology. It is also sometimes called "distortion" when it is generated as some geometric image-form changes to the picture or some waveform changes of the video signal at the point of observation. From this point of view, all deterioration factors in an image are noises. Usually, a narrower definition is used — namely, as the fluctuation

Evaluation of Image Quality

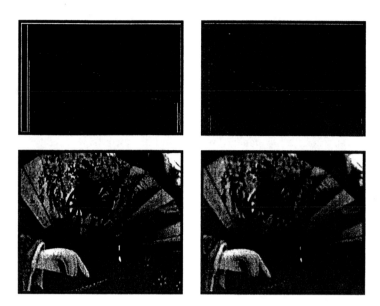

FIGURE 10.3 (See Color Figure 10.3 following page 178.) Difference of frequency response.

FIGURE 10.4 (See Color Figure 10.4 following page 178.) Normal (left) and noisy (right) images.

pattern (i.e., alternating signal element) overlapped with the output image when a flat pattern (i.e., input alternating signal element = 0) is input as a subject. The main sources of the noise are those generated during the photoelectric conversion process in the imaging device (such as shot noise, dark current, etc.) and those added or mixed in the signal processing, including interference of the circuit (such as amplifier noise, power supply switching noise, etc.).

Of course, less noise is preferable. Sometimes, a noise effect, such as simulated silver halide film grain, is added to give a realistic effect to a photograph. Nevertheless, it is always better for the camera to have less noise and allow such effects to be added by image processing. Figure 10.4 shows an example of the difference between normal and noisy images taken by the same camera under the normal and the highest gain.

To evaluate noise quantitatively, the deviation distribution of the recorded image signal that is a shot of a uniform white chart as a predetermined reference signal level (with the imaging lens desirably out-focused for the chart) is used as an indication of the noise. When the reference level is zero — that is, the imaging system is completely covered, the noise is called "dark noise." Standard deviation or a maximum deviation (peak-to-peak value) is generally used to indicate the magnitude of the noise. In the case of a standard television signal, the evaluation weighted with the spatial frequency distribution of luminous sensitivity is adopted as a standard way because the standard observation distance defined relative to the size of the screen is assumed. On the other hand, in the case of a digital still camera, weighting is not used generally because partial magnification (enlargement) is commonly supposed.

Signal-to-noise (S/N) ratio, which is the ratio of the noise level (N) calculated with respect to the reference signal level (S), is to be defined as an evaluation value. A logarithmic expression (decibel units) is usually used. ISO 15739 defines a more practical method of noise measurement for DSCs.

10.2.4 Gradation (Tone Curve, Gamma Characteristic)

Gradation is an item for estimating the curve of the input–output characteristic from the low-luminance part of the subject to the high. The curve is called a gradation curve or tone curve. A digital camera has a tone curve of the output digital signal to the input subject luminance and output devices such as a printer, a CRT display, an LCD, or a projector; each has a tone curve of the output luminance to the input signal. (Note that the reflectance of a print is equivalent to the luminance under the flat illumination.) Thus, apparently, total gradation (i.e., the curve of the displayed luminance to the subject luminance) is described by the total tone curve derived from the curves of the camera and the output device.

The tone curve of some of those image devices is sometimes approached by exponential approximation with the formula: $y = k \times x^\gamma$, where x is the input, y is the output, k is the linear constant, and γ (gamma) is the exponential index. Thus, gradation is sometimes called a "gamma characteristic."

Figure 10.5 shows an example of image enhancement by adjusting the tone curve. The tone curve attached to each picture shows the gradation characteristics with reference to the left image. (Note that the curves are just relative to one another; thus, the left linear graph does not mean linearity to the luminance of the input subject.) The quality of the picture is apparently changed as the tone curve changes. In this case, the picture becomes glittering (i.e., the contrast is enhanced), though it is difficult to compare the superiority of those.

The typical characteristic to be a reference for the total gradation is a linear characteristic. It is a condition of an ideal image reproduction that the intensity of the light output from the display (i.e., luminance of the display surface) is in proportion to the luminance of the original subject. However, the reproduction range would be quite insufficient if perfect linearity were required as a total gradation characteristic because the dynamic range of the display is extremely limited compared to the width of the luminance range of the subject. Therefore, it is common

Evaluation of Image Quality

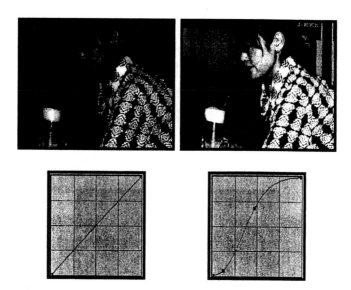

FIGURE 10.5 (See **Color Figure 10.5** following page 178.) Original (left) and tone-enhanced (right) images.

that the gradation characteristic is controlled properly with an actual image system (including the silver halide film system) and that the gradation is compressed at least in the high-luminance and low-luminance parts.

Moreover, this kind of limitation of the dynamic range occurs for the camera recording stage (in other words, signal interface between a camera and an output device) as well. Therefore, to make use of the dynamic range of the recording system effectively, a standard video signal is designed for recording with a gradation compression of $\gamma = 0.45$, and the corresponding expansion of $\gamma = 2.2$ is executed in the display system. (The original motivation for this in the early television system was the saving of hardware, which depends on the CRT's gamma value assumed to 2.2. Note that 0.45 is decided as 1/2.2 for total linearity.) Even when the total input–output characteristics are linear, the system contains a mechanism for such gradation compression recording-expansion regeneration.

As a reflection of the preceding circumstances on the evaluation of the gradation characteristic of the digital still camera, it should be taken into consideration with the next two viewpoints. The first concerns whether an input subject signal is processed to the gamma curve of the standard video signal ($\gamma = 0.45$), basically; the second is whether total tone reproduction is adequate, taking into account the presumed display (e.g., the printer or the video display). In other words, how the tone curve of a camera is shifted from the basic standard gamma curve to produce the best visual reproduction. There is no answer for what curve shape is commonly adequate; thus, it is one of the most important parameters of the camera design specification in terms of the output visual reproduction goals.

To measure a tone curve, a chart called a "gray-scale" chart, as shown in Figure 10.6(a), is commonly used. This chart has colorless patches (chips) with fixed reflectance (or transmittance). An image is taken of this chart, and the output level

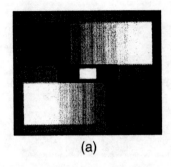

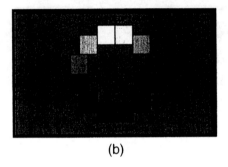

FIGURE 10.6 Test charts for measuring gradation: (a) common gray-scale chart; (b) ISO 14524 OECF chart.

for each chip is measured and plotted against the corresponding luminance value to give the tone curve. In addition, ISO 14524 defines the measurement method of the tone curve called OECF (optoelectronic conversion function) using an improved chart, which can avoid the interference of the lens shading shown in Figure 10.6(b).

10.2.5 Dynamic Range

The dynamic range (D-range) is defined as the ratio of the largest detectable signal to the smallest — that is, the highest luminance to the smallest. It is usually expressed as a logarithmic quantity (in decibels or EV). Especially in the case of digital still cameras, the concern is with input dynamic range (the width of the subject luminance range to be captured).

Consider the operating signal range divided into two parts: above and below an arbitrary boundary reference signal level (which the reference white level is usually selected as). The signal range above the boundary (high luminance) depends on the maximum signal level (saturation level). The signal range below the boundary (low luminance) depends on the noise (S/N ratio). The dynamic range is the logarithmic sum of these two ranges.

The input saturation level can be found as the subject luminance value of the point at which the output value reaches a maximum and stops changing (the saturation point) — that is, the point above which the gradation curve flattens off. Thus, the range above the boundary can be calculated as the ratio between the input saturation level and the subject luminance corresponding to the reference signal level.

Evaluation of Image Quality

FIGURE 10.7 (See Color Figure 10.7 following page 178.) Images with enough (left) and insufficient (right) D-range.

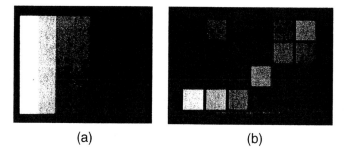

FIGURE 10.8 (See Color Figure 10.8 following page 178.) Test chart for measuring color reproduction: (a) color bar chart; (b) Macbeth's color chart.

Figure 10.7 shows an example of the effect of dynamic range difference. The right-hand picture taken by a camera with quite narrow range is clearly lower quality. Each deterioration phenomenon in the picture is mentioned later in this chapter (see Section 10.3.1.6).

Note that, in the field of photography, the term "latitude," which means the degree of an allowance of the exposure error, may sometimes be used as a counter-word for the meaning of the dynamic range. Quantitative measurement method of dynamic range is described in IEC61146-1 and ISO14524. In addition, ISO speed latitude similar to dynamic range is described in ISO12232.

10.2.6 COLOR REPRODUCTION

Needless to say, quality of color reproduction is determined by whether the color of the subject is reproduced faithfully and is conveyed to the observer properly. Quantitative evaluation of color reproduction uses a color chart like those shown in Figure 10.8, with specified color patches (color chips). A picture of the color chart is taken using the camera under test, and the picture is converted to give chromaticity values using the appropriate transforming formula. The differences can then be measured between the chromaticity values given by the camera and the expected values for each color. For evaluation of total color reproduction, using a display or a printed image, the chromaticity of the output light as represented on the display is measured and compared with the expected values.

On the other hand, various factors are involved in the "color" that a human being actually feels, e.g., the dependence on the brightness; the memory color; the influence of the lighting (background) color; the difference of the feeling between reflection and transmission; and, of course, the limited dynamic range of the system. Furthermore, though present chromatics and colorimetry are constructed based on the isochromatic theory by tristimulus value (three colors separation theory), actual human chromatic vision has some spectral dependency unmanageable with three colors.

For this reason, the smallness of the measured difference between the chromaticity of the image and the expected value does not simply indicate faithful color reproduction. Thus, visual observation using the color chart or an actual subject is commonly used as a final evaluation. With current techniques, therefore, the choice of color reproduction settings — that is, what color to display on the output system — is left to the designer's or evaluator's subjective sensitivity.

Nevertheless, numerical evaluation of color reproduction is still useful for an image system or the digital still camera. Quantitative estimates of various important properties of the image can be done on the chromaticity, for example; strength (saturation) of the color; width of the color reproduction range; the ability to distinguish delicate color differences; stability of color reproduction; the characteristic dispersion between the camera bodies, etc. Most importantly, recording the color characteristics of reproduction system quantitatively enables reproduction of the color decided by the designer's choice correctly. Color evaluation should be performed with a clear and limited purpose in mind, understanding the limitations of the measurements and their meaning.

10.2.7 Uniformity (Unevenness, Shading)

Uniformity is not an inherent evaluation item. Rather, it describes how the various properties of the image vary across the extent of image. Properly, "uniformity" can apply to any characteristic, such as resolution, that can vary across the image. However, the properties usually considered are luminance uniformity (luminance unevenness) and color uniformity (color unevenness). Additionally, it is usual to consider the converse — that is, a measure of nonuniformity, usually called "unevenness." In considering luminance unevenness, the term "shading" is sometimes used.

When a uniformly illuminated, flat pattern is recorded, the recorded value or the output value should be uniform without any variation across the whole image. However, with real cameras various factors are working against the output uniformity of the luminance or color — for example, the peripheral light intensity of the lens and irregularity in the sensor's responsivity to light. Usually, the unevenness is taken to be the peak–peak range of the luminance level of the recorded signal over the full image of a picture of a uniformly illuminated white chart. (Actually, it is rather difficult to keep the illumination flat for a reflected chart, so a transparent type chart for which it is comparatively easy to control the evenness of illumination is often used.)

For color unevenness, a simple method measures each peak-to-peak value of the two color-difference signals (R-Y and B-Y) over the full screen. A more precise measure would be to use the maximum color-difference, which is the maximum distance of the pixel data distribution on the color-differential plane. Other image properties can display nonuniformity, too. However, a property of which measured value has its own significance, such as resolution (that can be different (usually higher) at the center of the image than in the surrounding region), is not usually regarded as an evaluation subject.

10.3 DETAILED ITEMS OR FACTORS

Various factors and phenomena that affect final image quality will be explained next. Only some of these items have standard methods of evaluation and measurement. These have been selected because it is important to understand what can adversely affect the achievement of high image quality for digital cameras.

10.3.1 IMAGE SENSOR-RELATED MATTERS

10.3.1.1 Aliasing (Moiré)

Aliasing occurs when an image is acquired by discrete sampling and the sampling frequency of the imaging system is less than a spatial frequency component of the subject. A pattern of stripes or bands (called "moiré" from a term used by French weavers) is seen, such as might be seen when two fabric weaves are viewed one behind the other. The effect can arise in the image sensor and further down the image-processing chain, when, for example, the pixel resolution (the number of pixels) is reduced by using simple resampling. However, aliasing is most significant when it arises with the optoelectronic conversion at the image sensor. The reasons are as follow:

- The moiré arising in the post-capture image processing can be avoided by using adequate processing. On the other hand, that which arises at the sampling in the image sensor can never be removed from the image.
- For commonly used single color image sensors with a color mosaic filter, aliasing appears as color moiré because the color sampling frequency is lower than the luminance sampling frequency. It is highly visible, so it has a major detrimental effect on image quality.

Therefore, the quality criterion checked is usually color moiré, which tends to occur if a high-frequency component is in the subject; cyclicity is not necessary. Color moiré may also appear as a colored edge, which is the edge of the black-and-white subject pattern fringed with false color. In other words, color moiré is a typically observed as false coloring of edges.

Evaluation of color moiré is basically done by checking the false color generated when taking a high-frequency black-and-white subject. The most common chart is the CZP (circular zone-plate) chart whose pattern is a two-dimensional frequency sweep. This chart is an essential tool for designing and estimating the optical low-pass filter (OLPF), which is the optical device to avoid moiré creation by reducing the high-frequency components by blurring. The chart is used for measuring the trap

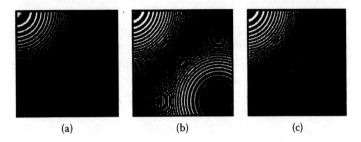

(a) (b) (c)

FIGURE 10.9 (See Color Figure 10.9 following page 178.) CZP pattern and examples of generated color moiré: (a) CZP pattern; (b) taken without OLPF; (c) taken with OLPF.

frequency of the OLPF, which is the spatial frequencies at which aliasing can be erased or suppressed enough.

Figure 10.9 shows simulated pictures of the measurement of color moiré. Figure 10.9(a) is the CZP pattern (only one-fourth area with the center of the CZP located at the upper left corner); Figure 10.9(b) is a picture taken by an image sensor equipped with an RGB Bayer color filter array without an OLPF, that is, the original imaging characteristic of the sensor. Quite an obtrusive circular color pattern is generated around the color sampling frequency point. Figure 10.9(c) shows that OLPF suppresses color moiré by trapping at the sampling frequency and decreasing the amplitude of the optical input image at the boundary frequency.

No method has been established to evaluate the effect on total image quality of color moiré that is still remaining in the output of a DSC that has an OLPF. The reasons are as follow:

- If the CZP chart is used, the occurrence of color moiré and its rough quantity can be detected. However, in case of residual moiré, the level is generally so low that the measured result is unstable under changes in the color or level due to variation in marginal camera angle because the sampling phase is shifted.
- When different imaging systems are compared, the color and the pattern of moiré are different. Also, the moiré is affected by small changes in the color such as saturation and hue; thus, it is not clear how the influence of the resulting color moiré components on the image quality should be compared.

Figure 10.10 shows an example with actual pictures. The color level of the pictures has been enhanced so that some color moiré can be observed in the right-hand image. In contrast, no color moiré is in the left image because it is taken out of focus and has no high-frequency component.

It is still possible that a camera, which would suppress color moiré enough under most conditions, could in some tests generate extremely distinct color moiré with certain test subject patterns. It is important, therefore, to check the performance of the camera with a photographic test after a certain level of evaluation test using a CZP or an edge pattern picture in the laboratory.

Evaluation of Image Quality

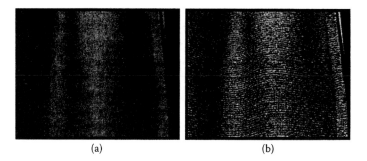

(a)   (b)

FIGURE 10.10 (See Color Figure 10.10 following page 178.) An example of color moiré in an actual picture: (a) moiré is not observed (out of focus); (b) color moiré generated (in focus).

10.3.1.2 Image Lag, Sticking

This is the phenomenon of a past image persisting in the sensor output. It is called image lag when the persistence period is short, and sticking when it is long or permanent. Though it was a significant item for evaluation in the days of imaging tubes, solid-state image sensors usually do not exhibit persistent images. However, an image lag sometimes appears even in a solid-state sensor if the transferring efficiency or reset of the pixel electric charge is not enough. A measurement method is basically the same as the one described in Section 6.3.8 in Chapter 6. The DSC under test is used, instead of an evaluation board for an image sensor.

10.3.1.3 Dark Current

An image sensor is a semiconductor device, so charge leaks due to thermal effects, even when no light falls on the sensor. This is called dark current. Spatially uniform dark current can be cancelled by subtracting the output from optical black pixels, i.e., pixels that are shielded from light. However, dark current varies from pixel to pixel and thus adds noise to the image. Moreover, the charge-accumulating parts of the charge transfer channels also generate dark current. This can be a source of noise if the time necessary to carry charge from pixels through the charge transfer channels is not constant, such as might happen if the charge transfer were suspended and the charges kept waiting.

The noise level is proportional to: (1) the accumulation time for the noise originating the pixels; and (2) the waiting time for the noise originating the transfer channels. Therefore, it is evaluated as the level of the noise in a still picture taken with dark input and the settings of accumulation time and waiting time, the same as in the actual use. Dark current is strongly dependent on the temperature, so this dependency should also be taken into account.

10.3.1.4 Pixel Defect

By intrusion of dust in the manufacturing process or the influence of cosmic radiation after manufacturing of the image sensor, some pixels have functional defects such as excessively large dark current, signal readout trouble, or excessively low pixel sensitivity, etc. Fixed white or fixed black signals are output from these pixels, so they are called "pixel defects."

Pixel defects are usually replaced with a neighboring pixel's data to prevent appearance of the pixel defect in the camera output. However, too many defects may affect the image quality even if corrected and may also make it impossible for the camera system to correct. Therefore, it is necessary to perform evaluation of pixel defects prior to installing the image sensor to a DSC. The method of evaluating the pixel defects is described in Section 6.3.9 in Chapter 6. The acceptability of the sensor is judged with consideration for the number of defects and the distribution (continuous defects particularly hinder the correction). It is common to control the pixel defect rate of the image sensor under specification acceptance criterion for purchase from the sensor manufacturer.

10.3.1.5 Blooming/Smear

These are abnormal signals caused by strong incident light into the imaging surface of the image sensor. Blooming occurs when the charge in a pixel saturates and leaks out to neighboring pixels. In CCD image sensors, smear occurs when long wavelength light penetrates deep into a silicon bulk. The generated charge leaks out into the transfer channel and is added to the pixel charges in transit on the transfer channels. Thus, it generates a characteristic vertical streak extended up and down from the bright spot. This white streak is commonly called smear.

Strictly speaking, overflowing charge from a saturated pixel often leaks into the transfer channel at the same time, so the white streak includes a blooming component. On the other hand, when the whole imaging surface is uniformly exposed high illumination and a high-speed electronic shutter operation is applied, smearing occurs over the whole image area almost uniformly and sometimes causes a lens flare-like effect (foggy image).

Figure 10.11 shows examples of smearing images taken by a camera without a mechanical shutter. The left part of the figure shows that a slight smear streak is extended from the flare of a cigarette lighter. The right part is a shot of the sun, showing a strong smear streak extended from the sun. In both images a bright area, which seems to be blooming, is spread around each luminescent spot.

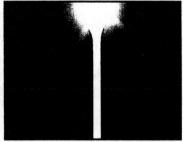

FIGURE 10.11 (See Color Figure 10.11 following page 178.) Examples of smear.

Evaluation of Image Quality

Detailed mechanisms of blooming and smear in CCD image sensors are described in Section 4.2.4.2 in Chapter 4. Also, the evaluation of blooming and smear is presented in Section 6.3.6 in Chapter 6. In the case of the DSC, a use of a mechanical shutter and an image sensor driving method specialized to a still image capture can suppress blooming/smear. However, this does not eliminate it completely. Thus, evaluation suitable for the still image capture should be performed.

10.3.1.6 White Clipping/Lack of Color/Toneless Black

White clipping, lack of color, and toneless black areas are observed partly because the image sensor does not have enough dynamic range and partly because the operating point of the image sensor is not well controlled.

White clipping occurs when an image sensor records a bright area of which luminance level is higher than the saturation level of the sensor. The saturated area is recorded as having the saturation level signal, resulting in a completely toneless white level area. Lack of color occurs when (using a color camera), just one or two color channels of the output color signals from the image sensor saturate. As a result, the color balance is severely degraded at the corresponding part of the subject area. This phenomenon affects image quality significantly, so suppressing or cutting the color signal is taken as a common countermeasure. The resultant image has some colorless areas, though they still maintain the tone as a luminance signal. In effect, some highest luminance parts of the subject become a monochrome photograph.

The toneless black phenomenon is quite similar to toneless white in that it is a loss of detail at the limits of the sensor's dynamic range, though the toneless black part does not always lose the gradation information completely. If the low-level gradation reproduction of the playback system is so poor that the slight gradation of the dark part is not recognized visually, the part is observed as toneless black. Thus, this phenomenon is closely related to the gradation characteristics of the playback system, and it tends to occur more often on active display systems susceptible to surrounding light (CRT, transparent type LCD, etc.) than on passive (surrounding light reflective) display systems (hard copy, reflective type LCD, etc.).

Thus, the primary cause of toneless black is the insufficient gradation characteristics of the image sensor and the playback system. In an actual camera, however, when the image is not noisy, the gradation characteristics can be set arbitrarily, allowing for emphasizing the gradation characteristics at the dark area to produce images with distinct tone. When the image is noisy, emphasizing the gradation characteristics at the dark area also emphasizes the noise, causing the image quality to deteriorate; a trade-off must be made between image tone and noise.

These phenomena are usually evaluated by recording actual images from the camera and a tone curve. The right-hand picture of Figure 10.7 shows the effect on the images of insufficient dynamic range. Both white clipping and toneless black can be observed. Other examples of these phenomena are shown in Figure 10.12. Even when a camera has normal dynamic range, these phenomena are often observed, depending on the dynamic range of the subject and the exposure. Figure 10.12(a) shows white clipping and Figure 10.12(c) shows toneless black. Figure 10.12(b) is the picture taken by an imaging system with especially wide dynamic range.

FIGURE 10.12 (See Color **Figure 10.12 following page 178.**) Example pictures: (a) with white clipping; (b) wide D-range; (c) with toneless black.

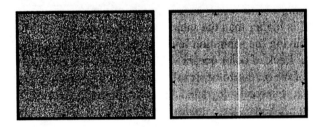

FIGURE 10.13 Examples of spatial random (left) and fixed pattern (right) noise.

10.3.1.7 Spatial Random Noise/Fixed Pattern Noise

The spatial distribution of noise consists of two elements: random noise, which is spread over the entire image without any regularity, and fixed pattern noise, which is noise with a certain pattern. Such fixed pattern noise as "white streak smear" and "flying-in noise" due to the driving signal to the image sensor, which have obvious linear patterns, are extremely noticeable in images. They are easier to detect than the random noise with the same peak level. Random noise, however, is less obtrusive than fixed pattern noise and tolerable to a certain degree, though, of course, excessive random noise also compromises image quality. The examples of these two types of noise are shown in Figure 10.13.

"Good" random noise — specifically, grainy noise — whose appearance is similar to that of silver halide film grain, can even be welcomed because it makes the image more "natural" when it is not excessive. Note that "bad" random noise, which is highly obstructive, also exists. The difference between good and bad random noise seems to be related to the frequency band of the noise along with other factors. Also, how random noise is felt is highly dependent on each individual user. This confuses the argument about it because evaluating image quality cannot be separated from discussing the preferences of the evaluator.

10.3.1.8 Thermal Noise/Shot Noise

An electrical amplifier generates a certain level of noise that is a function of temperature. Thus, noise is added to the image signal at the amplifier in the image sensor and in the preprocessing circuit receiving the output of the sensor. It is called thermal

Evaluation of Image Quality

noise. Thermal noise can be reduced by cooling, though it cannot be completely removed.

Quantum physics treats light (electromagnetic waves) and electric charge carriers (electrons or holes) as quanta, which display the properties of waves and of particles. The trend towards increasing the number of pixels and ever smaller pixel arrays inevitably means smaller and smaller photo pixels, decreasing the number of storage electrons generated by the photoelectric transformation down to several thousand or fewer per pixel — even at the saturation level, where quantum effects cannot be ignored. Consequently, the signal level of those image sensors fluctuates as a direct manifestation of the uncertainty of quanta.

This means that the output signal fluctuates randomly in time and in space even when the subject image is uniform (note: the concept of "uniform input" is hypothetical because the amount of photons coming from the subject fluctuates). The level of shot noise is known to be proportional to the square root of the amount of electrons (therefore to the output signal level). Accordingly, the ratio between the signal and the levels of shot noise increases proportionally to the square root of the signal level. Because each of these two noise sources arises from fundamental physics, they cannot be avoided but they can be estimated by theoretical calculation (see Section 3.3.3 in Chapter 3).

10.3.2 Lens-Related Factors

10.3.2.1 Flare, Ghost Image

Flare and ghost images are image disturbances caused by stray light rays falling into the imaging surface. They are due to inner reflections occurring at the surface of each single lens and also in the lens barrel. Flare refers to the state of image, where the entire image looks bleached and dull. This is caused by stray rays added over the entire image as low-frequency components. Ghost, on the other hand, is caused by the stray rays having a certain pattern added to the image. Stray rays can have a pattern when they come from the high-luminance parts of the subject, or when they are concentrated on particular areas after being reflected by some inner surfaces. Ghost images can also be interpreted as a kind of spatial fixed pattern noise.

Neither flare nor ghosting can be evaluated in a specific method because both phenomena manifest in various ways according to the specific arrangement of the camera, subject, and light source. An example evaluation method is to monitor how the image changes when a spot light source (e.g., a miniature light bulb) is placed just out of or within the subject area. When this test is performed, testing with the light source out of the subject area is more important than testing with the light source in the subject area. (See Figure 10.14b, which is the example of the former; Figure 10.14a is an example of the latter.) This is because when the cause of the disturbance is in the subject area and therefore noticeable for those who look at the image, they can see the cause of the problem and accept it. However, when the cause of disturbance is out of the subject area, they cannot see the cause and it would be difficult for them to accept the problem.

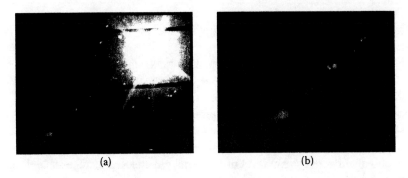

FIGURE 10.14 (See Color Figure 10.14 following page 178.) Ghost images with (a) a light source in the subject area and (b) a light source out of the subject area.

10.3.2.2 Geometric Distortion

Ideally, an image created by a lens system should be proportional to the subject in all dimensions. However, actual images have geometric distortion that changes the relative dimensions in the image. The two most common types of geometric distortion are barrel distortion and pincushion distortion. In a fixed focal length lens, the distortion is fixed; however, in a zoom lens system, the distortion phenomena manifest differently through the zoom range. Geometric distortion can be measured quantitatively (see Section 8.3.3 in Chapter 8). For general photography, it is desirable that geometric distortion be as small as possible, though distortion can be exploited as the special effect in artistic photography.

10.3.2.3 Chromatic Aberration

Strictly speaking, the light coming from one point on the subject does not focus on one point on the image because the refractive index of the lens differs according to the wavelength (dispersion of the refractive index). In a photographic lens system, chromatic aberration is corrected by using multiple lenses. Though the aberration can be significantly reduced, it cannot be completely eliminated.

Chromatic aberration manifests with two symptoms: tint (false color) on black-and-white edges in the image (due to positional shift of color) and deterioration of lens resolution. Tint on black-and-white edges can also be caused by aliasing. Tint caused by aliasing, however, is nearly uniform over the entire image because aliasing is related to the pixel pattern of the image sensor. On the other hand, chromatic aberration, which causes tint mainly on edges, is called "chromatic difference of magnification." Tint due to chromatic aberration is caused because the imaging magnification differs according to the color. Accordingly, especially at the lens edge, displacement occurs roughly proportional to the distance from the center, resulting in a particular type of tint. Therefore, tint due to aliasing and chromatic aberration can be distinguished clearly as follows:

- If the color tint occurs at the lens center and the color changes significantly with the position and direction of a black-and-white edge, the tint is due to aliasing.
- If tint is only eminent at the lens edge, the color is similar when the angle against the lens center is similar, and tint does not occur when the extension line of the edge penetrates the lens center, the tint is due to chromatic aberration.

10.3.2.4 Depth of Field

A recorded image is necessarily a flat image, that is, the stimulus intensity is distributed on a two-dimensional space. On the other hand, the subject is an object distributing in a three-dimensional space. Accordingly, the effect of an imaging lens system is to do a three-dimensional to two-dimensional conversion from the subject to the image. When this conversion is performed, the points on the subject that are at the exact focusing distance from the lens will be recorded with the best resolution; points not at the focusing distance will be blurred, and the further the distance, the more significant the blurring will be. "Depth of field" is defined as the range of distance within which the blurring is too small to be noticed.

Depth of field can usually be geometrically calculated using optical parameters, and the calculation results are accurate enough for practical use. It is well known that if the image frame size on the image sensor is the same, depth of field is deeper when the field angle and F-stop number are larger. (Approximately, depth of field is proportional to F-stop number [aperture] and inversely proportional to f^2, where f is the focal length; see Section 2.2.2 in Chapter 2.) Figure 10.15 shows the changes of the depth of field. The effects of field angle (focal length) and aperture (F-stop number) are demonstrated in (a), (b), and (c), (d), respectively.

On the other hand, if the field angle and F-stop number are the same, depth of field is inversely proportional to the image sensor frame size. Accordingly, general digital cameras, whose imaging frame is small due to the demand on the image sensorís yield, have deeper depth of field than silver halide cameras.

If image quality is viewed as accurately reproducing visual information, a greater depth of field is desirable because an imaging system with greater depth of field can record a subject with significant depth without compromising resolution much. In actual photography, however, the main subject is often emphasized by defocusing other subjects that are not at the focusing distance. In such cases, depth of field is controlled by changing the focal length or iris to bring the main subject into focus while defocusing other objects behind or in front of the main subject. From this standpoint, a necessary improvement for digital cameras is to make depth of field widely controllable. Though depth of field significantly affects the total quality of the recorded images, it is not usually measured because it is almost completely determined by optical parameters.

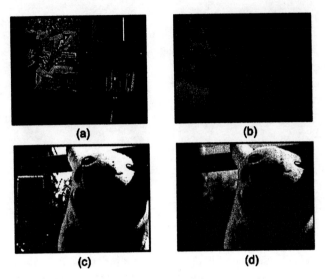

FIGURE 10.15 (See Color **Figure 10.15** following page 178.) Changes of depth of field: (a) wide angle (short focal length); (b) narrow angle (long focal length); (c) narrow aperture; (d) open aperture.

FIGURE 10.16 (See Color **Figure 10.16** following page 178.) Difference of perspective taken with (a) a narrow-angle lens and (b) a wide-angle lens.

10.3.2.5 Perspective

Perspective is one of the image characteristics related to the depth of the subject. It is a well-known fact in photography that a telephoto lens suppresses perspective, whereas a wide-angle lens emphasizes perspective, as is shown in Figure 10.16.

Essentially, perspective is the magnification change of the subject due to the distance. In other words, perspective is emphasized when the imaging magnification of the subject significantly differs as the distance changes in the depth direction, whereas perspective is suppressed when the magnification change is small. To be more specific, perspective (sense of depth) observed on a photograph is not due to the distance between the subject and camera, but rather to the relative change of the imaging magnification in relation to the relative distance in the depth direction from the main subject. As the subject distance that gives the same magnification to the main subject (the size of the main subject on the photograph), the reference distance when

Evaluation of Image Quality

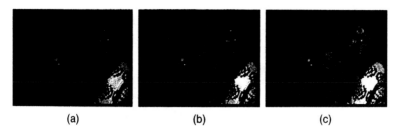

(a)   (b)   (c)

FIGURE 10.17 (See Color Figure 10.17 following page 178.) Difference of quantization bit number: (a) 8-b/color (full color); (b) 4-b/color; (c) 3-b/color.

considering the magnification change differs due to the focal length of the lens. Thus, it changes the rate of magnification change against the distance in depth direction. This is what the perspective effect of lens essentially is. Perspective is determined unambiguously only by the field angle; thus, it is not usually measured, though it may be considered when judging the total image quality of the recorded image.

10.3.3 SIGNAL PROCESSING-RELATED FACTORS

10.3.3.1 Quantization Noise

A digital camera records the luminance level of the subject after converting the luminance value to a digital value. During this analog-to-digital (A/D) conversion, the analog luminance levels, which can have any value within the range of the analog input, are assigned discreet digital bit values according to a set of predetermined threshold values. The A/D conversion inherently introduces some bit error or quantization error because the original values have been changed to one of the discrete digital values (the error is the difference between the input signal and the threshold to which it is changed). When interpreted as noise accumulated on the signal, such differences (namely, quantization errors) are called quantization noise. When the bit number of the digital recording is small, quantization noise significantly affects image quality, but when the bit number is large, it can be ignored.

Exif, which is currently used as the standard recording file format, employs an 8-b nonlinear quantization (the standard gamma: $\gamma = 0.45$) and is widely recognized to have enough bits for general purposes. However, 8-b resolution is not good enough when postprocessing is used to correct under- or overexposure problems with a recorded image. For example, when the recorded image has insufficient exposure by –2 EV (means 1/4), the effective quantization would be 6-b.

Quantization error becomes conspicuous at this low number of bits. Though the level of tolerance naturally differs depending on the purpose of the photograph, generally speaking, when the exposure deviation exceeds this level, the image quality deteriorates significantly, making the image unusable for actual use. Figure 10.17 shows quantization noise according to quantization bit number.

Another recently identified problem is one that arises when employing a low-saturation recording to secure reproduction range for colors, especially for such professional purposes as printing. In this case, quantization noise can have a considerable effect on the recorded image. It must be noted, however, that 8-b is quite

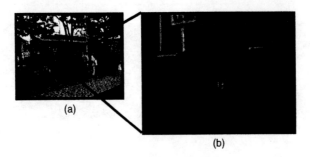

FIGURE 10.18 (See Color Figure 10.18 following page 178.) An example of compression noise (block distortion): (a) highly compressed picture; (b) expansion of a part.

appropriate currently because it is the standard specification for personal computers, which are the infrastructure for digital cameras. In the case in which quantum noise becomes a problem under current circumstances, the image quality deterioration caused by the image sensor system is usually more apparent, making quantization noise ignorable. Because the gamma processing is digitally applied in most digital cameras, linear 10- to 12-b A/D conversion for an image sensor output is widely employed in order to suppress the quantization error within the 8-b level after the gamma treatment.

10.3.3.2 Compression Noise

Exif, the standard digital camera file format, employs the JPEG compression, which is an irreversible (lossy) compression, in which a part of the subject information is abandoned to achieve a high compression rate. Raising the compression rate in JPEG compression increases the noise called compression noise because more of the original subject information will be lost and cannot be recovered when the image is restored from the compressed file. Because the compression rate can be determined relative to the information size of the subject, it is possible to suppress the compression noise below the noticeable level by optimizing the file size according to the information size of the subject. However, in actual digital cameras, if the image file size is changeable, it often causes problems in managing the system. Accordingly, the upper value of the file size is usually limited, thus making the compression noise explicit when recording a subject with a large amount of information.

Block distortion is caused by the fact that the JPEG compression processes the signal by an $8 \times 8 = 64$ pixel block unit. Luminance does not change smoothly at the border between two blocks, making the borderline conspicuous, as shown in Figure 10.18. This block distortion is especially prominent at the plain areas of the image when the image has plain and complicated patterns. Mosquito noise is a hazy and crispy noise pattern that arises at the edges of fine patterns in the recorded image. Block and mosquito noise are difficult to evaluate quantitatively because they are image-dependent noise. They are usually qualitatively evaluated from actual pictures.

10.3.3.3 Power Line Noise, Clock Noise

Power line noise and clock noise are so-called "flying-in noise." The switching regulator (DC/DC converter) circuit in the power supply and the timing pulse (clock) signals that activate the image sensor influence the analog part of the camera before the A/D conversion. These types of circuit activities can produce noisy bright dots or bright lines in the recorded image. One method of evaluating this kind of noise is to photograph a black subject with the highest imaging sensitivity setting, then replay the recorded image to see if the noise is visible and measure the peak level. This kind of noise is generated within the electronic circuitry, so the specific nature of the noise can vary with the environmental temperature and the self-generated heat from the camera.

10.3.4 System Control-Related Factors

10.3.4.1 Focusing Error

Even when the lens has ideal characteristics, if the focus is not right, the proper image quality cannot be achieved. If the subject is out of focus, the image will be blurred, and the degree of blurring increases as the deviation from the focus point increases. In other words, lens response will be lost from the high-frequency end, resulting in significant loss of resolution and sharpness. On the other hand, as a result of the decreasing resolution, aliasing decreases.

The previously stated phenomena occur at the main subject in the recorded image. However, in case the subject has significant depth, focusing error can unintentionally intensify lens response for such unimportant subjects as the background, de-emphasizing the main subject. Note that when relatively significant focusing error arises (when the subject is significantly far from the focus or when the subject has significant depth), the artifacts or such incongruous images as being blurred like double-lines, according to the lens characteristics and the subject pattern, can be observed.

10.3.4.2 Exposure Error

Although exposure error directly causes signal level error in the image, if the error is small, postprocessing with image-processing software and other methods can correct the error. On the other hand, if the exposure error is significant, satisfactory image quality cannot be achieved because the signal exceeds the dynamic range, making the lost information irretrievable. Also, exposure error will increase various kinds of noise including quantization noise.

Because the optimal image signal level will vary depending on the intended use of the image, one definition of an optimal exposure would be "the status in which the exposure (and therefore the level of the recorded image) is adjusted to give satisfactory image quality after postprocessing is applied." If the imaging system has sufficient dynamic range and sufficient exposure range, a second definition of optimal exposure would be "the circumstances in which the signal level has the highest probability that the images can be used without postprocessing."

These two definitions can sometimes contradict each other. The second definition has a significant meaning in marketing general-purpose digital cameras and is often prioritized over the first. However, the first definition should be considered as the primary definition because many digital cameras currently do not have sufficiently wide dynamic range capabilities to capture the desired details in fairly high-contrast scenes.

10.3.4.3 White Balance Error

If white balance is lost, colors are not recorded accurately and color reproduction image quality will be compromised. Basic performance of white balance can be measured by photographing an achromatic subject such as a uniform white chart and then evaluating the chromaticity of the recorded image. If the recorded image signal is not white, the white balance of the imaging system has a problem. However, when pictures are actually taken, subjects usually have various colored patterns. Although realistic scenes are more representative of actual camera usage, it is very difficult to make a generalized decision about the patterns and light sources to use for such tests. Currently, no standard evaluation method in this meaning has been established. Thus, the typical methods for evaluating white balance performance are based on qualitative assessments of actual photographs.

10.3.4.4 Influence of Flicker Illumination

Household electric power is alternating current (AC), which results in illuminating lights often having a pulsating component. This component is not noticeable for incandescent lamps because the heat accumulation in the lamps effectively damps the pulses. On the other hand, discharge lamps such as fluorescent lamps have a significant pulsating component that can affect the control system and recorded images.

In fluorescent lamps, cathode rays are emitted from each one of two electrodes alternately every half cycle, synchronous with the power supply cycle. The cathode rays excite ultraviolet rays, which in turn excite the fluorescent material to emit visible light. The emitting time of the light is shorter than the power supply cycle, thus pulsing the quantity and color of light. Accordingly, if the exposure time is short when a high-speed shutter is used, the exposure and color can differ significantly according to the timing of the visible light emission cycle. Typical users are likely to judge this as a problem even though it is an accurate representation of the actual scene at the moment at which the picture was captured.

10.3.4.5 Flash Light Effects (Luminance Nonuniformity/White Clipping/Bicolor Illumination)

Luminance nonuniformity arises particularly due to the nonuniformity in distribution of the strobe light. When using a built-in strobe, which requires saving the limited power supply, it is inevitable to darken the surrounding lighting to the acceptable limit. However, an improper setting of the acceptable limit can result in a photograph that looks as though it had been taken in a tunnel, although in fact it was taken in front of a wall. This phenomenon is also related to the dynamic range of the imaging system; the phenomenon becomes more conspicuous with a gradation characteristic that is likely to cause toneless black even if a strobe with the same light distribution characteristic is used. This means that luminance nonuniformity may not be managed only by the optical characteristic evaluation of the strobe.

When using a strobe, white clipping is more likely to happen than when photographing under the outdoor daylight. This is caused mainly by the depth-direction distribution of the subject. Because the subject illumination is inversely proportional to the square of the distance, the luminance dynamic range of a subject with depth is wider than that of a subject with less depth. The closest elements of the subject are most likely to have white clipping in the recorded image. This tendency becomes very conspicuous for short-distance photography because the distance ratio becomes large. Since digital photographs with white clipping tend to create an unpleasant impression (digital photographs often do not have enough gradation capacity and gradation steps tend to be merged), the white clipping should also be taken into consideration.

The phenomenon of bicolor illumination occurs when using a strobe under artificial lighting. If the strobe light irradiates only one part of the subject, the color at that part will be recorded correctly, but the white balance is lost on the artificially lighted parts, potentially resulting in unusual colors for those parts of the image. This is the so-called bicolor illumination. Like other two phenomena, bicolor illumination can also occur when using a silver halide camera. However, it should be noted that, for digital cameras, it can be corrected by partial image processing.

10.3.5 OTHER FACTORS: HINTS ON TIME AND MOTION

10.3.5.1 Adaptation

When dealing with a subject or an image, luminance adaptation and color adaptation of human visualization may need to be considered. These types of adaptation typically do not need to be considered in any detail for digital still cameras that handle still images. However, it should be noted that examiners of subjective image quality evaluation must be fully adapted to the environmental light before starting the test and that forcing them to gaze at one image for too long makes them insensitive to color contrast because of color adaptation.

10.3.5.2 Camera Shake, Motion Blur

It is well known that if the subject moves during exposure, the resultant image will have a trait of "motion blur." In other words, the image of the subject that moved is not sharp in a particular direction. Motion blur tends to be conspicuous with long exposure times. "Camera shake" occurs when the camera moves. Unlike motion blur, camera shake makes the entire image blur.

Motion blur is not evaluated because it is determined simply by the exposure time. On the other hand, probability of occurrence of the camera shake can differ based on the camera's physical design (e.g., grip or release action) even under the same exposure time. Camera shake is sometimes measured, though the measurement has a different implication from other image quality tests.

When a physical condition of the camera shake is the same, the extent of image blur is proportional to the image magnification. Therefore, a typical test would start with adjusting the lens focal length to a reference (e.g., its longest) position if the focal length is adjustable. The next step would be to photograph a small bright spot (e.g., a uniform black chart with a white spot at its center) with the predetermined exposure time. Typically, the test would need to be repeated multiple times or with multiple testers because the results of individual tests will vary with and between testers.

10.3.5.3 Jerkiness Interference

Digital cameras with movie recording functions or with electronic viewfinders deal also with movie signals. When dealing with movie signals in a camera, if the exposure time per image frame is too short compared to the frame rate (i.e., aperture ratio in time is low), the motion picture looks unnatural because the movement feels awkward or discrete. This is called jerkiness interference. It is conspicuous when a moving subject is recorded with a high-speed electronic shutter. Jerkiness interference is not usually problematic because it usually does not cause serious unpleasant feelings; on the other hand, longer exposure time enhances the motion blur.

10.3.5.4 Temporal Noise

Spatial random noise and fixed pattern noise are described in Section 10.3.1.7. Temporal noise fluctuating over time (see Section 3.3.1 in Chapter 3) is "frozen" as spatial noise when a still shot is taken; it is more or less filtered out by the human eye in video images. Thus, respective evaluation methods are needed for still images and movie images when a DSC can record movie images.

10.4 STANDARDS RELATING TO IMAGE QUALITY

In this section, standards related to image quality evaluation and digital camera evaluation are listed. Some of these standards are referred to in the main text. Image quality evaluation is difficult to perform quantitatively because it is quite subjective in principle. Also, because electronic imaging technologies regarding digital cameras have been developed as television/video technologies, proprietary standards for digital cameras are still limited. Accordingly, standards relating to video imaging technologies should also be referred to. IEC 61146 is a major one of these. ISO 20462, which is not referred to in the main text, explains the general methods for subjective image estimation. JEITA CP-3203 is described in Japanese only; nevertheless, it defines some useful test charts, some of which are referred by IEC61146. Finally, ISO12232, CIPA DC-002, and DC-004 can be helpful because they describe methods for measuring digital camera performance. However, these standards are not directly related to image quality.

Evaluation of Image Quality

ISO Standard

ISO 12232 Photography – Electronic still picture cameras – Determination of exposure index, ISO speed ratings, standarad output sensitivity, and recommended exposure index

ISO 12233 Photography – Electronic still picture cameras – Resolution measurements

ISO 14524 Electronic still picture cameras – Methods for measuring opto-electronic conversion functions (OECFs)

ISO 15739 Photography – Electronic still picture imaging – Noise measurements

ISO 20462-1 Psychophysical experimental method to estimate image quality
   Part 1: Overview of psychophysical elements

ISO 20462-2 Psychophysical experimental method to estimate image quality
   Part 2: Triplet comparison method

ISO 20462-3 Psychophysical experimental method to estimate image quality
   Part 3: Quality ruler method

IEC Standard

IEC61146-1 Video Cameras (PAL/SECAM/NTSC) – Methods of measurement
   Part 1: Non-broadcast single-sensor cameras

IEC61146-2 Video Cameras (PAL/SECAM/NTSC) – Methods of measurement
   Part 2: Two- and three-sensor professional cameras

CIPA Standard

CIPA DC-002 Standard procedure for measuring digital still camera battery consumption
   (http://www.cipa.jp/english/hyoujunka/kikaku/pdf/DC-002_e.pdf)

CIPA DC-003 Resolution measurement methods for digital cameras
   (http://www.cipa.jp/english/hyoujunka/kikaku/pdf/DC-003_e.pdf)

CIPA DC-004 Sensitivity of digital cameras
   (http://www.cipa.jp/english/hyoujunka/kikaku/pdf/DC-004_e.pdf)

JEITA Standard

JEITA CP-3203 Specifications of test-charts for video cameras (Japanese only)

11 Some Thoughts on Future Digital Still Cameras

Eric R. Fossum

CONTENTS

11.1 The Future of DSC Image Sensors 305
    11.1.1 Future High-End Digital SLR Camera Sensors 306
    11.1.2 Future Mainstream Consumer Digital Camera Sensors 308
    11.1.3 A Digital-Film Sensor 309
11.2 Some Future Digital Cameras 311
References 314

Thus far this book has covered the elements of DSCs and their functions. In this chapter, some thoughts on the improvements that might be made so that DSCs better meet customer needs and the new things that DSCs might do to make our future lives better are discussed. First, image sensors are discussed.

11.1 THE FUTURE OF DSC IMAGE SENSORS

Consumer "pull" demands ever improving performance from DSC image sensors. High-end digital SLR-type cameras (DSLRs) are usually designed to use lenses compatible with film SLR-type cameras (FSLRs). This is to encourage consumers to stay with a particular brand when they switch from film use to digital imaging. Consequently, these DSLRs can use sensors as large as 35-mm film (36 × 24 mm). Due to technological limitations having to do with semiconductor manufacturing, somewhat smaller sensor sizes are sometimes used. With such a large, fixed sensor size, pixels tend to be bigger, which enhances their light-gathering power and dynamic range. Pixel counts ranging from 5 to 16 million pixels are currently typical.

On the other hand, cost-sensitive compact cameras demand small physical size of the sensor to keep the sensor cost down and to keep optics commensurately small and inexpensive. Thus, to increase pixel count, it is necessary to shrink pixel size while maintaining a high level of image fidelity. Pixel counts are currently in the 1- to 5-million pixel range.

11.1.1 FUTURE HIGH-END DIGITAL SLR CAMERA SENSORS

The roadmap for DSLR cameras must include increasing the pixel count and (as always) maintaining or improving pixel performance. In addition to increasing image quality, pixel count has become an important marketing tool for differentiating products and for making old products obsolete. Practical consumer use of increasing pixel counts relies on concomitant progress in personal computing systems, including user interfaces, network speed, mass storage devices, displays, and printing engines. Recently, there is some indication that personal computing system throughput may begin to limit the adoption rate of higher resolution cameras.

The need for functional improvements in DSLR sensors is not a strong driver. Generally, the sensor's primary function is to record and output the photon flux incident on the sensor accurately. Some functionality, such as viewfinder mode, is already included in a few DSLR sensors. However, other basic functionality, such as on-chip A/D, is rare because lenses are relatively large, and camera body weight helps with balancing and with keeping the camera steady. Therefore, the lack of a strong volume and weight reduction "pull" means that the camera functionality need not be on-chip, especially if it compromises image quality.

For larger pixel counts, the readout of the sensor becomes increasingly important. The readout time of a sensor is determined by the sensor's pixel count and readout rate. A long readout time can complicate sensor operation because the dark signal will increase, and the opportunity for smear and other aberrations increases. Consumers will not likely accept a very large format sensor in a camera if the readout time becomes excessive because this limits the rapid taking of sequential pictures. Thus, for example, doubling the pixel count requires a doubling of the readout rate. Power dissipation and noise grow with readout rate, decreasing battery life, increasing sensor temperature and dark current, and reducing SNR.

Let us look at a hypothetical future DSLR camera sensor with 2^{26}, or 67 million, pixels with 10- to 12-b digital output. This is about four times the resolution of the highest resolution cameras in the public domain, and about eight times the resolution of current high-end consumer SLRs. In a 35-mm film format, the pixel count might be 10,000 × 6700 pixels, corresponding to a 3.6-μm pixel pitch. (To put this in perspective, if the sensor were as large as a soccer field, a pixel would be the size of a clover leaf.) A 3.6-μm pixel is nearly standard in 2004, so the timely readout of all those pixels, not small pixel size, is the real challenge. Even at a 66-MHz readout rate with ten parallel ports (100 to 120 pins), it will take a full 100 msec to read out a single image. In recent high-speed, digital-output, megapixel sensors, readout power (energy) was approximately 500 to 1000 pJ/pixel corresponding to this hypothetical sensor dissipating several hundred milliwatts during readout.[1] (Note that on-chip analog-to-digital conversion can actually reduce total chip power dissipation). Yield, packaging, optics, image processor engine, and storage memory I/O will be challenges for engineering such a prosumer product. However, we might expect such a product to be realizable within a decade or so.

In addition to increased pixel count, continued improvement in the pixel structure to improve image quality is expected. For almost any camera, the ideal goal is to have every photon that enters the lens add information to the image. No photon should go wasted. Yet, in fact, considering reflection losses; microlens efficiency (for sensors with microlenses); color filter transmission; detector quantum efficiency; and carrier collection efficiency, probably only one of every ten photons that come through the lens results in a collected photoelectron.

Color filter arrays are a large source of loss for photons and a source of spatial color aliasing. By definition, a color filter eliminates most of the photons in a broad spectrum image, and transmission within the passband is typically in the 70 to 80% range. The incomplete spatial sampling of any one color band can lead to color aliasing. The use of multiple sensor arrays and a color-splitting prism is now commonly used in video cameras (so-called 3-CCD cameras) to ameliorate these effects. For SLRs, though, the approach has not been commercially attractive.

One proposed solution is to use the characteristic absorption depth of silicon as a sort of color filter.[2] Because blue light is absorbed close to the surface and little red light is absorbed in the same region, a shallow detector would be more sensitive to blue photons than to red photons. The red photons are not discarded, however. A deeper detector buried vertically under the blue detector could be used to detect the red photons. Few blue photons would penetrate to this depth. Extending this concept, a third, intermediate green detector could also be used. These three vertically integrated detectors deliver three signals that contain enough information to recover the relative blue, green, and red signals at that pixel location. In addition to the more efficient use of photons, spatial color aliasing is also reduced by this technique. However, color reproduction is more complicated using the heavily convoluted absorption characteristics of each component detector, and other drawbacks such as increased dark current and reduced conversion gain limit this approach.

Another possible approach to improving performance is to use a multicolor stacked structure. On the top surface of the chip, three (or more) layers of silicon can be deposited to act as detector layers. By locating the detector(s) on the upper layer of the chip, higher fill factor and less color cross talk can be achieved, in principle. A single layer of a-Si:H has been explored in the past in order to achieve 100% fill factor, but due to lag and instability of the material, the technology has not yet been a commercial success.[3]

The stacked material need not be pure silicon. Materials that are like silicon can be used to vary the energy gap as a function of depth, making the layers more tuned to the colors that they are to detect. Examples include a-Si_xC_{1-x}, Si_xGe_{1-x}, or $GaIn_{x-}As_y$.[4] The main problem with these materials is dark current that is orders of magnitude larger than acceptable for image sensor performance. Material stability and lag will also be challenges for realizing a multicolor stacked structure.

As another avenue for avoiding color filter arrays, diffractive optics has been investigated.[5] In this approach, a micrograting is placed above each "pixel." This grating diverts light according to its spectral content into one of three or more receiving detectors. So far this approach only appears to be feasible for high F-number optical systems in which the light rays are nearly perpendicular to the surface of the sensor.

Color improvement can also be obtained through the use of additional colors besides red, green, and blue in the color filter array. For example, Sony has recently made use of a fourth color. However, sparser spatial sampling of any one color increases color aliasing problems. Randomizing the regular Bayer color filter array pattern can help reduce some color aliasing effects, as can micron-scale dithering of the sensor using a piezoelectric device.

Continued improvement in quantum efficiency (QE) can be expected over the coming years. The most obvious approaches for QE improvement are in materials modification, in which the energy gap is adjusted through strain or the addition of impurities. However, so far the trade-off between QE improvement and the increase in dark current has not been promising. Backside illumination in thinned, full-frame CCDs has been used by the scientific community for years to improve QE. Translation of this approach to interline transfer CCDs, especially those with vertical overflow drain and electronic shutter architectures, is more difficult. Backside illumination of CMOS sensors may be more promising for prosumer applications.

In addition to not wasting photons, improvements in the storage and readout of the photoelectrons can improve sensor performance. Increasing the storage capacity of pixels without increasing the electron-equivalent readout noise can improve the SNR of well-lit, shot-noise-limited images. It also improves the dynamic range of the sensor and allows greater exposure latitude. Reducing readout noise improves SNR in low-light conditions or in darker portions of an image.

11.1.2 FUTURE MAINSTREAM CONSUMER DIGITAL CAMERA SENSORS

Unlike digital SLR cameras, mainstream consumer digital cameras require small sensor size to maintain small optics and an affordable price point. The trend is towards increasing pixel count while maintaining a compact size. Subdiffraction limit (SDL) pixels are increasingly used for these cameras. We define an SDL pixel as a pixel smaller than the diffraction-limited Airy disk diameter of green light (550 nm) at F/2.8, which is 3.7 μm. Small pixels face several fundamental and technological challenges.

One problem with shrinking pixel sizes is the reduced number of photons that the pixel collects in an exposure. For a given scene lighting and lens F-number (both of which are nearly impossible to improve), a lens focuses a certain number of photons per square micron on the image sensor. The smaller the pixel is, the fewer photons are available to be collected. Thus, if a 5×5 μm pixel collects 50,000 photons, a 1×1 μm pixel would only collect 2000 photons. Due to photon shot noise, the signal-to-noise ratio (SNR) scales as the square root of the signal level. In this example, the photon SNR for the 5×5 μm pixel is 224:1 or 47 dB, whereas for the 1×1 μm pixel, the photon SNR is five times worse at only 45:1, or 33 dB. This is one of the penalties for increasing sensor resolution in a DSC with fixed sensor size.

A related problem has to do with the maximum number of photoelectrons a pixel can hold, known as full-well capacity. Because the capacitance of a pixel depends on its cross-sectional structure and the area of the pixel, pixel capacitance, and thus full-well capacity, generally scales linearly with area for a given technology and is of the order of 2000 $e^-/\mu m^2$. The maximum number of photoelectrons that a 1×1 μm pixel can store is approximately 1/25th the number that a 5×5 μm pixel can store. The maximum SNR attainable by a pixel (shot-noise limited) is determined by its full-well capacity, so the maximum SNR attainable by a 1 × 1 μm pixel is 1/5th the SNR of a 5 × 5 μm pixel. If one wishes to maintain a maximum SNR or, say, 40 dB with a 1 × 1 μm pixel, then the full-well capacity must be increased fivefold to 10,000 e$^-$/μm². Because increased resolution often goes with reduced operating voltage (to reduce readout power dissipation), the difficulty of realizing an increased full well becomes more evident because full well is often the product of capacitance and operating voltage. Three-dimensional device effects also become very important at such small lateral dimensions and are usually deleterious to device behavior.

In addition to the maximum SNR, where the noise is determined by shot noise, readout noise contributes to SNR at lower light levels. With smaller signals, a fixed readout noise level can result in rapidly deteriorating SNR because SNR then scales linearly with signal. For example, suppose the readout noise is 20 electrons r.m.s. A 5 × 5 μm pixel receiving 500 photoelectrons will have an SNR of 16, but a 1 × 1 μm pixel will receive 25 times less photoelectrons (i.e., 20) and have an SNR of about only 1. If the readout noise was only five electrons r.m.s., then the 5 × 5 μm pixel would have an SNR of 22 and the 1 × 1 μm pixel would have an SNR of 3, a 300% improvement. However, photography in this low SNR realm is generally not of interest to consumers.

In determining image quality, spatial fidelity is also important. In optics, there is a limit to how well one can focus something, as defined by the diffraction limit (see Chapter 2). For green light (0.55-μm wavelength) the smallest diameter spot that can be focused using F/2.8 optics is 3.7 μm. With a 5 × 5 μm pixel, this is not a problem, but with a 1 × 1 μm pixel, it is clear that edges will not appear sharply focused, even for perfectly focused perfect lenses. However, the use of color filter arrays generally requires the use of an optical antialiasing filter to avoid the introduction of spurious colors from the interpolation process. To a certain degree, diffraction effects will act as a low pass filter. A 1 × 1 μm pixel array with a Bayer pattern has a kernel that spans only 2 × 2 μm. Thus, spatial resolution for longer wavelengths (e.g., red) and at higher F-numbers will still be compromised by diffraction effects.

In addition to these fundamental challenges with SDL pixel sizes, many other technological challenges need to be overcome. These include

- Pixel-to-pixel uniformity, which becomes harder to control for SDL pixel sizes
- Microlens stack height, which needs to be reduced to minimize shading and cross-talk
- Pixel capacitance
- Quantum efficiency in thin active layers
- Dark current at higher doping concentrations and sharper edges
- Other scaling issues

11.1.3 A DIGITAL-FILM SENSOR

The first two discussions of future sensors were a linear extrapolation of current technology to a somewhat extreme level. It is interesting to contemplate alternative methods of imaging to stimulate new thought. Certainly, humans do not see with silicon pixels; their vision comprises biological detectors and neurons that "fire" signals periodically to other neurons. In the past 15 years, some exploration of neural-like imaging has been undertaken. So far, no practical application has been found for retinal-like image sensors that emulate biological imaging. Nevertheless, this avenue is still intriguing.

We can expect that microelectronics feature sizes will continue to shrink, as predicted by Moore's law, at least for a while. Thus, it will be possible to build very small pixels. One is then faced with the question of what one can do with deep-SDL pixels. One possibility is to take advantage of spatial oversampling to reduce aliasing effects and improve signal fidelity. However, analog pixels, as described previously, do not scale easily into the deep-SDL size regime.

Another avenue for possible exploration is the emulation of film. In film, silver halide (AgX) crystals form grains in the submicron to the several-micron size range. A single photon striking the grain can result in the liberation of a single silver atom. This grain is effectively tagged as "exposed" and constitutes the latent image. In the subsequent wet chemical development process, the one silver atom results in a "runaway" feedback process that chemically liberates all the silver atoms in the grain. This leaves an opaque spot in the film because the grain has been converted to silver metal. Unexposed grains are washed away. The image intensity is thus converted to a local density of silver grains.

The probability that any particular grain is exposed under illumination grows linearly at first and only eventually approaches unity. (Mathematically, this is equivalent to calculating the probability that a particular patch of ground is hit by a rain drop after a certain total amount of rain has fallen upon the ground). The quantitative calculation is beyond the scope of this chapter, but this process gives rise to film's special D-logH contrast curve, in which D is density and H is light exposure. The smaller the grain size is, the lower the probability that the grain will be struck by a photon in a given exposure, and the "slower" the film speed will be because more light is required to ensure a high probability that all grains are struck by photons. However, the spatial resolution of the image is determined by grain size, with smaller grain sizes and slower film having higher image resolution.

Binary-like picture elements are found in a film image because they are exposed or not exposed. The local image intensity is determined by the density of exposed grains, or in digital parlance, by the local spatial density of "1s."

It is easy to transform this concept into a digital-film sensor by emulating this process. Consider an array of deep-SDL pixels in which each pixel is just a fraction of a micron in size. The conversion gain needs to be high and the readout noise low so that the presence of a single photoelectron can be determined. (Actually, several photoelectrons could contribute to pushing the output signal above some threshold, but ultimately single photoelectron sensitivity would be desired). From the preceding discussion, it is evident that a pixel that only needs to detect a single photoelectron has much lower performance requirements for full-well capacity and dynamic range than an analog pixel in a conventional image sensor. Upon readout, the pixel is set to a "0" or a "1." (This is reminiscent of using memory chips as image sensors except that much higher sensitivity is being sought).

Due to the single-bit nature of the "analog-to-digital" conversion resolution, high row-readout rates can be achieved, allowing scanning of a "gigapixel" sensor with perhaps 50,000 rows in milliseconds. The readout binary image could be converted to a conventional image (a sort of digital *development*) of arbitrary pixel resolution, thus trading image intensity resolution for spatial resolution. Because of its D-logH exposure characteristics, the sensor would have improved dynamic range and other film-like qualities.

It is also possible to consider multiple scans during a single exposure that can be logically OR'd or added together to improve intensity and/or spatial resolution. Although it is unclear whether this digital film approach would have compelling performance advantages over an extrapolation of current technology, it is important to explore new paradigms for imaging.

As a last thought for this section, we can consider future hybrid sensors that depart substantially from the current technology. The current technology relies on nonequilibrium silicon semiconductor physics and excellent results have thus far been obtained. Silicon is used as the detector and for the readout electronics. However, by operating in nonequilibrium, dark current (the equilibrium restorative process) results in limiting exposure times and forcing fast readout rates. Perhaps some new photosensitive material that is less susceptible to dark current, such as an organic semiconductor or an electronic plastic, will be invented. When deposited on a stacked silicon readout chip, perhaps the change in this material's properties upon exposure to light could be read out using silicon technology and, equally importantly, reset to the dark condition. Certainly between where technology is today and this concept, many inventions have yet to be invented.

Now that we have discussed some ideas about future sensors for future DSCs, let us look at what those DSCs might be doing.

11.2 SOME FUTURE DIGITAL CAMERAS

A major purpose of a camera is to share "being there" with others or one's self at a future date (neglecting strictly artistic photographs; see Figure 11.1). It is no wonder that consumer DSCs are adding features that include short movie mode and audio recording. We can expect these features to be improved in the future. For example, movie mode may move to HDTV-compatible format and audio may become stereo. One might argue that the trend to increasing pixel count is to improve the sensation of "being there." Certainly, in the trend of consumer television, "ultra-definition" TV, or UDTV, with 32-Mpixel or higher resolution (using four 8-Mpixel sensors), is being considered for future applications.[6]

Two different paths to "being there" can be considered. One is a sort of surround vision in which 2-π steradians (hemispherical) or more of solid angle viewing is recorded and displayed. Already, researchers are experimenting with such imaging for video telepresence. For applications such as real-estate marketing, some surround-vision cameras already exist. A conventional sensor is used in these cameras, but a special lens maps the solid viewing angle into a two-dimensional image area. To recreate the scene at a later time, algorithms are used to take the recorded data and make a two-dimensional image that represents what one would see by gazing

FIGURE 11.1 (See Color Figure 11.1 following page 178.) An illustration of "being there."

in a particular direction. Pan and tilt are readily achieved in this process. Higher resolution sensors would help this surround-vision mapping process. It is also possible to imagine a hemispherical projection process using an LCD display and the optics used in recording, except in reverse. However, this would require a special display surface or viewing room outside the current purview of consumer applications.

A second path to creating the sensation of "being there" is to record not only the reflected light from objects, but also the range of the objects. In this case, it is possible that more faithful three-dimensional reproduction of the original scene can be achieved to increase the sensation. Recording the range of objects using an active ranging mechanism, such as pulsed IR light and time-of-flight measurement, has been investigated for several years. However, this approach is probably not feasible for most scenes or for consumer application.

Stereo vision is another possibility for obtaining range information. Because this method of obtaining range information is most closely related to the way in which people obtain a sensation of range, it is probably the most attractive for this application. Technically, it is not difficult to envision a consumer-level stereo-vision system. One could almost fasten two DSC cameras together and obtain an instant stereo vision system. More sophistication is needed for the display of the stereo image. The use of red and blue color coding for stereo display has been used often, but the color content of the image suffers. Polarization techniques have been successfully used, but this requires special display and viewing devices. Quantification of range data from stereo imagery is computationally intensive; however, once it is obtained, possibly a more satisfying display can be made of the data.

One might be able to argue successfully that digital imaging did not really succeed until the widespread availability of suitable display devices. The creation of a better sensation of "being there" will depend on the availability of high-resolution and physically larger displays. Perhaps, "virtual-reality" glasses that project the image locally (or even directly onto the retina) will also promote more advances in digital imaging. Many perceptual problems remain with personal virtual-reality displays, including nausea and headaches. As these problems are solved, we might expect a window of opportunity for expanding the applications afforded by digital imaging.

Digital imaging, in addition to allowing us to "be there" where we might have been, also allows us to be in places where no human could go. Examples of this include the robotic exploration of space and sea, voyaging inside the human body, and seeing things that happen so fast that they can be missed "in the blink of an eye."

Digital imaging might also allow us to go to places that we may be unable to go ourselves. For example, one might be able to imagine a "briefcase body" that someone else carries around. The briefcase body is equipped with stereo vision and audio (bidirectional) so that, using virtual reality glasses and ear phones, one can be "virtually there" at important meetings. Handicapped people might be able to enjoy the sights and sounds in real time of being some place to which their bodies do not allow them to travel. For that matter, one can imagine a "robotic body" that sits in one's office at work and can be moved around by remote control to interact with other robotic bodies at meetings, etc. One would interact through virtual reality glasses and ear phones and "talk" to other robotic bodies — except that the display could be modified to make it look like the actual person behind the robotic body. One might dispense with the robotic body altogether and create future virtual offices where one could still interact with other workers but do so in cyber space, as was portrayed (in a different way) in the movie *The Matrix*. In this case, digital imaging may not be so necessary.

Another emerging concept is for personal continuously recording devices. In the future, someone might wear a visual/audio recorder that records all daily activities. Using advanced compression techniques, low-resolution imaging, and mass data storage systems, recording a full day of visual information is becoming technically feasible. These data could be archived, but reviewed to improve the sensation of memory (e.g., what was I to get at the store?). More advanced future functions could include high-level automatic summarization of the events of the day and indexing for future retrieval.

Now, as with many new technologies, new opportunities arise for the use of the technology and many new social and ethical questions arise as well. Is it ethical to capture, analyze, and archive high-resolution images of people in public or semi-private places without their knowledge in the name of security? Is it ethical to record every detail of one's interactions with other people or is a slightly fuzzy memory kinder for the purposes of social interactions? Would someone really want every interaction with someone else recorded for posterity by those other people? Of course, human memory does that already, but data that do not "lie" or become forgotten may not be so desirable. Therefore, although we may be able to envision and create these new technologies, as responsible engineers we also have a duty to be sure that we are, in fact, improving society.

REFERENCES

1. A. Krymski, N. Bock, N. Tu, D. Van Blerkom, and E. R. Fossum, A high-speed, 240 frames/second, 4.1 megapixel image sensor, *IEEE Trans. Electron Devices*, 50(1), 130–135, 2003.
2. R. Merrill, Color separation in an active pixel cell imaging array using a triple-well structure, U.S. Patent No. 5,965,875.
3. S. Bentheim et al., Vertically integrated sensors for advanced imaging applications, *IEEE J. Solid-State Circuits*, 35(7), 939–945, 2000.
4. J. Theil et al., Elevated pin diode active pixel sensor including a unique interconnection structure, US Patent No. 6,018,187.
5. Y. Wang, JPL, private communication.
6. I. Takayanagi, M. Shirakawa, K. Mitani, M. Sugawara, S. Iversen, J. Moholt, J. Nakamura, and E.R. Fossum, A 1-1/4-inch 8.3-Mpixel digital output CMOS APS for UDTV application, *ISSCC Dig. Tech. Papers*, 216–217, February 2003.

Appendix A

Number of Incident Photons per Lux with a Standard Light Source

Junichi Nakamura

Radiometry describes the energy or power transfer from a light source to a detector and relates purely to the physical properties of photon energy. When it comes to visible imaging in which the response of the human eye is involved, photometric units are commonly used.

The relationship between the radiometric quantity $X_{e,\lambda}$ and the photometric quantity X_v is given by

$$X_v = K_m \cdot \int_{\lambda_1}^{\lambda_2} X_{e,\lambda}(\lambda) \cdot V(\lambda) \cdot d\lambda \tag{A.1}$$

where $V(\lambda)$ is the photopic eye response, K_m is the luminous efficacy for photopic vision and equals 683 lumens/watt; $\lambda_1 = 0.38 \mu m$; and $\lambda_2 = 0.78 \mu m$. Table A.1 shows the photopic eye response and it corresponds to the response of $\bar{y}(\lambda)$ shown in Figure 7.2 in Chapter 7.

A standard light source with a color temperature T can be modeled using Planck's blackbody radiation law (see Section 7.2.5. in Chapter 7). The spectral radiant exitance of an ideal blackbody source whose temperature is T (in K) can be described as

$$M_e(\lambda, T) = \frac{c_1}{\lambda^5} \cdot \frac{1}{\exp\left(\dfrac{c_2}{\lambda T}\right) - 1} \left[\frac{W}{cm^2 - \mu m}\right] \tag{A.2}$$

$$c_1 = 3.7418 \times 10^4 \text{ watt-}\mu m^4/cm^2$$

$$c_2 = 1.4388 \times 10^4 \text{ }\mu m\text{-K}$$

TABLE A.1
Photopic Eye Response $V(\lambda)$

Wavelengh (nm)	Photopic $V(\lambda)$	Wavelength (nm)	Photopic $V(\lambda)$
380	0.000039	590	0.757
390	0.00012	600	0.631
400	0.000396	610	0.503
410	0.00121	620	0.381
420	0.0040	630	0.265
430	0.0116	640	0.175
440	0.023	650	0.107
450	0.038	660	0.061
460	0.060	670	0.032
470	0.09098	680	0.017
480	0.13902	690	0.00821
490	0.20802	700	0.004102
500	0.323	710	0.002091
510	0.503	720	0.001047
520	0.710	730	0.000520
530	0.862	740	0.000249
540	0.954	750	0.00012
550	0.99495	760	0.00006
560	0.995	770	0.00003
570	0.952	780	0.000015
580	0.870	–	–

Without a color filter array, the number of photons is represented by

$$n_{ph\_blackbody}(T) = \frac{\int_{\lambda_3}^{\lambda_4} M_e(\lambda,T) \cdot \left(\frac{hc}{\lambda}\right)^{-1} \cdot d\lambda}{K_m \int_{\lambda_1}^{\lambda_2} M_e(\lambda,T) \cdot V(\lambda) \cdot d\lambda} \quad [\text{photons}/\text{cm}^2 - \text{lux} - \text{sec}] \quad (A.3)$$

where λ_3 and λ_4 are the shortest and longest wavelengths, respectively, of the light that hits a detector. If an IR cut filter is used, then $\lambda_4'' \lambda_2$. Figure A.1 shows the number of photons per cm²-lux-sec with $\lambda_3 = \lambda_1$ and $\lambda_4 = \lambda_2$, as a function of color temperature of the black body light source.

With an on-chip color filter array and an IR cut filter, the numerator has to be modified to

$$n_{ph\_blackbody}(T) = \frac{\int_{\lambda_1}^{\lambda_2} M_e(\lambda,T) \cdot \left(\frac{hc}{\lambda}\right)^{-1} \cdot T(\lambda) d\lambda}{K_m \int_{\lambda_1}^{\lambda_2} M_e(\lambda,T) \cdot V(\lambda) \cdot d\lambda} \quad (A.4)$$

Appendix A

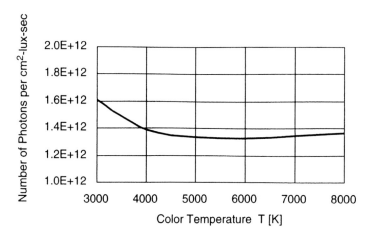

FIGURE A.1 Number of incident photons per cm²-lux-sec as a function of color temperature of the blackbody light source.

where $T(\lambda)$ is the total spectral transmittance.

REFERENCE

G.C. Holst, *CCD Arrays Cameras and Displays*, 2nd ed., JCD Publishing, Winter Park, FL, 1998, chap. 2.

Appendix B

Sensitivity and ISO Indication of an Imaging System

Hideaki Yoshida

The sensitivity of a photo detector is usually represented as the ratio of an output signal level to a received illuminance level. In other words, it is a proportionality constant of the linear curve of the output signal level to the input illuminance. When the detector is the charge integrating (accumulating) type, the term "illuminance" in the above sentence should be replaced by "exposure," which is a product of illuminance and the charge integration time.

The above-mentioned "sensitivity" can be defined and measured for both DSCs and image sensors since they both can be considered types of photo detectors. However, it is not a practical "sensitivity" for DSC users because it never identifies which exposure is adequate for taking a picture. The most important parameter for users is the level of exposure needed to generate an adequate output level to produce a good picture.

The ISO indicator of sensitivity in a photographic system represents this "adequate exposure," which is represented in the equation below:

$$\text{ISO indication value } S = K / H_m \qquad (\text{B.1})$$

where K is a constant and H_m is the exposure in lux-seconds.

According to the standard ISO 2240 (ISO speed for color reversal films) and ISO 2721 (automatic controls of exposure), a numeric 10 is to be used for K in electronic imaging systems.

Thus, the equation can be written as

$$S = 10 / H_m \qquad (\text{B.2})$$

For example, ISO 100 means that the adequate average exposure H_m (= 10/S) of the imaging system is 10/100 = 0.1 (lux-s).

Hereupon, there are some parameters that can be regarded as indicating the sensitivity of a DSC described in ISO 12232. Two of them, "ISO saturation speed" and "ISO noise speed," are described in the present text published in 1998. Two new parameters, "SOS" (standard output sensitivity) and "REI" (recommended exposure index), have been added to the first revision, which is to be published in 2005. Differences among these four parameters are only caused by what is regarded as the adequate exposure in Equation B.2.

1. *ISO saturation speed* is the S value when the exposure level generates a picture with image highlights that are just below the maximum possible (saturation) camera signal value. The adequate average exposure H_m is regarded as 1/7.8 of the exposure level at the saturation point (saturation exposure), where 7.8 is the ratio of a theoretical 141% reflectance (which is assumed to give the saturation exposure with 41% additional headroom, which corresponds to 1/2 "stop" of the headroom (= $\sqrt{2}$) to an 18% reflectance (the standard reflectance of photographic subjects). Thus, Equation B.2 can be changed into

$$S_{sat} = 78 / H_{sat} \qquad (B.3)$$

   where H_{sat} is the saturation exposure in lux-s. The saturation speed only shows the saturation exposure as a result. Suppose there are DSCs that have the same sensitivity at low-to-medium exposure levels, meaning that their tone curves at those levels are identical. If a tone curve of one of them has a deeper knee characteristic near the saturation exposure, the saturation speed of that DSC becomes lower. Thus, it is preferable to use the saturation speed to indicate the camera's overexposure latitude.

2. *ISO noise speed* is the S value when the exposure generates a "clear" picture of a given S/N value. An S/N value of 40 is used for an excellent image, while 10 is used for an acceptable image.

   This parameter seems to be a good indicator for taking a picture in that it shows the necessary exposure to obtain a certain low-noise image. However, an actual camera often has various image capture settings, such as the number of recording pixels, compression rate, and noise reduction. In these cases, even if the same tone curve and exposure control are used, the S/N changes as the camera settings change. Thus, ISO noise speed does not fit a camera set directly.

3. *SOS* is the S value when the exposure generates a picture of "medium" output level corresponding to 0.461 times the maximum output level (digital value of 118 in an 8-bit system). H_m in Equation B.1 corresponds to the exposure that produces 0.461 times the maximum output level. The numeric 0.461 corresponds to the relative output level on the s-RGB gamma curve for the 18% standard reflectance of photographic subjects.

   SOS gives an acceptable exposure because the average output level of the picture becomes "medium." Thus, it is convenient for a camera set. However, there is no guarantee that the exposure indicated by SOS is the

best. Also, it is not suitable for an image sensor, whose output characteristic is linear.
4. *REI* is the S value when the exposure generates a picture with an "adequate" output level that a camera vendor recommends arbitrarily. According to this definition, it is apparent that REI can apply only to a camera set and that the exposure indicated by REI would be adequate only if the vendor's recommendation is appropriate.

According to above considerations, the author of this appendix suggests the following to designers or manufacturers for communicating with the next users (consumers in the case of DSC manufacturers, and camera designers in the case of image sensor manufacturers):

- Use SOS or REI (or both) for indicating the sensitivity of a camera set. They are both effective for users when choosing an exposure level or finding a usable subject brightness.
- Use ISO speeds for an image sensor. Report noise speed at each noise level (S/N = 40 or S/N = 10, the preference is 40) as basic information. In this case, it is most important to address the signal processing algorithms and an evaluation method of the S/N. Preferably, they are as simple as possible. Also, some kind of standardization is needed, but regretfully there is nothing now.

Also, report the ratio of the saturation speed to noise speed as added information indicating the upper dynamic range.

Index

A

Abbe number, 40–41
Aberrations, 30–31
Active pixels, 145–146
Active-pixel sensors, 61
Adaptation, 301
Advanced color interpolation, 248–250
Airy disc, 36
Algorithms, *See* Camera control algorithm; Image-processing algorithms
Aliasing
  artifacts, 173
  description of, 287–288
  spatial color, 307
  tint due to, 294–295
Aliasing noise, 226
Amplifiers
  analog, 238
  column, 165
  column parallel programmable gain, 165
  digital gain, 238
  floating diffusion, 75
  line-amplifier, 165
  programmable gain
    column parallel, 165
    description of, 164
    serial, 165
Amplitude response, 200–201
Analog amplifier, 238
Analog electronic still cameras, 2
Analog front end
  analog-to-digital conversion, 267–269
  correlated double sampling, 266–267
  description of, 164–165
  device example, 269–271
  diagram of, 267
  function of, 266
  future designs for, 274–275
  optical black level clamp, 267
Analog front-end chip, 137
Analog signal capture, 258
Analog-to-digital conversion, 267–269, 297
Analog-to-digital converter, 149
Angle of exit light, 46–47
Angular response, 193–194
Antiblooming, 114–117
Applications
  network use, 19
  newspaper photographs, 18
  printing press, 18–19
  real estate, 311
Arrays
  imaging sizes for, 88–89
  optical format and, 88
  performance of, 85–87
Artifacts
  digital, 227
  flare-like, 202
  light sources as cause of, 202–203
Aspherical lenses, 41–43, 50–51
Astigmatism, 30
Astrophotographers, 19
Auto exposure
  changes in, 174–175
  description of, 172, 238–239
Auto focus
  changes in, 174–175
  description of, 172
  digital integration, 240–241
  filter and output, 241
  measurement methods for, 239–240
Auto white balance, 238–239
Automatic gain control, 17

B

Backside illumination, 308
Band-pass filters, 183
Bayer primary color filter array, 62–63, 124
Bicolor illumination, 301
Bicubic interpolation, 246
Bilinear color interpolation, 230
Bilinear interpolation, 230, 244–245
Birefringency, 34
Black body locus, 212
Black defects, 120
Black level calibration, 269
Block distortion, 298
Blooming, 77, 114, 119, 290
Blurring, 237, 299
Boltzmann's distribution factor, 116
Boosting, 158

Buried channel CCD, 101–105
Buried photodiode, 116–117

C

Camera(s)
    analog electronic, 2
    characterization of, 214–215
    classification of, 3
    digital still, *See* Digital still camera
    purpose of, 311–312
    SLR, *See* SLR cameras
Camera back type digital still cameras, 14
Camera control algorithm
    auto exposure, 238–239
    auto focus
        digital integration, 240–241
        measurement methods for, 239–240
    auto white balance, 238–239
    data
        compression of, 242–243
        storage of, 243
    video mode, 241–242
    viewfinder, 241–242
Camera shake, 301–302
Camera spectral sensitivity, 212–213
Casio QV-10, 9–10
CCD(s)
    analog signal capture, 258
    buried channel, 101–105
    charge transfer in
        description of, 97
        fringing field effect, 97–99
        mechanism of, 97–99
        self-induced drift, 97–98
        thermal diffusion, 97–98
    concept of, 95–97
    description of, 17, 32, 55
    floating diffusion charge detector, 107–108
    four-phase driving, 105–107
    horizontal
        description of, 58, 121
        selective charge transfer mechanism, 134–135
    load capacitance of, 122
    metal-oxide-semiconductor capacitor, 96
    noise reduction in, 107–110
    output circuitries, 107–110
    readout architectures of, 57–58
    surface channel, 99–101
    two-phase driving, 105–106
    vertical, 58, 116, 122, 128
    vertical transfer, 105
CCD image sensors
    antiblooming, 114–117
    black defects, 120
    blooming, 119, 290
    characteristics of, 117–122
    charge transfer efficiency, 120
    CMOS image sensor vs., 90, 146–147
    dark current noise, 119–120
    digital still camera application requirements, 122–123
    frame transfer, 110–112, 172
    frame-interline transfer, 111–112
    future of, 138–139
    horizontal charge transfer, 110
    incident light on, 224
    interline transfer
        description of, 110–112, 121, 171
        full frame signal sequences of, 134
        interlace scan type of, 123–125
        Nyquist limits of, 131
        progressive scan type of, 124–125
    operation of, 121
    photoelectric conversion characteristics, 117–119
    pixel interleaved array
        block diagram of, 133
        description of, 129–131
        full frame signal sequences of, 134
        Nyquist limits of, 131
        resolution characteristics of, 131
    pixel sizes for, 89
    power consumption by, 121–122
    p-substrate structure of, 112–114
    p-well structure of, 114
    shading in, 72
    smear, 119, 290
    system solution for, 137
    types of, 110
    vertical offset drift of, 225
    vertical transfer, 111
    white blemishes, 120
CCD registers, 72
Cellular phones with cameras, 15–16, 274
Charge collection efficiency, 65–66
Charge detector
    floating diffusion, 107–108
    sensitivity of, 108–109
Charge diffusion, 83–84
Charge packets
    description of, 97
    horizontal mixing of, 136–137
    mixing process for, 136
    vertical mixing of, 136
Charge transfer
    description of, 56–58
    efficiency of, 120

Index

low-voltage, 158
reset noise and, 157
selective, 134–135
Charge-coupled device, *See* CCD(s)
Charge-coupled device image sensors, *See* CCD image sensors
Charge-to-voltage conversion factor, 84
Chroma clipping, 248–249
Chromatic aberration
    description of, 30–31, 294–295
    tint due to, 294–295
Chromatic adaptation, 216–217
Chromaticity, 207–210, 286
CIE, 206–207
CIPA standards, 303
Circle of least confusion, 33–34
Circular zone-plate chart, 200, 287–288
C-800L, 10
C-1400L, 10
Clipping
    chroma, 248–249
    description of, 247
Clock noise, 298–299
CMOS image sensors
    active-pixel, 62
    analog signal, 224
    benefits of, 145
    CCD image sensor vs., 90, 146–147
    concepts of, 144–147
    configuration of, 147
    dark current from transistor in an active pixel of, 72
    description of, 54, 144
    digital signal from, 224–225
    fill factor in, 62
    fixed pattern noise suppression, 148–149
    future of, 175
    global shutter, 151, 171
    memory-in-pixel scheme in, 171–172
    noise in
        column-wise, 167
        description of, 165–166
        pixel-to-pixel random, 166–167
        row-wise, 167
    output stage of, 149
    peripherals used in, 149–150
    pixel serial readout architecture, 150
    pixel size for, 89
    power consumption of, 151–153
    rolling shutter, 151
    shading in, 167–168
    system-on-a-chip, 145
    X–Y pixel addressing, 148
Color, lack of, 291
Color adaptation, 301
Color aliasing, 173
Color constancy, 217
Color conversion
    description of, 216–218
    filters for, 183
Color correction
    RGB, 232
    YCbCr, 232–233
Color difference, 210–211
Color filter array
    Bayer, 230
    description of, 62–63, 227–228
    RGB, 248–249
    spatial color aliasing caused by, 307
Color interpolation
    advanced, 248–250
    definition of, 229
    description of, 226–227, 229–231, 243
    quincunx grid sampling, 229
    rectangular grid sampling, 228–229
Color management
    colorimetric definition, 218–219
    description of, 218
    image state, 219–220
    profile approach, 220
Color moir, 287–289
Color phase noise, 227
Color rendering index, 213
Color reproduction, 285–286
Color spot diagram, 31
Color temperature, 211–212
Color theory
    chromaticity, 207–210
    color-matching functions, 206–207
    human visual system, 206
    light sources, 211–212
    overview of, 205–206
    tristimulus values, 206–207
    uniform color spaces, 207–210
Color unevenness, 286–287
Colorimetry, 220
Color-matching functions, 206–207
Column amplifiers, 165
Column correlated double sampling, 162–163
Column parallel programmable gain amplifier, 165
Column parallel readout architecture, 150–151
Column-wise noise, 167
Comatic aberration, 30
Compact film cameras, 47
Complementary color filter pattern, 227
Composite aspherical lenses, 42
Compression noise, 298
Cones, 206
Conversion factor, 198

Conversion gain
    description of, 60–61, 108
    estimation of, 81–82
Convolution kernel, 235
Correlated color temperature, 212
Correlated double sampling
    analog front end and, 266–267
    column, 162–163
    description of, 67
    diagram of, 268
Cosine fourth law, 27
Cropping of image, 247
Cross color noise, 227
Crosstalk, 83–84
Cubic interpolation, 245–246
Cyclic noise injection, 167

D

Dark characteristics
    average dark current, 195
    description of, 194
    fixed-pattern noise at dark, 196–197
    temporal noise at dark, 195–196
Dark current
    activation energy of, 119
    average, 195
    from CCD registers, 72
    definition of, 103, 289
    description of, 67–68
    diffusion current, 69–70
    generation current in the depletion region, 68–69
    pinning effects on, 104
    PN photodiode, 154
    shot noise, 76–77
    surface generation, 70
    temperature dependence of, 71, 289
    total, 70
    white spot defects, 71–72
Dark current noise, 119–120
Dark fixed-pattern noise, 196–197
Dark noise, 282
Dark noise floor, 168
Data compression, 242–243
Data storage, 243
Decenter sensitivity, 48
Defect pixel, 202
Depth of field, 32–33, 295–296
Depth of focus, 33–34
Differential delta sampling, 163–164
Differential nonlinearity, 269
Diffraction, 24–25, 36–37
Diffractive optics, 307

Diffusion current, 69–70
Digital artifacts, 227
Digital back end
    description of, 271–272
    future designs for, 275–276
Digital gain amplifier, 238
Digital imaging, 313
Digital signal processors
    architecture of, 262
    description of, 17
    frame rate of, 264–265
    general-purpose, 262, 263–266
Digital signal-processing block, 240
Digital SLR cameras
    description of, 169, 305
    future of, 306–308
    high-end, 306–308
    multicolor stacked structure used with, 307
    spatial color aliasing, 307
Digital still cameras
    analog circuit of, 17
    applications of
        CMOS image sensors for, 175
        description of, 18–19, 122–123
    block diagram of, 16
    camera back type, 14
    cellular phones with, 15–16
    definition of, 2–3
    digital circuit of, 17
    digital signal-processing block of, 225
    frame rates for, 259
    future types of, 311–314
    history of, 3–10
    imaging devices for, 17
    imaging optics of, See Imaging optics
    middle-class compact, 169
    optics of, See Optics
    performance trends, 273
    point-and-shoot type of, 10–13
    power consumption by, 261–262
    structure of, 16–17
    system control of, 17
    toy, 14–16
    trends of, 272–274
    variations of, 10–16
Digital video cameras
    description of, 224
    frame rates for, 259
Digital-film sensor, 309–311
Discrete cosine transform, 242–243
Distortion, 30
Drain
    lateral overflow, 114
    vertical overflow, 114–116
Dummy pixels, 87–88

Index

Dynamic random access memory, 65
Dynamic range, 79, 284–285

E

Edge detectors, 250
Edge illumination, 27
Electrical components, of crosstalk, 83
Electrical shading, 167
Electronic global shutter, 171–172
Electronic shutter, 126–127
Electronic shutter pulse, 121
Electronic viewfinders, 11
Electronic zoom lenses, 243–246
Exif, 297
Exposure error, 299–300

F

F2.8 lens, 26
F5.6 lens, 26
Face-plate exposure, 186
Face-plate illuminance, 185
False color noise, 227
Fast Fourier transform algorithm, 235
Feature flexibility, 257–258, 263–264
Fill factor, 61–62
Film, 256
Film direct transmitter, 18
Filter(s)
    band-pass, 183
    color conversion, 183
    description of, 235
    finite impulse response, 236–237
    infinite impulse response, 236–237
    low-pass, *See* Low-pass filters
    unsharp mask, 237–238
Filter kernel, 235
Finite impulse response filters, 236–237
Fixed-pattern noise
    causes of, 227
    dark current
        from CCD registers, 72
        definition of, 103
        description of, 67–68, 196–197
        diffusion current, 69–70
        generation current in the depletion region, 68–69
        pinning effects on, 104
        surface generation, 70
        temperature dependence of, 71
        total, 70
        white spot defects, 71–72

    definition of, 103
    description of, 66–67, 292
    under illumination, 199
    shading, 72
    suppression of
        circuit for, 162
        description of, 148–149
Flare, 293
Flare-like artifacts, 202
Flash light effects, 300–301
Flicker illumination, 300
Flicker noise power spectral density, 166
Floating diffusion amplifier, 75
Floating diffusion charge, 61
Floating diffusion charge detector, 107–108
Fluorite, 41
"Flying-in" noise, 292, 298
F-number
    description of, 25–26, 36
    modulation transfer function and, 37
Focal length
    description of, 23
    lens configurations and, 23–24
Focusing errors, 299
Forward-based diode, 68
Four-phase driving CCD, 105–107
Frame rate, 258–261, 264–265
Frame transfer CCD, 110–112
Frame-interline transfer CCD, 111–112
Frequency response, 279–280
Fringing field effect, 97–99
F-settings, 89
Fuji DS-200F, 8
Fuji DS-1P, 8
Full-frame transfer CCD, 111
Full-well capacity
    definition of, 308
    description of, 66
    estimation of, 82–83

G

G1, 96
G2, 96
Gamma characteristic, 282
Gamma correction, 267
Gamma curves, 233–234
Gate electrode potential, 102
Gaussian distribution, 73
Generation current in the depletion region, 68–69
Geometric distortion, 294
Ghost image, 293–294
Gigapixel sensor, 311
Glass

imaging optics use of, 40–41
reflectance of, 45
refractive index of, 45
ultrahigh refractive index of, 41
Glass molded aspherical lenses, 42
Global reset synchronization with mechanical shutter, 170–171
Global shutter
    description of, 151
    electronic, 171–172
Gradation, 282–284
Gray-scale chart, 283–284
Grid sampling
    quincunx, 229
    rectangular, 228–229
Ground glass aspherical lenses, 42

H

Hard reset, 155
Hardwired ASIC
    description of, 262
    feature flexibility of, 264
    frame rate of, 265
    general-purpose digital signal processor vs., 263
    power consumption by, 265
    time-to-market considerations for, 265
High resolution, 123
High-frame-rate movie, 132–138
High-resolution still picture, 132–137
Horizontal CCDs, 58
Horizontal charge mixing, 136–137
Horizontal charge transfer CCD, 110
Huffman coding, 242–243
Hybrid camera, 275
Hyperfocal distance, 33

I

ICC, *See* International Color Consortium
IEC 61146, 302–303
Illumination
    conversion factor, 198
    fixed-pattern noise under, 199
    temporal noise under, 198
Image
    clipping of, 247
    compression of, 242–243
    cropping of, 247
    description of, 2
    reproduced, of natural scenes, 202–203
    resize of, 246–247

smear, 290
storage of, 2
Image acquisition board, 184–185
Image capture
    description of, 2
    high-definition, 274
Image capture function, 2
Image lag
    characteristics of, 201–202
    description of, 77, 155, 289
    effects of, 201
    evaluation of, 182, 201–202
    measurement of, 202
Image quality
    definition of, 278
    evaluation parameters for, 180–181
    factors that affect
        adaptation, 301
        aliasing, 287–288
        bicolor illumination, 301
        blooming, 290–291
        camera shake, 301–302
        chromatic aberration, 294–295
        clock noise, 298–299
        color adaptation, 301
        color reproduction, 285–286
        compression noise, 298
        dark current, 289
        depth of field, 295–296
        dynamic range, 284–285
        exposure error, 299–300
        flare, 293
        flash light effects, 300–301
        flicker illumination, 300
        focusing error, 299
        frequency response, 279–280
        geometric distortion, 294
        ghost image, 293–294
        gradation, 282–284
        image lag, 289
        jerkiness interference, 302
        lack of color, 291
        luminance adaptation, 301
        luminance nonuniformity, 300–301
        motion blur, 301–302
        noise, 280–282
        perspective, 296–297
        pixel defect, 289–290
        power line noise, 298–299
        quantization noise, 297–298
        resolution, 278–279
        signal processing-related, 297–299
        smear, 290–291
        sticking, 289
        system control-related, 299–301

Index 329

temporal noise, 302
toneless black, 291–292
uniformity, 286–287
white balance error, 300
white clipping, 291, 301
standards regarding, 302–303
Image sensor(s)
alignment of, 183
with analog output, 196
array performance of, 85–88
CCD, *See* CCD image sensors
characterization of, 180
charge collection and accumulation, 56
charge detection by, 59–61
CMOS, *See* CMOS image sensors
conversion gain by, 60–61
definition of, 54, 289
description of, 256
development of, 180
functions of, 54–61
future of, 305–311
imaging array scanning, 56–59
interlaced scan, 59–60
jig for alignment of, 183
lack of color, 291
modulation transfer function of, 86–87
noise
  definition of, 66
  fixed-pattern, 66–72
  $1/f$, 74, 76
  random, 66
  schematic diagram of, 67
  shot, 74
  temporal, *See* Temporal noise
  thermal, 73
  "white," 74
performance of, 181–182
photoconversion by, 55–56
progressive scan, 59–60
readout of, 306
toneless black, 291–292
ultrahigh-definition television, 175
white clipping, 291
Image sensor evaluation
angular response, 193–194
dark characteristics
  average dark current, 195
  description of, 194
  fixed-pattern noise at dark, 196–197
  temporal noise at dark, 195–196
defect pixel, 202
environment for
  band-pass filters, 183
  color conversion filters, 183
  configuration of, 181–185

description of, 181
evaluation board, 184
image acquisition board, 184–185
imaging lens, 183
IR-cut filters, 183
light sources, 182–183
optical filters, 183
temperature control, 183–184
illuminated characteristics
  definition of, 198
  temporal noise, 198
image lag characteristics, 201–202
light intensity measurements, 185–186
optical axis adjustment, 186
parameters, 180–181
photoconversion characteristics
  description of, 186–188
  linearity, 189
  saturation exposure, 189–190
  sensitivity, 189
preparation of, 185–186
purposes, 180
quantum efficiency, 193
resolution characteristics, 200–201
smear characteristics, 199–200
software for, 185
spectral response data for, 190–193
Image state, 219–220
Image-processing algorithms
color correction
  RGB, 232
  YCbCr, 232–233
color interpolation, 226–228
  definition of, 229
  description of, 226–227, 229–231
  quincunx grid sampling, 229
  rectangular grid sampling, 228–229
description of, 224–225
gamma curves, 233–234
noise reduction
  aliasing noise, 226
  description of, 225
  offset noise, 225–226
  pattern noise, 226
tone curves, 233–234
Image-processing engine
architecture of, 262
components of, 271
feature flexibility, 257–258, 263–264
frame rate, 258–261
hardwired ASIC, 264
imaging functions, 257
imaging performance, 258
operations performed by, 257
power consumption, 261

semiconductors, 261
time-to-market, 261
Imaging array scanning, 56–59
Imaging lens
    configuration of, 32
    description of, 32, 183
    modulation transfer function for, 35
Imaging optics
    aspherical lenses, 41–43
    characteristics of, 31–37
    coatings, 43–46
    configuration of, 32
    depth of field, 32–33
    depth of focus, 33–34
    design process for, 38–40
    diffraction effects, 36–37
    glass materials used in, 40–41
    mass production of, 47–49
Incident light angle, 193–194
Incident photons per lux, 315–317
Infinite impulse response filters, 236–237
Infrared cut filters, 32
Input referred noise, 77
Interlaced scan, 59–60, 123–125
Interline transfer CCD
    description of, 110–112, 121, 171
    full frame signal sequences of, 134
    interlace scan type of, 123–125
    Nyquist limits of, 131
    progressive scan type of, 124–125
International Color Consortium, 220
International Commission on Illumination, *See* CIE
Interpolation
    bicubic, 246
    bilinear, 244–245
    color
        advanced, 248–250
        definition of, 229
        description of, 226–227, 229–231, 243
        quincunx grid sampling, 229
        rectangular grid sampling, 228–229
    cubic, 245–246
    linear, 244–245
IR-cut filters, 183
Irreversible compression, 242
ISO 12233, 279
ISO noise speed, 320
ISO saturation speed, 320
ISO standards, 303

J

JEITA CP-3203, 302–303

Jerkiness interference, 302
JPEG image compression, 242–243, 298
Junction field effect transistor, 160–161

K

Kodak DCS-1, 8–9
kTC noise, 74–75, 154–155

L

Lack of color, 291
Lateral buried charge accumulator and sensing transistor, 161
Lateral overflow drain, 114
Latitude, 285
LBCAST, 161
LCD monitor
    Casio QV-10, 9
    in SLR cameras, 13
Lens
    aberrations that affect, 30
    aperture mechanisms for, 183
    aspherical, 41–43, 50–51
    coatings on, 43–46
    F-number of, 25
    retrofocus, 24
    schematic diagram of, 22
    telecentric, 47
    telephoto, 24
    thickness of, 23
    type of, 38
    zoom
        angle of light exiting from, 46–47
        electronic, 243–246
        multigroup moving, 50
        short, 50–51
        video type, 49
Lens distortion correction, 251
Lens shading correction, 251–252
Light box, 182
Light intensity measurements, 185–186
Light receptors, 206
Light sources
    artifacts caused by, 202–203
    description of, 211–212
Line noise, 227
Line shift, 110
Line-amplifier, 165
Linear interpolation, 230, 244–245
Linear saturation exposure, 189
Linearity, 83, 189
Low-pass filters

Index

nonseparable, 236
optical
  birefringency of, 34
  characteristics of, 34–36
  description of, 32, 231
  materials for, 35
  modulation transfer function for, 35
  reasons for using, 34–35
  separable, 236
Low-voltage charge transfer, 158
L-shaped transfer gate, 158
Luminance, 298
Luminance adaptation, 301
Luminance nonuniformity, 300–301
Luminance unevenness, 286
Luther condition, 213, 220
Lux-seconds, 79

M

Magnetic video camera, 5
Mechanical shutter
  description of, 128
  global reset synchronization with, 170–171
  rolling reset synchronization with, 170
Memory-in-pixel scheme, in CMOS image sensors, 171–172
Metal-oxide semiconductor capacitor
  description of, 96
  interactions between, 96–97
Metal-oxide semiconductor diode, 56, 58
Metal-oxide semiconductor field effect transistor, 144–145
Metal-oxide semiconductor transistor
  noise model of, 75–76
  reset noise from, 74
Micrograting, 307
Microlens array, 63–64, 84
Middle-class compact digital still cameras, 169
Modulation transfer function
  description of, 27–30
  F-number and, 37
  frequency response and, 280
  of image sensors, 86–87
  measuring of, 85
  for optical low-pass filters, 35
  resolution and, 28
  spatial frequency characteristics, 28–29
  with uniform detector sensitivity, 86
Moir, effect, 34–35, 87, 287–288
Moore's law, 310
Mosquito noise, 298
Motion blur, 301–302
Movie mode
  description of, 172
  mode changes in, 174–175
  subresolution readout, 173
MTF_{IMAGER}, 86–87
Multicolor stacked structure, 307
Multigroup moving zooms, 50

N

Nearest neighbor interpolation, 244
Networks, 19
Newspaper photographs, 18
Noise
  aliasing, 226
  blooming, 77
  clock, 298–299
  CMOS image sensors
    column-wise, 167
    description of, 165–166
    pixel-to-pixel random, 166–167
    row-wise, 167
  color phase, 227
  column-wise, 167
  compression, 298
  cross color, 227
  dark, 282
  definition of, 280–281
  distortion vs., 280
  false color, 227
  fixed-pattern, *See* Fixed-pattern noise
  "flying-in," 292, 298
  image quality and, 280–282
  input referred, 77
  kTC, 74–75, 154–155
  line, 227
  mosquito, 298
  offset, 225–226
  output referred, 77
  pattern, 226
  pixel-to-pixel random, 166–167
  power line, 298–299
  quantitative evaluation of, 282
  quantization, 297–298
  read, 75–76
  reduction of, 107–110
  reset, 74–75, 155–156
  row-wise, 167
  shot
    dark current, 76–77
    definition of, 74
    description of, 292–293
    equation for, 166
    photon, 76–77
  smear, 77

spatial random, 292
suppression of
    column correlated double sampling for, 163
    differential delta sampling for, 163–164
temperature effects on, 184
temporal
    causes of, 227
    at dark, 195–196
    definition of, 72
    description of, 66–67
    under illumination, 198
    pixel, 162
    reset noise, 74–75
    thermal noise, 73
    variance of, 72–73
thermal, 73, 292–293
white, 74, 196
Noise charge, 75
Noise effect, 281
Noise equivalent exposure, 83
Noise floor, 75–76, 79, 155
Noise injection, 167
Nonlinearity, 269
Nonseparable filter structure, 235
Nonseparable low-pass filters, 236
n-substrate, 114, 158–160
n-well/p-type substrate, 154
Nyquist frequency, 35, 87, 200, 226, 228–229

O

Odd field, 123
Offset noise, 225–226
Olympus C-800L, 10
On-chip integration, 168
On-chip microlens array, 63
On-chip reference generators, 149–150
On-chip signal processing, 152
On-chip timing generator, 149–150
One-dimensional filter, 235–236
1/f noise, 74, 76
Optical axis, 186
Optical black level clamp, 267
Optical black pixels, 87
Optical components, of crosstalk, 83
Optical filters, 183
Optical format, 88–89
Optical low-pass filters
    birefringency of, 34
    characteristics of, 34–36
    description of, 32, 231, 287
    materials for, 35
    modulation transfer function for, 35

reasons for using, 34–35
Optical shading, 167
Optics
    description of, 16–17, 22
    fundamentals of, 22–27
    modulation transfer function, 27–20
Optoelectronic conversion function, 284, 287
Output referred noise, 77
Output-referred image data, 219

P

p layer, 116
Passive pixels, 145–146, 148
Pattern noise, 226
PC displays, 89
Periscopes, 43
Personal continuously recording devices, 313
Perspective, 296–297
Photoconversion
    characteristics of, 78–85, 186–190
    conversion gain
        description of, 60–61
        estimation of, 81–82
    crosstalk, 83–84
    description of, 55–56
    dynamic range, 79
    linearity, 83
    noise equivalent exposure, 83
    quantum efficiency
        definition of, 78
        estimation of, 81
    responsivity, 78
    sensitivity of, 84
    signal-to-noise ratio
        description of, 79–80
        increasing of, 84–85
        sensitivity and, 84
Photodetectors
    in pixels, 61–66
    sensitivity of, 319
Photodiode pixels
    description of, 139
    pinned
        boosting, 158
        configuration of, 156–157
        description of, 156–157
        low-voltage charge transfer, 158
    PN
        description of, 153–154
        hard reset, 155
        reset noise suppression, 155–156
        soft reset, 155
        structure of, 154

Index

reset noise, 154–155
vertical integration, 161–162
Photodiode voltage, 147
Photoelectric conversion, 117–119
Photon flux absorption, 55–56
Photon shot noise, 76–77, 80
Photopic eye response, 316
Pinned photodiode pixels
    boosting, 158
    configuration of, 156–157
    description of, 156–157
    low-voltage charge transfer, 158
Pixel(s)
    active, 145–146
    addressing of, 150–151
    for analog-to-digital conversion, 267
    boosting, 159
    charge collection efficiency, 65–66
    color filter array, 62–63
    configuration of, 158–159
    dark current in, 68
    defect, 202
    description of, 10, 54–55
    dummy, 87–88
    fill factor, 61–62
    full-well capacity, 66, 82–83
    isolation of, 160
    microlens array, 63–64
    optical black, 87
    overflowing charge from, 290
    passive, 145–146, 148
    photodetector in, 61–66
    photodiode, *See* Photodiode pixels
    of point-and-shoot digital cameras, 11
    SIO_2/SI interface, 64–65
    structure of, 61, 125, 306–307
    subdiffraction limit, 308
    temperature dependence of, 183–184
Pixel binning, 173–174
Pixel counts, 305
Pixel defect, 289–290
Pixel follower circuit, 166
Pixel interleaved array CCD
    block diagram of, 133
    description of, 129–131
    Nyquist limits of, 131
    resolution characteristics of, 131
Pixel parallel readout architecture, 151
Pixel resolution, 287
Pixel selective transistor, 159
Pixel serial readout, 150, 162
Pixel size
    considerations for, 89–90, 122, 130
    shrinking of, 308
Pixel skipping, 173–174

Pixel temporal noise, 162
Pixel-to-pixel random noise, 166–167
PIXOUT, 161
Planck's blackbody radiation law, 315
Planck's constant, 55
Plastic aspherical lenses, 42
p–n junction capacitance, 109, 112
PN photodiode pixels
    description of, 153–154
    hard reset, 155
    reset noise suppression, 155–156
    soft reset, 155
    structure of, 154
Point-and-shoot cameras, 10–13
Poisson equations, 99
Polarization, 312–313
Polaroid, 3
Poly-silicon electrode technology, 123
Power consumption
    by CCD image sensors, 121–122
    by CMOS image sensors, 151–153
    by digital still cameras, 261–262
    by hardwired ASIC, 265
    by image-processing engine, 261
Power line noise, 298–299
PowerShot A5 Zoom, 48
PowerShot S100 DIGITAL ELPH, 42, 49
Printing press, 18–19
Programmable gain amplifier
    column parallel, 165
    description of, 164
    serial, 165
Progressive scan, 59–60, 124–125
p-substrate, 112–114, 159–160
Pulse code modulation, 257
p-well structure, 114

Q

Quantization error, 268
Quantization noise, 297–298
Quantum efficiency
    definition of, 78
    estimation of, 81
    improvement in, 308
    procedure for obtaining, 193
Quincunx grid sampling, 229
QV-10, 9–10

R

Radiometry, 315
Random noise, 66, 292

Rayleigh limit, 36–37
Read noise, 75–76
Recommended exposure index, 320–321
Rectangular grid sampling, 228–229
Recursive conversion technique, 214
Red, green, blue primary color filter array, 63, 218
Reflectance, 44–45
Reflection, 24–25
Refraction, 36
Refractive power, 23
REI, 320–321
Reproduced images of natural scenes, 202–203
Reset gate, 108
Reset noise
    active feedback for correction of, 156
    charge transfer and, 157
    description of, 74–75, 155
    nondestructive readout for reduction of, 156
    suppression of, 155–156
Resize of image, 246–247
Resolution
    characteristics of, 200–201
    description of, 138, 183
    high, 123
    image quality, 278–279
    measurement of, 279
    modulation transfer function and, 28
    Sparrow, 37
Responsivity, 78–79
Retrofocus lens, 24
Reverse biased metal-oxide semiconductor diode, 58
Reverse biased photodiode, 58
Reversible compression, 242
RGB color correction, 232
Rods, 206
Rolling reset synchronization with mechanical shutter, 170
Rolling shutter, 151
Row-wise noise, 167

S

Saturation charge, 66
Saturation exposure, 189–190
Scene-referred image data, 219
Schwarzschild aberrations, 30
Seidel aberrations, 30
Selective charge transfer mechanism, 134–135
Self-induced drift, 97–98
Semiconductors, 261
Sensitivity, 189, 319–321
Sensitivity metamerism, 213
Separable low-pass filters, 236
Serial programmable gain amplifier, 165
Shading
    description of, 72, 167–168
    lens, 251–252
Shockley–Read–Hall theory, 69
Short zoom lenses, 50–51
Shot noise
    dark current, 76–77
    definition of, 74
    description of, 292–293
    equation for, 166
    photon, 76–77, 80
Shutter speed, 127
Signal binning, 173
Signal charge transfer, 124
Signal-to-noise ratio
    calculation of, 117–118
    definition of, 282
    description of, 79–80, 117–118, 308–309
    increasing of, 84–85
    maximum, 309
    sensitivity and, 84
Silicon, 68
Silver halide crystals, 310
Silver halide photography system, 2
Silver halide SLR cameras, 17
Single lens
    retrofocus, 24
    schematic diagram of, 22, 26
    telephoto, 24
    thickness of, 23
Single lens reflex cameras, *See* SLR cameras
SIO_2/SI interface, 64–65
SLR cameras
    CMOS image sensors for, 169
    cost of, 8
    description of, 13–14
    diffraction use by, 24–25
    digital
        description of, 169, 305
        future of, 306–308
        high-end, 306–308
        multicolor stacked structure used with, 307
        spatial color aliasing, 307
    film, 305
    LCD display in, 13
    silver halide, 17
    structure of, 14
Smear
    definition of, 290
    description of, 77, 119
    examples of, 290

Index

frame transfer CCD, 110
   image sensor evaluation and, 199–200
   mechanical shutter advantages for, 128
   suppression ratio, 200
   "white streak," 292
Smear-to-signal ratio, 110–111, 119
Soft reset, 155
Sony Mavica, 4–5
SOS, 320–321
Sparrow resolution, 37
Spatial color aliasing, 307
Spatial frequency response, 280
Spatial random noise, 292
Spectral reflectance, 207
Spectral response, 190–193
Spherical aberration, 30
Standard output sensitivity, 320–321
Stereo vision, 312
Sticking, 289
Still images, 169–170
Still video cameras
   description of, 5–7
   failure of, 7–8
   price of, 8
Still video floppy, 6–7
Subdiffraction limit pixels, 308
Subresolution readout, 173
Surface channel CCD, 99–101
Surface reflectance, 44
System-on-a-chip, 261

T

Telecentric lens, 47
Telephoto lens, 24
Temporal noise
   causes of, 227
   at dark, 195–196
   definition of, 72
   description of, 66–67
   under illumination, 198
   image quality affected by, 302
   pixel, 162
   reset noise, 74–75
   thermal noise, 73
   variance of, 72–73
Thermal diffusion, 97–98
Thermal noise, 73, 292–293
Thin film transistor, 241
Three-dimensional space, 295
Threshold voltage, 148–149
Time-to-market, 261
Tint, 294–295
Tone curves, 233–234, 282–283, 291

Toneless black, 291–292
Total dark current, 70
Total noise power, 73
Toy cameras
   CMOS image sensors for, 168–169
   description of, 14–16
Transfer gate pulse, 126–127
Tristimulus values, 206–207
Two-dimensional filter, 235–236
Two-dimensional space, 295
Two-phase driving CCDs, 105–106

U

Ultrahigh-definition television image sensor, 175
Uniform color spaces, 207–210
Uniformity, 286–287
Unsharp mask filters, 237–238

V

Vacuum deposition, 43
Valence band pinning, 103
Vertical charge mixing, 136–137
Vertical charged-coupled devices, 58, 116
Vertical integration photodiode pixels, 161–162
Vertical offset drift, 225
Vertical overflow drain, 114–116
Vertical transfer CCDs, 105, 111
Very large-scale integrated circuits, 54
Video mode, 241–242
Video zoom lens, 49
Viewfinder, 241–242, 306
Visual system, 206
Voltage buffer, 147–148

W

Wafer baking, 63
Wave optics, 36
Wedge, 279
White balance
   color conversion, 216–218
   description of, 215
   error, 300
   loss of, 300
   white point, 215–216
White blemishes, 120
White clipping, 291, 301
White noise, 74, 196
White spot defects, 71–72

X

X-Y address, 56–58, 145
X-Y address readout, 152
X–Y pixel addressing, 148

Y

YCbCr color correction, 232–233

Z

Zoom lenses
    angle of light exiting from, 46–47
    electronic, 243–246
    multigroup moving, 50
    short, 50–51
    video type, 49